Contributions to Survey Sampling and Applied Statistics

Papers in Honor of H. O. Hartley

H. O. HARTLEY

Contributions to Survey Sampling and Applied Statistics

Papers in Honor of H. O. Hartley

Edited by

H. A. DAVID

IOWA STATE UNIVERSITY
AMES, IOWA

ACADEMIC PRESS New York San Francisco London 1978

A Subsidiary of Harcourt Brace Jovanovich, Publishers

ACADEMIC PRESS, INC.
111 Fifth Avenue, New York, New York 10003

United Kingdom Edition published by
ACADEMIC PRESS, INC. (LONDON) LTD.
24/28 Oval Road, London NW1

Library of Congress Cataloging in Publication Data

Main entry under title:

Contributions to survey sampling and applied statistics.

Bibliography of H. O. Hartley: p.
CONTENTS: Cochran, W. G. Laplace's ratio estimator.–Kempthorne, O. Some aspects of statistics, sampling, and randomization.–Sampford, M. R. Predictive estimation and internal congruency. [etc.]
Includes bibliographies.
1. Mathematical statistics–Addresses, essays, lectures. 2. Sampling (Statistics)–Addresses, essays, lectures. 3. Hartley, H.O. I. Hartley, H. O. II. David, Herbert Aron.
QA276.16.C56 519.5 77-6593
ISBN 0–12–204750–8

PRINTED IN THE UNITED STATES OF AMERICA

Contents

Sampling in Two or More Dimensions

ROBERT S. COCHRAN

Part II THE LINEAR MODEL

The Analysis of Linear Models with Unbalanced Data

RONALD R. HOCKING, O. P. HACKNEY, AND F. M. SPEED

Nonhomogeneous Variances in the Mixed AOV Model; Maximum Likelihood Estimation

WILLIAM J. HEMMERLE AND BRIAN W. DOWNS

Concurrency of Regression Equations with k Regressors

A. M. KSHIRSAGAR AND VIOLETA SONVICO

Robustness of Location Estimators in the Presence of an Outlier

H. A. DAVID AND V. S. SHU

The Ninther, a Technique for Low-Effort Robust (Resistant) Location in Large Samples

JOHN W. TUKEY

Completeness Comparisons among Sequences of Samples

NORMAN L. JOHNSON

Part V MATHEMATICAL PROGRAMMING AND COMPUTING

Absolute Deviations Curve Fitting: An Alternative to Least Squares

ROGER C. PFAFFENBERGER AND JOHN J. DINKEL

Tables of the Normal Conditioned on t-Distribution

D. B. OWEN AND R. W. HAAS

List of Contributors

Numbers in parentheses indicate the pages on which the authors' contributions begin.

G. E. P. Box (203), Department of Statistics, University of Wisconsin, Madison, Wisconsin

Edward C. Bryant (41), Westat, Inc., Rockville, Maryland

Robert S. Cochran (113), Department of Statistics, University of Wyoming, Laramie, Wyoming

William G. Cochran (3), Department of Statistics, Harvard University, Cambridge, Massachusetts

H. A. David (235), Statistical Laboratory, Iowa State University, Ames, Iowa

John J. Dinkel (279), Department of Statistics, The Pennsylvania State University, University Park, Pennsylvania

Brian W. Downs (153), Department of Computer Science, and Experimental Statistics, University of Rhode Island, Kingston, Rhode Island

James E. Gentle (223), Statistical Laboratory, Iowa State University, Ames, Iowa

R. W. Haas (295), Department of Statistics, Southern Methodist University, Dallas, Texas

O. P. Hackney (133), Department of Computer Science and Statistics, Mississippi State University, Mississippi

Morris H. Hansen (57), Westat, Inc., Rockville, Maryland

William J. Hemmerle (153), Department of Computer Science, and Experimental Statistics, University of Rhode Island, Kingston, Rhode Island

Ronald R. Hocking (133), Department of Computer Science and Statistics, Mississippi State University, Mississippi

Norman L. Johnson (259), Department of Statistics, The University of North Carolina, Chapel Hill, North Carolina

Oscar Kempthorne (11), Statistical Laboratory, Iowa State University, Ames, Iowa

A. M. Kshirsagar (173), Institute of Statistics, Texas A & M University, College Station, Texas

D. B. Owen (295), Department of Statistics, Southern Methodist University, Dallas, Texas

Roger C. Pfaffenberger (279), Management Science and Statistics Department, College of Business and Management, University of Maryland, College Park, Maryland

J. N. K. Rao (69), Department of Mathematics, Carleton University, Ottawa, Ontario, Canada

M. R. Sampford (29), Department of Computational and Statistical Science, University of Liverpool, Liverpool, England

D. M. Scott† (191), Institute of Statistics, Texas A & M. University, College Station, Texas

S. R. Searle (181), Biometrics Unit, Cornell University, Ithaca, New York

V. S. Shu (235), Statistical Laboratory, Iowa State University, Ames, Iowa

W. B. Smith (191), Institute of Statistics, Texas A & M University, College Station, Texas

Violeta Sonvico‡ (173), Institute of Statistics, Texas A & M University, College Station, Texas

F. M. Speed (133), Department of Computer Science and Statistics, Mississippi State University, Mississippi

Benjamin J. Tepping (57), Silver Spring, Maryland

G. C. Tiao (203), Department of Statistics, University of Wisconsin, Madison, Wisconsin

John W. Tukey (251), Princeton University, Princeton, New Jersey and Bell Laboratories, Murray Hill, New Jersey

W. H. Williams (89), Mathematics and Statistics Research Center, Bell Laboratories, Murray Hill, New Jersey

† Present address: U.S. Steel, Monroeville, Pennsylvania.

‡ Present address: Dt. Estadistica, INTA–Castelar–Pcia Bs. As., Argentina.

Preface

This volume is a tribute to H. O. Hartley on the occasion of his 65th birthday. HOH's career as a statistician started immediately after he left Nazi Germany for England in 1934, with a fresh Ph.D. in mathematics from Berlin University. There followed postgraduate research in statistics under John Wishart at Cambridge, resulting in 1940 in a Ph.D. in mathematical statistics. During part of this period HOH took a post as statistician at the Harper Adams Agricultural College, Shropshire, where he met his wife, Grace. From 1938 to 1946 he was heavily engaged in war work at the Scientific Computing Service Ltd., London. His academic career started in full force with an appointment as Lecturer in Statistics at University College, London. Glimpses of this and later periods are provided in the Greetings introducing this volume: E. S. Pearson on the UCL years, followed by T. A. Bancroft on HOH's decade as Professor at Iowa State University, and R. J. Freund on HOH's tenure as first Director of the Institute of Statistics, Texas A & M University. The association with Texas A & M, begun in 1963, will continue after HOH relinquishes administrative duties during 1977.

Such a skeletal career outline omits a great deal. Wherever HOH has worked he has given unsparingly of his talents, and always with enthusiasm, humanity, and charm. As a result he has not only achieved recognition as a leading statistician but has also acquired innumerable friends and admirers. The contributors to this Festschrift have all had close contacts with HOH, many as former students (myself included) or as past and present colleagues. Only limitations of space have precluded enlarging the list. It is natural that the papers appearing here should mirror many of HOH's diverse interests; even where no direct reference to his work is made, the Hartley influence is clear. We wish HOH many more good and fruitful years, and look forward to his continued lively presence at statistical gatherings.

I greatly appreciate the ready cooperation of all contributors to this volume. It is a pleasure to record with thanks the early encouragement I received from Egon Pearson, the repeated help of Rudi Freund, and the advice on the choice of contributors given by J. N. K. Rao and J. E. Gentle. Each paper submitted was read by at least one reviewer and the refereeing assistance of the following is gratefully acknowledged: B. C. Arnold, B. A. Bailar, H. T. David, W. A. Fuller, J. E. Gentle, C. P. Han, D. A. Harville, R. J. Jessen, N. L. Johnson, A. M. Kshirsagar, R. E. Lund, R. C. Pfaffenberger, J. N. K. Rao, S. R. Searle, V. A. Sposito, S. M. Stigler, B. V. Sukhatme, R. D. Warren, and W. H. Williams. Expert secretarial help was provided by Avonelle Jacobson, Kathy Shaver, and Dallis Sonksen.

Ames, Iowa
June 1977

H. A. DAVID

Greetings to HOH for 1977

FROM EGON PEARSON

I do not seem now to have any stand-by paper to contribute to your *Festschrift* which is worthy of the occasion, but looking back from beyond 80 years I feel that I must send you a message of good cheer at this time. You came to me at University College in the fall of 1947 when, after what had been effectively seven years of wartime break, I had great need to build up the Department of Statistics at University College again. But it was perhaps not mainly as a lecturer that I needed your help, but as a colleague in the production of a new *Biometrika Tables for Statisticians*, Volume I, to replace the long obsolescent *Tables for Statisticians and Biometricians*, Part I, first issued as a pioneer volume by K. P. in 1914 and little altered since then.

You had already been supervising, under L. J. Comrie at the Scientific Computing Service, the computation of those fundamental tables of "normal theory" which I had seen, prewar, must be included to bring our volume up-to-date: namely extensive tables of percentage points of χ^2 and of the beta-variable x, which was readily transformed into a table for the inverted beta or variance ratio F. No matter that Amos, with the powerful equipment of the Sandia Corporation has since found several errors in our "primitive" computations.

But as soon as you came, with an office across the corridor from my room on the first floor of the old Department of Statistics, it was easy to begin to discuss together (a) which of the old Part I tables should be retained, (b) which should be enlarged or modified, and (c) what new forms of table, useful in relatively straight-forward statistical analysis, should be added. It was you primarily with your natural "computational sense", inborn but sharpened by Comrie, who suggested a good deal of the fresh work, how it should be carried out and how best fitted to the page. Some of the new calculation was performed by contract outside, but much of it was done within the Department, either by our two selves or by such hardworking helpers as Joyce May. Others formed part of students' research programmes. You devised ingenious methods of interpolation and suggestions of how to proceed when the limits of a tabled function had been reached. As Karl Pearson remarked long ago, the table-maker is always having to balance the user's convenience against cost of typesetting, paper, and general unwieldiness.

There will, I fear, have been a tendency for the user to give most credit to myself for this volume as apparently the main author and editor, but in the case of *Tables*, Volume I, it was truly a joint effort, which neither of us could have completed without the other. It should be remembered that the work was done before the coming of the electronic computer, and in this sense followed the style of those pioneer table-makers, Karl Pearson and L. J. Comrie.

The scope and form of Volume II was scarcely planned when Volume I went to press in 1953. Unfortunately, there was still no money in the British universities available for mathematical statistics and no prospect yet for new posts—or at any rate I had not the drive which secures funds. So that just as in the case of Jerzy Neyman fifteen years before, I felt that I must advise you to look for possibilities in the United States. I did this with special reluctance, not only because it effectively meant the breaking up of our table-making partnership which had been one of the cornerstones on which I had hoped to build in the postwar years, but also because it involved exile for Grace and your children from their West of England homeland. I was grateful that, whatever you may have felt, you never let me see any bitterness about my advice.

From 1954, the year of publication of Volume I, work on Volume II proceeded by starts and jerks; we were too busy for either of us to be good correspondents and our short half-day meetings on your visits to England never provided time for proper discussion, although it was remarkable how on each occasion when we met, some small problem broke through a blockage with the help of your quick wit and always refreshing outlook.

Heavily occupied with the Department and *Biometrika*, and wishing to get on with some work of my own, there was not much to encourage me to get on with Volume II alone. However, enough of a skeleton plan and a number of new tables which had already appeared in print in *Biometrika*, at last justified my short visit to you in Texas in 1969, and as a follow-up a two-week visit in 1970. Then and there we drove to completion a number of sections of the Introduction and afterwards it was relatively straightforward for me to fill in the gaps, either in California or at home in the year which followed, ending in publication of Volume II in 1972.

I remember vividly that two-week visit to your home a mile or so off the campus of Texas A & M University in 1970, and how touched I was by the way in which Grace with no more than humorous protest, allowed her dining room table and large parts of the floor to be covered continuously with a spread of papers. It showed, I felt, that I must have been forgiven for having failed to prevent her exile from England 17 years before. We took a day off to visit the little estate in the wilds where you were able to retire to a log cabin and escape from office-piling surges of administrative business. I remember in particular that large murky pond from the depths of which small alligators emerged at certain hours like the Loch Ness Monster to see the light of day.

May official retirement now give you a little more time to visit your daughter near Paris and perhaps to stay a few days longer in England!

FROM T. A. BANCROFT

I am pleased to give an account of the second episode of the Hartley story following the previous account given by Professor Pearson for the 1947–1953 period.

HOH, you joined the staff of our Iowa State University statistical center (Statistical Laboratory and Department of Statistics) on July 1, 1953 and left to become Director of a newly created Institute of Statistics at Texas A & M beginning July 1, 1963. Your first twelve-month period here was as a visiting professor; but, fortunately for us, you accepted our invitation to continue as a regular faculty member and remained so for nine more years.

You came to Ames at a critical point in time in the development of the teaching, consulting, and research programs of our statistical center. While the Statistical Laboratory had been well-established by Professor George W. Snedecor as its first Director, twenty years earlier, in 1933 and its on-campus consulting and campus-wide computing services were flourishing, both were primarily limited to agriculture and certain biological sciences and related areas. Also, although the Department of Statistics had been formally established in 1947, some courses listed in the initial departmental offerings had not been activated, even when you arrived in 1953, due to the exodus of some key faculty members just before 1950, the year I happened to have been made Director and the first permanent head. This was due primarily to the scarcity of available statisticians; in particular, for our needs, statisticians who were well trained, interested, and experienced in both applications and theory.

Beginning in 1950, plans were laid and subsequently implemented to add on-campus consulting and the teaching of related service courses in the areas of the physical sciences and engineering, the social and behavioral sciences, and the humanities, while maintaining the already well-established work with agriculture and related areas. In addition, the campus-wide computing service, initiated in the 1920's by G. W. Snedecor and A. E. Brandt with the assistance of Henry A. Wallace, was equipped with an IBM 650 digital computer early in 1957. The campus-wide computing service remained as part of our Statistical Laboratory until the rapid expansion of digital-computing necessitated the establishment of a separate University Computation Center. On April 1, 1962 the hardware and general software of our computing service were transferred to the University Computation Center, while statistical numerical analysis and data processing remained a responsibility of the Statistical Laboratory. Plans were also initiated to strengthen and expand our graduate training program for the M.S. and Ph.D. programs in statistics as well as joint graduate degree programs and minor programs.

In the above connection, your support and assistance were most valuable. Being well-trained, interested, and experienced in statistical applications and theory, and in addition having experience in computing with L. J. Comrie at the Scientific Computing Service, London, you welcomed responsibilities at Iowa State in four important areas: (1) teaching of the most advanced courses in survey sampling theory, and research and direction of M.S. and Ph.D. theses in this area, (2) a responsibility similar to (1) in statistical numerical analysis and digital computing software, (3) the preparation of lectures and the initiation of our most advanced general methods course, and (4) on-campus consulting and cooperative research on problems related to (1), (2), and (3). In addition, your established competence in statistical research attracted many off-campus supported research grants and contracts, providing financial support for graduate students and some other staff members. At Iowa State you were truly "a statistician for all seasons!"

Our Annual Report shows that while here you published some 39 research papers either singly or coauthored, and directed 17 Ph.D. theses, including three joint majors, and ten M.S. theses. In addition, the book which started much earlier in England with E. S. Pearson and H. O. Hartley, editors: *Biometrika Tables for Statisticians,* Volume I. Cambridge University Press, June 1954, was released at the end of your first year in Ames, and some planning was subsequently started on Volume II, published in 1972. Further the book, already begun in London and continued in Princeton, J. A. Greenwood and H. O. Hartley, *Guide to Tables in Mathematical Statistics*, Princeton University Press, 1961, and the 'Section on Analysis of Variance' in *Mathematical Methods for Digital Computers*, edited by F. Ralston and S. Wilf, Wiley, 1960, were both published during your Iowa State period. While at Iowa State, in recognition of your outstanding scientific contributions, you were awarded the D.Sc. by University College, London.

While we were not able to lure you back to Iowa State upon my retirement from administrative duties in 1972, the faculty in statistics here did succeed in employing as my successor, Professor H. A. David, an able former Ph.D. graduate student with you while you were still at University College, London.

Finally, we trust that your official retirement will provide you with more time to visit more often your many friends at Iowa State and elsewhere in the United States and abroad. Also, you and Grace should have more time to visit your children, Mike and Jennifer.

FROM RUDI FREUND

It is claimed that you are now retiring. Having watched you closely over the last fourteen years I know that retiring is not the proper word. Nevertheless, at this important juncture in your life, it is appropriate to reminisce about those years.

I had, of course, known about you before 1962; both of us ran IBM 650's in the early days and I was much impressed with your innovative work on what we now consider to be a very small and limited computer. We also had a brief interlude at one of the SREB Summer Sessions at Oklahoma and again I thoroughly enjoyed your insight into things statistical and computational.

Then came the offer from Texas A & M where they indicated that they had hoped to hire you as Director; this certainly was a wonderful opportunity to be associated with someone I had long admired and whose general philosophy of statistics was so appealing. So for these past years we have been neighbors, both at the office and at home, and I must say that I have enjoyed both aspects tremendously.

Our first official contact was at an airport motel in Chicago in the Fall of 1962 where we began to prepare the National Science Foundation development grant proposal which we ultimately obtained and which was so instrumental in getting the Institute off to such a good start. Then you arrived at Texas A & M in 1963, and we were quartered in a subdivided room in the Animal Sciences Building. In typical Hartley fashion your first step was to visit every department head to inform them of the creation of this new Institute of Statistics and what it would try to do for the university as a whole and their departments in particular. Tireless efforts such as that resulted in the growth of the sizeable department which has prospered under your leadership and now ranks nationally as one of the well-established departments in the country.

Your associates have never ceased to be amazed at the new ideas you have, your insights into problems, and your devotion to your work which keeps you working many times into the night (much to Grace's discomfiture). Obviously the shortcomings of your associates have caused you much extra work, but you have never complained. In fact, even if you wish to comment unfavorably on someone's work, you will always start by complimenting on that part of the job that was well done. I have never heard you become angry; for as trying a job as you have had, this is certainly an amazing accomplishment.

In spite of your excessively long working hours you have managed to mix socially with the department, something that has helped to create and maintain the spirit of unity among both students and faculty of the department. The Continental Breakfasts that you and Grace have sponsored every Fall are now an Aggie tradition. Among the many times we have enjoyed your company at other events, I particularly remember the 10th anniversary celebration of the Institute where one of our students (fearfully shaking in his boots) did a takeoff on your performance in one of your classes. Also, none of us will forget Grace's and your terpsichorean performances at some of our parties.

So as you "retire" we all look back on the past years and hope that we shall continue our association with you in the coming years, on whatever basis you choose.

Published Works of H. O. Hartley

(a) Books

[with D. W. Smithers and K. M. H. Branson] *The Royal Cancer Hospital Mechanically Sorted Punched Card Index System*, published by the Royal Cancer Hospital, London, 1945.

Section on "Statistics" for *Civil Engineering Reference Book*, Butterworth Scientific Publications, London, 1951.

[with E. S. Pearson] *Biometrika Tables for Statisticians, Vol. I,* Biometrika Publications, Cambridge University Press, Bentley House, London, 1954.

Section on "Analysis of Variance" in *Mathematical Methods for Digital Computers* (F. Ralston and S. Wilf, eds.), Wiley, New York, 1960.

[with J. A. Greenwood] *Guide to Tables in Mathematical Statistics*, Princeton University Press, Princeton, New Jersey, 1961.

[with E. S. Pearson] *Biometrika Tables for Statisticians, Vol. II,* Biometrika Publications, Cambridge University Press, Bentley House, London, 1972.

(b) Papers

1935†

A connection between correlation and contingency. *Proceedings of the Cambridge Philosophical Society* **31** 520–524.

1936

Continuation of differentiable functions through the plane. *The Quarterly Journal of Mathematics,* Oxford Series **7** 1–15.

A generalization of Picard's method of successive approximation. *Proceedings of the Cambridge Philosophical Society* **32** 86–95.

[with J. Wishart] A theorem concerning the distribution of joins between line segments. *Journal of the London Mathematical Society* **11** 227–235.

1937

The distribution of the ratio of covariance estimates in two samples drawn from normal bivariate populations. *Biometrika* **29** 65–79.

1938

Statistical study to Dr. Hunter's paper: Relation of ear survival to the nitrogen content of certain varieties of barley. *Journal of Agricultural Science* **28** 496–502.

† The first five papers are in Hartley's former name, H. O. Hirschfeld.

Studentization and large sample theory. *Supplement to the Journal of the Royal Statistical Society* **5** 80–88.

1940

Testing the homogeneity of a set of variances. *Biometrika* **31** 249–255.

1941

[with L. J. Comrie] Table of Lagrangian coefficients for harmonic interpolation in certain tables of percentage points. *Biometrika* **32** 183–186.

1942

[with E. S. Pearson] The probability integral of the range in samples of *n* observations from a normal population *Biometrika* **32** 301–310.

The range in random samples. *Biometrika* **32** 334–348.

1943

[with E. S. Pearson] Tables of the probability integral of the studentized range. *Biometrika* **33** 89–99.

1944

Studentization. *Biometrika* **33** 173–180.

1945

Note on the calculation of the distribution of the estimate of mean deviation in normal samples. *Biometrika* **33** 257–265.

[with H. J. Godwin] Tables of the probability integral and the percentage points of the mean deviation in samples from a normal population. *Biometrika* **33** 252–265.

1946

The application of some commercial calculating machines to certain statistical calculations. *Supplement to the Journal of the Royal Statistical Society* **8** 154–183.

Tables for testing the homogeneity of estimated variances. (Prefatory note to tables prepared with E. S. Pearson) *Biometrika* **33** 296–304.

1947

[with S. H. Khamis] A numerical solution of the problem of moments. *Biometrika* **34** 340–351.

1948

Approximation errors in distributions of independent variates. *Biometrika* **35** 417–418.

[with C. A. B. Smith] The construction of Youden Squares. *Journal of the Royal Statistical Society*, Series B, **10** 262–263.

The estimation of nonlinear parameters by "Internal Least Squares." *Biometrika* **35** 32–45.

1949

Tests of significance in harmonic analysis. *Biometrika* **36** 194–201.

1950

[with H. H. Stones, F. E. Lawton, and E. R. Bransby] Dental caries and length of institutional resistance, etc. *British Dental Journal* **89** 199–203.

The maximum *F*-ratio as a short-cut test for heterogeneity of variance. *Biometrika* **37** 308–312.

A simplified form of Sheppard's correction formulae. *Biometrika* **37** 145–148.

The use of range in analysis of variance. *Biometrika* **37** 271–280.

[with E. S. Pearson] New Tables of Statistical Functions, Table of the probability integral of the *t*-distribution. *Biometrika* **37** 168–172.

Table of the χ^2-integral and of the cumulative Poisson distribution. *Biometrika* **37** 313–325.

1951

[with E. R. Fitch] A chart for the incomplete beta-function and the cumulative binomial distribution. *Biometrika* **38** 423–426.

The fitting of polynomials to equidistant data with missing values. *Biometrika* **38** 410–413.

[with E. S. Pearson] Moment constants for the distribution of range in normal samples. *Biometrika* **38** 463–464.

[with E. S. Pearson] Charts of the power function for analysis of variance tests, derived from the non-central F-distribution. *Biometrika* **38** 112–130.

[with H. H. Stones, F. E. Lawton, and E. R. Bransby] Time of eruption of permanent teeth and time of shedding of deciduous teeth. *British Dental Journal* **90** 1–7.

1952

Second order autoregressive schemes with time-trending coefficients. *Journal of the Royal Statistical Society*, Series B **14** 229–233.

Tables for numerical integration at non-equidistant argument steps. Proceedings of the Cambridge Philosophical Society **48** 436–442.

1953

[with S. S. Shrikhande and W. B. Taylor] A note on incomplete block designs with row balance. *The Annals of Mathematical Statistics* **24** 123–126.

Appendix—Some theoretical aspects to Estimating variability from the differences between successive readings (Joan Keen and Denys J. Page, authors) *Applied Statistics, II* **13** 20–23.

1954

[with H. A. David and E. S. Pearson] The distribution of the ratio, in a single normal sample, of range to standard deviation. *Biometrika* **41** 482–493.

[with E. L. Kozicky, G. O. Hendrickson] A proposed comparison of fall roadside pheasant counts and flushing rates. *Proceedings of the Iowa Academy of Science* **61** 528–533.

[with E. C. Fieller] Sampling with control variables. *Biometrika* **41** 494–501.

[with A. Ross] Unbiased ratio estimators. *Nature* **866** 270.

[with H. A. David] Universal bounds for mean range and extreme observation. *The Annals of Mathematical Statistics* **25** 85–99.

1955

Some recent developments in analysis of variance. *Communications on Pure and Applied Mathematics* **8** 47–72.

1956

A plan for programming analysis of variance for general purpose computers. *Biometrics* **12** 110–122.

[with Helen Bozivich and T. A. Bancroft] Power of analysis of variance test procedures for certain incompletely specified models, I. *The Annals of Mathematical Statistics* **27** 1017–1043.

1957

[with J. F. Jumber, E. L. Kozicky, and A. M. Johnson] A technique for sampling mourning dove production. *The Journal of Wildlife Management* **21** 226–229.

[with E. C. Fieller and E. S. Pearson] Tests for rank correlation coefficients, I. *Biometrika* **44** 470–481.

Equipping a University Laboratory to Satisfy the Computational Demand. *The Computing Laboratory in the University* (Preston C. Hammer, ed.) The University of Wisconsin Press, Madison 175–179.

1958

Changes in the outlook for statistics brought about by modern computers. *Proceedings of the Third Conference on the Design of Experiments in Army Research Development and Testing*, Washington, D. C. 345–363.

[with Laurel D. Loftsgard] Linear programming with variable restraints. *Iowa State College Journal of Science* **33** 161–172.

Maximum likelihood estimation from incomplete data. *Biometrics* **14** 174–194.

[with Leo Goodman] The precision of unbiased ratio-type estimators. *Journal of the American Statistical Association* **53** 491–508.

[with H. O. Carter] A variance formula for marginal productivity estimates using the Cobb–Douglas functions. *Econometrica* **26** 306–313.

1959

Analytic studies of survey data. Volume in honor of Corrado Gini, issued by the Institute of Statistics of the University of Rome 1–32.

The efficiency of internal regression for the fitting of the exponential regression. *Biometrika* **46** 293–295.

Smallest composite designs for quadratic response surfaces. *Biometrics* **15** 611–624.

1960

[with Edward C. Bryant and R. J. Jessen] Design and estimation in two-way stratification *Journal of the American Statistical Association* **55** 105–124.

1961

The modified Gauss Newton method for the fitting of nonlinear regression functions by least squares. *Technometrics* **3** 269–280.

Nonlinear programming by the simplex method. *Econometrica* **29** 223–237.

Statistics, a reprint from *Civil Engineering Reference Book*, second edition. (J. Comries, ed.) Butterworth & Company, London 125–139.

1962

Foreword, Contributions to Order Statistics (Sarhan & Greenberg, authors) Wiley, New York, vii–ix.

Multiple frame surveys. *Proceedings of the Social Statistics Section* American Statistical Association 203–206.

[with J. N. K. Rao] Sampling with unequal probabilities and without replacement. *The Annals of Mathematical Statistics* **33** 350–374.

[with J. N. K. Rao and W. G. Cochran] On a simple procedure of unequal probability sampling without replacement. *The Journal of the Royal Statistical Society*, Series B **24** 482–491.

1963

The analysis of design of experiments with the help of computers. *Applications of Digital Computers* (Freiberger and Prager, ed.) Ginn & Company, Boston 179–194.

[with R. R. Hocking] Convex programming by tangential approximation. *Management Science* **9** 600–612.

In Dr. Bayes' consulting room. *The American Statistician* **17** 22–24.

[with D. L. Harris] Monte Carlo computations in normal correlation problems. *Journal of the Association for Computing Machinery* **10** 302–306.

1964

Exact confidence regions for the parameters in nonlinear regression laws. *Biometrika* **51** 347–353.

Linear programming in the feed industry. *Proceedings of the 19th Annual Texas Nutrition Conference* 27–43.

Some small sample theory for nonlinear regression estimation. *Proceedings of the Ninth Conference on the Design of Experiments in Army Research Development and Testing* 655–660.

1965

[with A. W. Worthan] Assessment and correction of deficiencies in PERT. *Proceedings of the Tenth Conference on the Design of Experiments in Army Research Development and Testing* 375–400.

Multiple purpose optimum allocation in stratified sampling. *Proceedings of the Social Statistics Section of the American Statistical Association* **8** 258–261.

[with Aaron Booker] Nonlinear least squares estimation. *Annals of Mathematical Statistics* **36** 638–650.

1966

Maximum likelihood estimation for unbalanced factorial data. *Proceedings of the Eleventh Conference on the Design of Experiments in Army Research Development and Testing* 597–606.

[with A. W. Worthan] A statistical theory for PERT critical path analysis. *Management Science*, Series B **12** 469–481.

Systematic sampling with unequal probability and without replacement. *Journal of the American Statistical Association* **61** 739–748.

1967

[with R. P. Chakrabarty] Evaluation of approximate variance formulas in sampling with unequal probabilities and without replacement. *Sankhyā: The Indian Journal of Statistics*, Series B **29** 201–208.

Expectations, variances, and covariances of ANOVA mean squares by "synthesis". *Biometrics* **23** 104–114.

Industrial applications of regression analysis. *Technical Conference Transactions*, Chicago 143–149.

[with R. R. Hocking and W. P. Cooke] Least squares fit of definite quadratic forms by convex programming. *Management Science* **13** 913–925.

[with J. N. K. Rao] Maximum-likelihood estimation for the mixed analysis of variance model. *Biometrika* **54** 93–108.

[with R. J. Freund] A procedure for automatic data editing. *Journal of the American Statistical Association* **62** 341–352.

1968

[with J. N. K. Rao] Classification and estimation in analysis of variance problems. *Review of the International Statistical Institute* **36** 141–147.

[with L. D. Broemeling] Confidence region estimation of response surface ordinates. *Metron*, XXVII–N. 1–2 59–68.

The future of departments of statistics. *The Future of Statistics*, Academic Press, New York, 103–137.

[with L. D. Broemeling] Investigations of the optimality of a confidence region for the parameters of a nonlinear regression model. *Annals of the Institute of Statistical Mathematics* **20** 263–269.

[with J. N. K. Rao] A new estimation theory for sample surveys. *Biometrika* **55** 547–557.

[with W. B. Smith] A note on the correlation of ranges in correlated normal samples. *Biometrika* **55** 595–596.

1969

[with Paul G. Ruud] Computer optimization of second order response surface designs. *Statistical Computation*, Academic Press, New York, 441–462.

[with S. R. Searle] A discontinuity in mixed model analyses. *Biometrics* **25** 573–576.

[with J. N. K. Rao and Grace Kiefer] Variance estimation with one unit per stratum. *Journal of the American Statistical Association* **64** 841–851.

1971

[with R. R. Hocking] The analysis of incomplete data. *Biometrics* **27** 783–823.

[with J. N. K. Rao] Foundations of survey sampling (A Don Quixote Tragedy). *The American Statistician* 21–27.

[with W. K. Vaughn] A computer program for the mixed analysis of variance model based on maximum likelihood. *Snedecor Memorial Volume*, Iowa State University Press, Ames, Iowa, (T. A. Bancroft, ed.) 129–144.

[with R. C. Pfaffenberger] Statistical control of optimization. *Optimizing Methods in Statistics*, Academic Press, New York, 281–300.

[with J. H. Matis] Stochastic compartmental analysis: Model and least squares estimation from times series data. *Biometrics* **27** 77–102.

1972

[with D. Lurie] Machine-generation of order statistics for Monte Carlo computations. *The American Statistician* **26** 29–30.

[with R. C. Pfaffenberger] Quadratic forms in order statistics used as goodness-of-fit criteria. *Biometrika* **59** 605–611.

1973

[with K. S. E. Jayatillake] Estimation for linear models with unequal variance. *Journal of the American Statistical Association* **68** 189–192.

[with L. J. Ringer, and O. C. Jenkins] Root estimators for the mean of skewed distributions. Social Statistics Section *Proceedings of the American Statistical Association* 413–416.

[with O. C. Jenkins, L. J. Ringer] Root estimators. *Journal of the American Statistical Association* **68** 414–419.

[with R. L. Sielken, Jr.] Two linear programming algorithms for unbiased estimation of linear models. *Journal of the American Statistical Association* **68** 639–641.

[with W. J. Hemmerle] Computing maximum likelihood estimates for the mixed A. O. V. model using the *W* transformation. *Technometrics* **15** 819–831.

1974

Multiple frame methodology and selected applications. *Sankhyā: The Indian Journal of Statistics* Series C, Pt. 3, **36** 99–118.

1975

[with J. E. Gentle] Data monitoring criteria for linear models. *A Survey of Statistical Design and Linear Models*, North-Holland Publishing Company, Amsterdam.

[with R. L. Sielken, Jr.] A "super-population viewpoint" for finite population sampling. *Biometrics* **31** 411–422.

[with C. E. Gates, K. S. E., Jayatillake, K. E. Gamble, and T. L. Clark] Texas mourning dove harvest survey. *The Texas Journal of Science* vol. **XXVI**, Nos. 3 and 4.

1976

The impact of computers on statistics. *On the History of Statistics and Probability* (D. B. Owen, ed.) Marcel Dekker.

The impact of mathematical modeling and computers in biomedical research. *Proceedings of Conference of the Society for Advanced Medical Systems* Houston, Texas, 159–168.

1977

[with R. L. Sielken, Jr.] Estimation of "safe doses" in carcinogenic experiments. *Biometrics* **33**.
Solution of Statistical Distribution Problems by Monte Carlo Methods. *Volume III of Mathematical Methods for Digital Computers* (Kurt Enslein, Anthony Ralston and Herbert S. Wilf, eds.) John Wiley and Sons.
[with J. N. K. Rao] Some comments on labels: A rejoinder to a section of Godambe's Paper. *Sankhyā Series C* (to be published shortly).

Part I

SAMPLING

Laplace's Ratio Estimator

William G. Cochran

HARVARD UNIVERSITY

This paper describes the sample survey from which Laplace estimated the population of France as of September 22, 1802 by means of a ratio estimator (ratio of population to births during the preceding year). His analysis of the distribution of the error in his estimate used a Bayesian approach with a uniform prior for the ratio of births to population. He assumed that the data for his sample and for France itself were both drawn at random from an infinite superpopulation. He found the large-sample distribution of his error of estimate to be approximately normal, with a small bias and a variance that he calculates. He ends by calculating the probability that his estimate of the population of France is in error by more than half a million. To my knowledge this is the first time that an asymptotic distribution of the ratio estimate has been worked out.

1. INTRODUCTION

Our friend HO Hartley has contributed to so many important areas in statistics that authors of this Festschrift should not find it difficult to select a topic related to some of HO's work. Assuming that we are not expected to rival his originality and brilliance, I would like to write about an early use of the ratio estimator in a sample survey. This is the well-known survey from which Laplace estimated the population of France as of September 22, 1802.

This survey interests me for at least two reasons. It is a survey taken under the direction of the French government at the request of a private citizen.

Key words and phrases Laplace, ratio estimator, sample survey, error of estimate, superpopulation.

However, as Dr. Stephen Stigler has reminded me, by 1802 Laplace already had previous government service and political prestige having been Minister of the Interior for six weeks in 1799, President of the Senate in 1801, and been made a Grand Officer of the Legion of Honor by Napoleon in 1802. Thus the sample survey may have had a more official origin than I at first thought.

Secondly, Laplace not only gave the formula for the ratio estimate, but worked out the large-sample distribution of the error in his estimate, showing this distribution to be normal with a bias and a variance that he gives, and noting that in large samples the effect of the bias becomes negligible. He ends by calculating the probability that his estimate of the population of France is in error by more than half a million. His work is the first known to me in which the sampling distribution of the ratio estimate has been tackled.

His survey is described with the arithmetical estimates but without any details of the error theory in Chapter 8 of the English translation: "A philosophical essay on probabilities," (Laplace, 1820, English translation, 1951). His assumptions and mathematical strategy plus the resulting normal approximation are given in Chapter 6 of the "Théorie Analytique des Probabilités." For the details of the analysis the reader is referred to an earlier chapter in which Laplace worked on a similar problem.

2. THE SURVEY AND THE ESTIMATE

In France, registration of births was required, and Laplace assumed without comment that he could obtain the numbers of births in a year for France as a whole or for any administrative subdivision. In this he may have been a little naïve. The fact that some action is compulsory in a country, even in a dictatorship, does not ensure that everyone takes this action. Actually, Laplace's theoretical assumptions do not demand completeness of birth registration, but something less—namely, that any percent incompleteness should be constant throughout France, apart from a binomial sampling error.

From such data as he could get his hands on, Laplace had noticed that the ratio of population to births during the preceding year was relatively stable. He persuaded the French government to take a sample of the small administrative districts known as communes, and count the total population y in the sample communes on September 22, 1802. From the known total number of registered births during the preceding year in these communes, x, and in the whole country X the ratio estimate $\hat{Y}_R = Xy/x$ of the population of France could be calculated.

The method—registered births × estimated ratio of population to births—was not original with Laplace. It dates back at least to John Graunt in 1662. In his "Observations on the bills of mortality," Graunt suggested it as one method of estimating the population of a country in which the number of annual births was known. In fact, Graunt went one better than Laplace, and estimated the key ratio y/x without any sample at all, by what we might call crude guesswork. He argued that since women of childbearing age seemed to be having a baby about every second year, the number of women of childbearing age could be estimated as $2X$, where X was the number of annual births. Then he guessed that the number of women aged between 16 and 76 was about twice the number of women of childbearing age and was therefore $2 \times 2 \times X$. (The number of live women over 76 was, I suppose, negligible.) He regarded every woman over 16 as belonging to a family, and, by a step which I do not fully understand, counted $2 \times 2 \times X$ as the number of families. The final step was to estimate the average size of a family as 8—husband, wife, three children, and three servants or lodgers. Thus to estimate population from births, multiply by $2 \times 2 \times 8 = 32$.

Commenting on this arithmetic, Greenwood (1941) thought that Graunt had made two big errors. As an average, a baby every two years was much too often; Greenwood thought the first multiplier should have been 4 instead of 2. On the other hand, 8 looked much too large as the average size of a family. Since these two errors were in opposite directions, perhaps Graunt was lucky. For what it is worth, Greenwood quotes data from the British Census of 1851, nearly 200 years after Graunt, which gave a ratio of population to births as 31.47.

Laplace's use of the ratio estimate based on a sample was not even original in France, two other Frenchmen having used the method in the late eighteenth century, as Stephan (1948) has noted.

Laplace's sample was what we now call a two-stage sample. He first selected 30 departments—the large administrative areas—chosen to range over the different climates in France. In each sample department a number of communes were selected for complete population counts on September 22, 1802. The criterion for selection was that the communes should have zealous and intelligent mayors: Laplace wanted accurate population counts. The combined population of the sample communes as of September 22, 1802 was 2,037,615 (I guess about a 7% sample).

At this point Laplace did two things that puzzle me slightly. As regards births, he totaled the sample births for the three-year period September 22, 1799 to September 22, 1802, finding a value 215,599 so that his sample x is 215,599/3. No explanation is given for the use of the three-year period. One might guess that the objective was to obtain higher precision from a larger sample with little effort since the births were already on record. But the device might not accomplish this if the birth rate varied from year to year. In his theoretical development of the sampling distribution of the error in $\hat{Y}_R$, he acts as if x was counted for only the year preceding September 22, 1802. In his numerical estimate, however, the sample ratio y/x was taken as $3(2{,}037{,}615)/(215{,}599) = 28.852345$.

Secondly, having been responsible for this important survey, he might be expected to present and use as X the actual number of registered births in France in the year preceding September 22, 1802. Instead, he writes casually: "supposing that the number of annual births in France is one million, which is nearly correct, we find, on multiplying by the preceding ratio (y/x), the population of France to be 28,352,845 persons."

3. THE SAMPLING ERROR: STANDARD METHODS

Suppose that we had the population and birth data x_{ij}, y_{ij} for the jth sample commune in the ith department and wanted to try to estimate the sampling error of Laplace's $\hat{Y}_R$. The sample was a two-stage sample, but with no use of probability sampling. We might first judge whether the sample of departments could be regarded as a random sample, or perhaps better as a geographically stratified random sample. Secondly, we might judge whether the distribution of selected communes with zealous and intelligent mayors within a department seemed to approximate a random selection. Depending on these judgments we might decide to use the textbook formula for the large-sample variance of the ratio estimator in two-stage sampling. In this formula, $V(\hat{Y}_R)$ depends on the variances and the covariance of the y_{ij}, x_{ij} between departments and between communes in the same department.

Laplace simplified his mathematical problem by a sweeping assumption. He considered an infinite urn consisting of white and black balls and representing a superpopulation of French citizens alive on September 22, 1802. The white balls represented those born in the preceding year and registered. The ratio p (proportion of white balls) is unknown. He regarded the ratio x/y from the sample of communes as a binomial estimate of p. Choice of this model seems at odds with his deliberate choice of sample departments scattered throughout France, which presupposes a judgment that the birth rate p varies from department to department.

If we adopt Laplace's assumption and assume the sampling fraction y/Y small, we write $\hat{Y}_R = Xy/x = X/\hat{p}$, where $\hat{p}$ is a binomial estimator of p based on a sample of y persons. By Taylor's approximation,

$$\hat{Y}_R = \frac{X}{\hat{p}} \cong \frac{X}{p} - \frac{X(\hat{p}-p)}{p^2} + \frac{X(\hat{p}-p)^2}{p^3}. \tag{1}$$

With $Y = X/p$, it follows that with binomial sampling $\hat{Y}_R$ has a bias whose leading term is $Xpq/yp^3 = Xq/yp^2$. On substituting $\hat{p} = x/y$, $\hat{q} = (y - x)/y$, the sample estimate of this bias is $X(y - x)/x^2$.

From (1), the approximate variance of $\hat{Y}_R$ is

$$V(\hat{Y}_R) \cong X^2 V(\hat{p})/p^4 = X^2 pq/yp^4 = X^2 q/yp^3. \tag{2}$$

The ratio of the bias to the s.e. of $\hat{Y}_R$ is approximately $(q/yp)^{1/2}$, becoming negligible for y large.

If we drop the assumption that y/Y is small, and use the hypergeometric, the variance (2) is multiplied by the factor $(Y - y)/(Y - 1)$.

The sample estimate of the variance (2) is

$$\tilde{v}(\hat{Y}_R) = X^2(y - x)y^3/y^2x^3 = X^2 y(y - x)/x^3. \tag{3}$$

The preceding analysis assumes in effect that y and $E(x) = yp$ are both large.

Laplace's analysis finds that $\hat{Y}_R$ is approximately normally distributed in large samples, with the same leading term $X(y - x)/x^2$ in the bias, but with a variance slightly larger than the estimate in (3), namely

$$v(\hat{Y}_R) = X(X + x)y(y - x)/x^3. \tag{4}$$

There is also the difference that since his approach was Bayesian, he would regard (4) as the correct asymptotic variance given the sample information rather than as a sample estimate of the variance. If y/Y is small, the difference between (3) and (4) is minor, since x/X will nearly always be small so that $X(X + x) \cong X^2$.

4. LAPLACE'S ANALYSIS OF THE SAMPLING ERROR

When we examine Laplace's analysis more carefully, we note that he assumes that France itself and the sample communes are *both* binomial samples from the infinite superpopulation or urn with unknown p (ratio of births to population). Thus he regards X as a random binomial variable from a sample of unknown size Y, the population of France. So far as I know, this use of an infinite superpopulation in studying the properties of sampling methods was not reintroduced into sample survey theory until 1963, when Brewer (1963), followed by Royall (1970), applied it to the ratio estimator with the following model in the superpopulation

$$y = \beta x + \varepsilon \qquad \varepsilon \sim (0, \lambda x), \tag{5}$$

this being the model under which the ratio estimator is expected to perform well.

If we follow Brewer and Royall in obtaining results conditional on the known value of X, we write $\hat{Y}_R = X/p_y$, $Y = X/p_Y$, where p_y and p_Y are estimates of p obtained from binomial samples of sizes y, Y. Then

$$\hat{Y}_R - Y = X\left(\frac{1}{p_y} - \frac{1}{p_Y}\right) \cong X(p_Y - p_y)/p^2. \tag{6}$$

Averaging over repeated selections of France itself and of the random subsample of communes drawn from France, we get

$$E(\hat{Y}_R - Y)^2 = EV(\hat{Y}_R) \cong \frac{X^2pq}{p^4}\left(\frac{1}{y} - \frac{1}{Y}\right) \tag{7}$$

$$= \frac{X^2q}{yp^3}\frac{(Y-y)}{Y}. \tag{8}$$

This is practically the same result for $V(\hat{Y}_R)$ as we obtained when we replaced the binomial by the hypergeometric in (2).

The reason why Laplace's variance (4) differs from (8) and from the sample estimate of (8) does not lie in his Bayesian approach, but in his assumption that France and the sample of communes were *independent* samples, whereas the latter is a subsample drawn from France.

To proceed with Laplace's analysis, he begins by assuming that the unknown p follows a uniform prior dp $(0 \leq p \leq 1)$. A comment here is that, assuming almost no advance knowledge, Laplace might have tightened his prior by assuming, say, $p < 0.2$. For even if the women of childbearing age work overtime, the denominator of p contains many women of nonchildbearing ages as well as men.

Given the binomial sample data from the communes (x successes out of y trials), the posterior distribution of p is then

$$p^x(1-p)^{y-x} \Big/ \int_0^1 p^x(1-p)^{y-x}\,dp. \tag{9}$$

Then he assumes that he has a second independent binomial sample, France itself, which gave X successes (births) out of an unknown number of trials Y (the population of France). He writes $Y = \hat{Y}_R + z$ and concentrates on the distribution of z, the error in the ratio estimate of the population of France. The probability of X successes in $(Xy/x) + z$ trials is

$$\frac{\left(\dfrac{Xy}{x} + z\right)!}{X!\left[\dfrac{X(y-x)}{x} + z\right]!}\, p^X(1-p)^{[X(y-x)/x]+z} \tag{10}$$

In order to find the ordinate of the distribution of z in terms of the known quantities X, y, z, Laplace first multiplies (10) by (9). This is where he assumes that his two samples, France and the communes, are independent. He then integrates this product with respect to p from 0 to 1. He multiplies the resulting expression by dz and divides by its integral with respect to z from $-\infty$ to ∞ in order to obtain a distribution function $f(z)$ whose integral is unity. He proceeds to find an approximate form of this distribution when x, X, and y are all large.

Multiplication of (10) by (9) followed by integration with respect to p yields

$$P(z \mid X, y, x) \propto \frac{\left(\dfrac{Xy}{x} + z\right)!}{X!\left[\dfrac{X(y-x)}{x} + z\right]!} \frac{\int_0^1 p^{X+x}(1-p)^{[(X+x)(y-x)/x]+z}\, dp}{\int_0^1 p^x(1-p)^{y-x}\, dp} \tag{11}$$

$$= \frac{(y+1)!\,(X+x)!\left(\dfrac{Xy}{x} + z\right)!\left[\dfrac{(X+x)(y-x)}{x} + z\right]!}{X!\,x!\,(y-x)!\left[\dfrac{X(y-x)}{x} + z\right]!\left[\dfrac{(X+x)y}{x} + 1 + z\right]!}. \tag{12}$$

In finding an approximation to this multiple of the distribution of z, we may discard all factorials not involving z which will cancel when (12) is divided by its integral over z. In the four large factorials in (12) involving z, Laplace first applied Stirling's approximation to $n!$, then expanded in a Taylor series up to terms in z^2. Ignoring the terms in $\sqrt{2\pi}$ we have

$$\ln[(\phi + z)!] \cong (\phi + z + \tfrac{1}{2}) \ln \phi + (\phi + z + \tfrac{1}{2}) \ln[1 + (z/\phi)] - \phi - z. \tag{13}$$

Discarding terms in (13) not involving z and using Taylor's expansion and noting that all four ϕ terms in (12) are large, we have

$$\ln(\phi + z)! \cong z \ln \phi + \left(\phi + z + \frac{1}{2}\right)\left(\frac{z}{\phi} - \frac{z^2}{2\phi^2}\right) - z \tag{14}$$

$$\cong z \ln \phi + \frac{z^2}{2\phi} + \frac{z}{2\phi}. \tag{15}$$

In (12) there are four ϕ terms, Xy/x, etc. The leading term in the sum of the four $\ln \phi$ terms in (15) works out as $-\ln[1 + x/(X + x)y] \cong -x/(X + x)y$. The leading term in the sum of the four $(1/\phi)$ terms in (15) is

$-x^3/X(X+x)y(y-x)$. Hence, from (12) and (15), if $f(z)$ is the frequency distribution of z,

$$\ln f(z) \cong \frac{-xz}{(X+x)y} - \frac{x^3(z^2+z)}{2X(X+x)y(y-x)} \tag{16}$$

$$f(z) \cong \exp\left\{-\frac{1}{2}\frac{x^3}{X(X+x)y(y-x)}\left[z^2+z-\frac{2X(y-x)}{x^2}z\right]\right\} \tag{17}$$

(apart from the multiplier needed to make the integral unity).

Thus Laplace's analysis showed the large-sample distribution of the error z in the ratio estimate to be normal, with a bias whose leading term is $X(y-x)/x^2$ if x/X is negligible, and a variance $X(X+x)y(y-x)/x^3$.

Given Laplace's binomial assumption and his data, the standard error of z is calculated to be 107,550. This makes the odds about 300,000 to 1 against an error of more than half a million in his estimate.

It is unfortunate that Laplace should have made a mistake in probability in a book on the theory of probabilities. In his application, however, the mistake was of little consequence. His working out of the large-sample distribution of the ratio estimator and his concept of the superpopulation as a tool in studying estimates from samples are pioneering achievements.

REFERENCES

BREWER, K. W. R. (1963). Ratio estimation in finite populations: Some results deducible from the assumption of an underlying stochastic process. *Austral. J. Statist.* **5** 93–105.

GRAUNT, J. (1662). Natural and political observations made upon the bills of mortality. Reprinted by Johns Hopkins Press, Baltimore, Maryland, 1939.

GREENWOOD, M. (1941). Medical statistics from Graunt to Farr. I. *Biometrika* **32** 101–127.

LAPLACE, P. S. (1820). A philosophical essay on probabilities. English translation, Dover, N.Y., 1951.

ROYALL, R. M. (1970). On finite population sampling theory under certain linear regression models. *Biometrika* **57** 377–387.

STEPHAN, F. F. (1948). History of the uses of modern sampling procedures. *J. Amer. Statist. Assoc.* **43** 12–39.

Some Aspects of Statistics, Sampling, and Randomization

Oscar Kempthorne

IOWA STATE UNIVERSITY

In this essay the general nature of conventional mathematical statistics is discussed. It is suggested that Neyman–Pearson–Wald decision theory is ineffective and that the making of terminal decisions must involve some sort of Bayesian process. Bayesian theory is, however, deficient with respect to obtaining of a prior distribution which seems to be a data analysis problem and to be the basic inference problem in a decision context. The relevance of parametric theory to the finite population inference problem is questioned. It is considered that intrinsic aspects of the logic of inference are given by the cases of populations of size one and two, and these are discussed. The role of labeling is considered, and some work of Hartley and Rao discussed. It is suggested that there are difficulties in unequal probability sampling with respect to the ultimate inferences, that is, beyond the matters of estimation and variance of estimators. It is suggested that admissibility theory is ineffective. Absence of attention to pivotality is deplored.

1. INTRODUCTION

It is an honor to contribute to this volume. Hartley's contributions run through general mathematical statistics, experimental statistics, sampling

Key words and phrases Inference, decision theory, Bayesian theory, finite populations, admissibility, pivotality.

Journal Paper No. J-8673 of the Iowa Agriculture and Home Economics Experiment Station, Ames, Iowa, Project 890.

statistics, numerical analysis; indeed, there seems not to be a branch of statistics to which Hartley has not made significant contributions. He has the ability to react intelligently and with considerable creativity to almost any problem in pure or applied statistics. Hartley was a colleague for about a decade; one could not find a better colleague on both professional and personal levels. To contribute to the volume is also, then, a great personal pleasure.

Because of the very broad efforts of Hartley, there are many topics that could be discussed which relate to his own contributions. From the time of his joining the Iowa State University faculty in 1953, Hartley has worked extensively in the area of sampling with many original contributions. The general problem of making inferences about a finite population from observation of a subset is one that has crucial importance to the whole of humanity in its political and economic processes. It has stimulated a very large literature; on one hand, the variety of actual situations and needs encountered has led to a huge literature on suggestions of design and analysis of samples; on the other hand, the problem poses the general problem of inference, and there is now a very large literature for even the finite population case, to which Hartley has made considerable contributions.

2. THE GENERAL NATURE OF CONVENTIONAL MATHEMATICAL STATISTICS

I take the view that in this arena the data D are *assumed* to be the realization of a random variable X from some probability distribution, say F, which is a member of a class $\mathscr{F}$ of distributions, and that the great bulk of conventional mathematical statistics is concerned with functions of X and F and their probability distributions. This is, I suppose, a mere truism, but I shall use it later. It is remarkable, perhaps, that a large area of human creative activity can be characterized so briefly. One may be concerned with functions of the random variable only, or with functions of the random variable and one or more of the parameters, as in the development of statistical tests or of pivotal quantities. It may be that the distribution functions are arbitrary within a class $\mathscr{F}$, and one wants to find functions of the data alone which have a known distribution.

It is relevant to later discussion to note that "almost all" distributional problems are insoluble with a discrete sample space, notwithstanding the fact that our elementary texts are replete with finite space problems that are soluble. The pure mathematician can perhaps take the view that the finite space problems are all soluble and that they are merely a matter of computation. But this view is too facile by far. One merely has to consider the question of the exact distribution of, say, the χ^2 criterion for a binomial or a multinomial. The conventional route then is by way of asymptotic approximation. These remarks have relevance because the bulk of conventional mathematical statistics deals with continuous distributions. I have often taken the view that real data are necessarily discrete and that the consequences of this are inadequately appreciated. To counter the view that this is just a "hobby horse" to which I am addicted, I give some quotations from a paper by Wolfowitz (1969) which, interestingly enough, was a contribution to a volume honoring S. N. Roy:

> In actuality the observations which occur in any statistical experiment take values which belong to a (finite) set of integral multiples of some unit. Thus, the chance variables in the mathematical formulation of the problem should all take their values on a (finite) lattice. When the unit of measurement is small the problem can often be simplified and a good approximation obtained by replacing the distributions of the "actual" chance variables by continuous approximations.
>
> In some problems involving chance variables with continuous distributions there arise measure theoretical difficulties not present when the chance variables are discrete. . . . For example, many of the difficult and delicate problems in the study of sufficient statistics arise *solely* for this reason. The trouble is that most, if not all, of these problems are of no particular interest to the measure theorist and really have no enduring place in mathematics except possibly in mathematical statistics. In mathematical statistics they owe their existence to the idealization described above. Is there not something basically wrong with an idealization which creates difficult problems rather than serves to avoid them? And is there not something basically wrong with a subject if difficult problems arise from a supposedly simplifying idealization?

Similar views on the use of continuous approximating distributions have been given by D. Basu. The previous quote, however, represents only part of the whole picture, for the reason that most problems with finite sample spaces are impossible to solve, and one has to be content with an approximation. However, the approximation may lead to a result that is clearly absurd with possible observations. The real question is whether results from continuous approximation, coupled with the assumption that observations are possible of indefinitely fine precision, are approximately valid with observations one really can obtain. Obviously, for instance, moments and cumulative distribution functions of grouped continuous random variables can be so approximated, while asymptotic distributions of sample spacings break down if there is a limit to the fineness of observation.

It is also relevant to note that even if argumentation is confined to finite sample spaces with infinite parameter spaces, there are very difficult problems. For amusement, I give one which is trivial to state: let X be a binomial random variable with parameter θ, $0 < \theta < 1$. How is one to estimate θ given, say, 0, successes in n trials? The point of this "little" problem, is that a vast amount of literature discusses estimation of a parameter θ which is specified to lie in a space Θ by means of a mapping from the data space to a space which includes points not in Θ. The "corny" joke about this that was being circulated some decades ago is that an estimate of a survival probability equal to a negative number represents "the fate worse than death." I think the view must be taken that this "little" problem is not well-formed.

Later in the same paper, Wolfowitz (1969) gives an extensive criticism of the now conventional theory of testing statistical hypotheses. I must here confine myself to two quotations:

> Yet, first rate statisticians, . . . , write papers about admissibility, stringency, etc., of these tests. . . . (These papers) do not contribute to the essential development of statistics or to the solution of realistic problems, nor are they of permanent mathematical interest per se.

> [The book by Lehmann (1959)] is a distinguished book . . . But I would consider it a disaster for statistics if this book should determine the direction of research for any appreciable period of time.

I think that these remarks may have some implications with respect to much of the literature of the past two decades on the theory of sampling. The ideas of the Neyman–Pearson–Wald theory of statistical inference as a decision process, with the natural ideas for a decision process of admissibility, for instance, have dominated highly theoretical work in sampling, but with negligible impact, it would seem, on the practice of sampling. It would seem to be an unavoidable corollary of this that the direction of effort is ill-chosen. It is natural to expect that a theory of the baking of cakes would have a little impact on actual processes of baking cakes. I have the view that part of the literature of the theory of sampling is cluttered with an abstract (but very elegant) formulation in terms of σ-algebras, measurability considerations (even the standard nonmeasurable set appears!) that do not add at all to the real problem of drawing inferences (whatever that may be) about a finite population from a finite subset. The dilemma is that such theory has a seductive beauty.

3. WHAT IS INFERENCE?

Throughout all of statistical theory and applications is this basic question. The situation is very simple and has been known now for centuries. In conventional presentations we have by some fantastic intellectual feat, it is posited, the knowledge that our dare are realizations of a random variable (a frequency random variable or a belief random variable, it does not matter which) with Probability (Data | θ) known for all possible data. Our task then is to obtain what we may term: Inference (θ | Actual Data). I have hazarded the view that this is not a well-formed problem of the real world. We never know $P(\text{Data} \mid \theta)$ except in one case, that of finite frequency–probability sampling from a finite population, and that is why I find this problem fascinating. Whether we know this function, on a product space $D \times \Theta$, to be sure, even in this case depends on whether we are entitled to take for granted that we know that our randomizer is in fact a randomizer. In the following, I am prepared to make this assumption.

The difficulty of formulating inference as a well-formed problem led to the replacement of the "inference problem" by the decision problem in the hands of the Neyman–Pearson–Wald school. The failure, and I consider it to be an outright failure, of this school to give a solution to the problem of actually making a decision, led to the Bayesian theory of decision, the summit of which is, I judge, the remarkable effort of Savage (1954). The "problem of inference" then became the problem of personalistic decision. I find it interesting that I have been making terminal decisions for all of my professional life, and before, of course, without really giving serious thought to how I did so. I had regarded, indeed, the making of terminal decisions as something which lay outside the discipline of statistics. Curiously enough, I could develop a rationale, even now, for this view, but I could also develop what I regard as a powerful argument against that view, and I now give the preponderance of weight to the contrary view.

The matter is worth discussing at a little length, I feel. When faced with a terminal decision, as, for instance, buying an auto battery, or buying a common stock, I find, as I am sure we all do, that I have to use something vaguely like a prior distribution. I do not go through a formal Bayes calculation but I am quite sure that I follow some process which has Bayesian features that are intrinsic, and I am sure everyone does. I see no possibility of making any terminal decision without using some sort of Bayesian process. The critical question is then, of course, how "some sort of Bayesian process" should be formulated, and Savage made a fine effort to do this. I do not wish to support the view, which has been taken "lock, stock and barrel" by some, that Savage's formulation is perfect, but on the contrary, suggest considerable objections to it can be well-formulated. I take the view that the formal Bayesian computation is a mathematically trivial process which, indeed, we teach in a first course on probability. The real problem, which has not been faced adequately, is how one is to get a prior distribution for a decision analysis. This problem of "decision making" I do regard as a problem of statistics and one which has not been discussed with other than superficiality. The Savage prescription is that one "introspects," and this appears to be the prescription of current Bayesianism. I believe the view must be taken that this is no prescription at all. One has data and the basic problem for the Bayesian decision maker is to "analyze" the data and to reach a prior distribution by some process which can be sustained. Introspection has led to segments of humanity having very strong beliefs which other equally "rational" segments reject as completely irrational.

Just as Neyman–Pearson theory of testing of hypotheses was and still is presented as a solution of the "inference problem," without any discussion at all of its cornerstones, so the Bayesian approach is now being presented in a high proportion of teaching texts without any discussion of its cornerstones. It is inevitable, I think, that it will be rejected. at least in its common form. In the very fine exposition of I. R. Savage (1968) it seems that the matter is one of beliefs, rather than coherent preferences. We are supposed to have a belief probability for every element of the product space $\mathscr{D} \times \Theta$, that satisfies the usual probability axioms. The calculus of belief probabilities is trivial mathematics for finite spaces but does present mathematically interesting problems otherwise. I suggest that the remarks I quoted from Wolfowitz (1969) have deep import and that none of the mathematics has bearing on the real problem of inference, being merely an avenue for curious mathematical problems. But how does one get the basic belief probability space? If one follows the belief route, it would seem that this is a real statistical problem.

Another point is that while each exposition is a coherent string of well-formed formulae, to apply this string to a real-world population, as, in general, to apply any purely formal theory to reality, one must establish epistemic correlations [see Northrop (1948)]. One has a mathematical argument using, say, P, V, and T; but for this to be useful in interpreting the real world, one must have a means of convincing oneself that the reading of a manometer corresponds to the mathematical variable, P, in our theory, and so forth. Such epistemic correlations cannot be made with perfect assurance. Even in our best modeling of a portion of the real world, physics, the establishing of perfect epistemic correlations cannot be achieved. There is the feature that the actual sequence is something like: observation → intuitive epistemic correlation → theory → prediction by theory → validation using the epistemic correlation in the reverse direction. The reason the conventional sampler uses randomization is to establish what the sampler considers to be an irrefutable epistemic correlation. The purpose of forcing the individual to choose his personal probability by consideration of preferences among wagers is, of course, to try to establish beyond question epistemic correlation between the "belief probabilities" of the mathematical sentences of the theory, and some measurable property of the real world.

It almost seems that if one is to validate the theory of compounding beliefs, one has to see if the beliefs given by the theory are in agreement with the beliefs without the theory. But this then makes the whole exercise rather silly. There are shades in this discussion of Hartley's interesting essay (1963). I hazard the view that the epistemological underpinnings of the Bayesian process need a lot of consideration, and I say this in spite of agreeing or even asserting that terminal decision making that I do has Bayesian features.

We now have, I am told, the phenomenon of statisticians who have accepted the Bayesian process going to distant countries that are in deep need of reliable state statistics and running the gamut of Bayesian Design (whatever that is) and analysis. The summit of such activity in the political arena is, perhaps, the choice of one small community to forecast a presidential election. Why not indeed? If one believes that Cherokee, Iowa, say, will give a proportion favoring candidate A which deviates from the nation by less than 1%, one would be foolish to look anywhere else. The problem with the Bayesians is that "they believe their beliefs." That this statement is a linguistic monstrosity is beside the point, I believe. Either one understands it or one does not, and no amount of rhetoric or carefully formulated verbiage can really clarify the essential content—which, I surmise, simply cannot be verbalized. The answer is, *of course*, to incorporate your disbelief of your beliefs to reach a final belief, which you are then to accept unequivocally and totally. This is reminiscent of the antique infinite regress problem, which is avoidable only by postulating a first cause, i.e., an unquestionable belief in the present context. I believe the application of the full Bayesian process by a Bayesian will generate most unfortunate consequences. I believe, furthermore, that the particular Bayesian must honestly declare his conclusions in the form "my belief probability is 95% that $270 < \theta < 330$," and that when he acts in full *honesty and openness*, his prescriptions will be rejected, unless he has established a record of performance.

What, then, does the conventional "frequentist" sampler do about inference? It is not at all clear, and I suggest that the widely-used texts are curiously quiet on the matter. It seems that the "problem of inference" is solved by making a choice of an estimator on an informal Bayesian basis, and then quoting the realized value of the estimator and its "standard error" computed by using some formula which has a plausible basis. If pressed, it would seem that the final output is a 95% confidence interval. In this, he seems to be mimicking the naïve theoretical frequentist who constructs a function from the data space to the space of all subsets of the parameter space such that the probability that the image subset under the mapping contains the true parameter point is, say 0.95, this number having been prechosen. This naïve frequentist view is exposited excellently by Lehmann (1959) with all the panoply of measure theory to which Wolfowitz (1969), previously quoted, alludes, and with the absurdity, claimed by many to be just a minor technical irrelevancy, that to achieve the desired probabilities one fudges the data by adding, say, the realized value of an extraneous uniform (0, 1) random variable. The claim that this is an irrelevancy is, I believe mere sophistry. It seems, rather, to be intrinsic to the theory with its prechosen α which appears from nowhere. It leads to "fuzzy" confidence sets with definite probabilities, which many regard as absurd. If one follows a route of this general type, it seems rather that one should contemplate definite subsets with "fuzzy" probabilities. Or to be less naive, as Cox and Hinkley (1974) or Kempthorne and Folks (1971) suggest, one should consider mapping into a nested class of subsets, there being associated with each member of the nested class of subsets a number equal to the prior probability that the subsets given by the mapping rule covers the true parameter point. In the case of continuous underlying random variables with discrete observation, the "fuzziness" of the subsets decreases with finer accuracy of observation. The relevance of this to the finite population problem is, it seems, quite moot, because the finite population problem is "strongly" discrete.

Subject to the problems of graininess, I regard the less naive process of the previous paragraph as a partial solution to the problem of inference. I insist that it is only a partial solution, in that the problem of inference is given a particular form, or perhaps one should say that it has been changed to a different problem. The failure to exposit that a different problem is "solved" is one of the basic reasons for the attacks of the Bayesians, who would claim, it seems, that the solution of the altered problem is irrelevant to inference. The less naïve confidence-type inference does not lead, obviously, to probabilities of hypotheses, but it is considered by some to give a rational measure of the tenability of hypotheses. It may be suggested also that probabilities of hypotheses are not needed except in the terminal decision context.

4. THE FINITE POPULATION PROBLEM

The preceding views have deep relevance to the problem of inference about a finite population. I would first like to claim that the logical nature of the problem may be stripped out by considering the simplest cases. Given a mode of approach, say, the frequentist one, a large amount of theoretical work on approximation for large finite populations may be, and has been, undertaken. But I take the view that one can see the logical problems merely by considering the case of populations of size one and of size two. These cases cannot be misleading.

In the former case, we may suppose that some particular individual has been instructed to write a number on a piece of paper and place it in an envelope. My task as a statistician is to make an inference about that number, without seeing it, of course. As a person who claims to be a practising statistician, I am quite unable to do anything about the problem. Bayesians, however, have no difficulty, it seems. Anyone of them will, I judge, engage in an introspective process and obtain a prior distribution on what the number is. The absurdity of the Bayesian position is conveyed by the claim that the Bayesian can make an inference; indeed, he will be able to state his probability that the number lies in any (measurable, according to some σ-algebra) set of the real line. The example is interesting, of course, because the Bayesian could say: "You are to state whether the number is greater than 5 (or to make things cute, greater than e), and if you are incorrect, you will be shot at dawn tomorrow." Faced with this, I would draw on the existential literature and reply: "Life really is absurd."

Now turn to the case of a finite population consisting of two members. We have to be most careful because we encounter the matter of labels, on which there has been so much discussion. Suppose that we have the two numbers on paper inside thick envelopes and the envelopes are chosen very carefully so as not to be distinguishable to me. This would require of course, some prior empirical work on my envelope-recognizing ability. I assume for the purposes of my argument that this is possible. I call this the unlabeled case. In contrast is the case in which each envelope is given a distinctive mark, so the marks are isomorphic to the set $\{1, 2\}$.

I consider the unlabeled case first. Let us suppose that I am permitted to look at one envelope, and am then asked to make an inference about the parameter $\{\theta_1, \theta_2\}$. I deliberately use set notation here. What am I to do? A procedure is to pick one of the envelopes "at haphazard" to use a phrase of D. V. Lindley. I see that the value is 7, say. My problem is now to make an inference about the number in the envelope that I have not seen. Then, surely, I am back in the previous case. I have to make an inference about the number in the other envelope without any information. It is obvious, I think, that I have no recourse but to refuse to make an inference. This situation has a curious aspect with regard to repeated "sampling." Suppose that the piece of paper is returned to the envelope, and I have not marked it in any way. I then again take an envelope "at haphazard." Suppose I find that the number is 2003. Then from my two observations I know that the two numbers are 7 and 2003. Suppose, however, that the number I see on the second view is again 7. What am I now to do? I have to take the view that there are two possibilities with regard to the two numbers $\{7, 7\}$ or $\{7, ?\}$, but I am unable to distinguish between them, let alone make a judgment about ? if the second possibility holds.

This situation interests me because it is not a regular case of statistical estimation. The information given by two observations is *not* the sum of the information given by the separate observations. The reader may ask here what I mean by "information". I shall not endeavor to explain, except to state that the situation is entirely different in logical nature from having one and two observations on, say, a normal (0, 1) random variable. Conventional basic mathematical statistics is not helpful. Fisherian ideas of information do not, as D. Basu has often said, help us at all. The Fisherian information ideas tell us only about reduction of data and about the prior informational content of an unknown observation. They do not tell us what information is given by the actual observations, e.g., that the number in the examined envelope is 7.

I now turn to the frequentist outlook. The frequentist will insist that the sampler pick one of the envelopes at random with a proven randomizer. He will then be given the number in the chosen envelope. With this background what "inferences" can he make? If trained in conventional inference, he will consider the problem of estimation. Given the history, he will consider unbiased estimators of the population total. I believe it is clear that in the absence of labels, we can select one envelope with probability $\frac{1}{2}$, and not any other probability. It does not make sense to pick one envelope with probability $\frac{3}{4}$. Suppose then we pick an envelope with probability $\frac{1}{2}$. Then we observe a random variable X equal to θ_1 with probability $\frac{1}{2}$ and to θ_2 with probability $\frac{1}{2}$. Our estimator of the population total is some $g(X)$. For unbiasedness we must have

$$\tfrac{1}{2}g(\theta_1) + \tfrac{1}{2}g(\theta_2) = \theta_1 + \theta_2$$

and this must hold for all $(\theta_1, \theta_2) \in R^2$. Obviously, there is only one answer for $g(X)$, namely, $g(X) = 2X$. We are in the familiar situation that if we have just one unbiased estimator, then it is the best. The variance of the estimator is clearly $(\theta_1 - \theta_2)^2$, but we can have no idea what this is.

Can we make any partial inferences from our sample of 1? We can, indeed, in the framework of frequency statistics. There are three possibilities $\theta_1 < \theta_2$, $\theta_1 = \theta_2$, $\theta_1 > \theta_2$. Consider the assertion $\theta_1 + \theta_2 \geq 2x$, where x is the observed number. Then the frequency probability that this assertion is true is $\geq \frac{1}{2}$. Some, including Basu, I think, would take the view that this is useless. I do not think so.

It is here that we reach a fundamental dilemma. Suppose I have observed one number "at haphazard." I cannot make such an assertion. If, however, I have observed one number with probability $\frac{1}{2}$, I can make the assertion. This leads to a situation that seems ludicrous. Suppose individuals A and B are given the same task. Individual A chooses an envelope at haphazard and individual B by a perfect-coin-flip. They get the same envelope, and yet B can make an assertion that A cannot make. Is the situation indeed ludicrous? I believe not. B can make the assertion because he has introduced a probability and his assertion is about the particular instance of a family of possible instances and he is attributing to the particular instance the frequency associated with the family. It is interesting that the argument here is an example of the primitive fiducial argument of R. A. Fisher, which almost the whole profession rejects.

5. THE LABELED CASE

Some philosophical difficulties that I think I see in tye unlabelled case do not arise in the case that our two envelopes are given, say 1 and 2. I can now talk about observing envelope 1 with probability p and envelope 2 with probability $1 - p$. Then our observation is a random variable X, which equals θ_1 with known probability p_1 and θ_2 with known probability p_2. Now consider unbiasedness. We shall, indeed we must, use as estimator of $\theta_1 + \theta_2$, a function that depends *only* on X. So we shall consider $g_1(\theta_1)$ and $g_2(\theta_2)$ as estimators for the two cases. Then we must have

$$p_1 g_1(\theta_1) + p_2 g_2(\theta_2) = \theta_1 + \theta_2$$

and this must hold for all $(\theta_1, \theta_2) \in R^2$. It is obvious that we must have $g_1(\theta_1) = \theta_1/p_1\, g_2(\theta_2) = \theta_2/p_2$. This is, of course, the Midzuno or Horvitz–Thompson estimator, and it is the only unbiased estimator and is therefore by the familiar argument the best, whatever criterion of optimality is used. The notion of repeated sampling leads to complexities here. If we repeat the sampling, we may observe the same envelope, which obviously tells us nothing more than observing the first envelope did. Or we may observe the second envelope and then we know everything that was unknown.

In this latter case, we should obviously take $\theta_1 + \theta_2$ as our estimate of the population total, and not, for example, $(\theta_1 + \theta_2)/p_1 p_2$. It then seems that one cannot define an unbiased estimation which has this property. One can get an unbiased estimator only by being quite stupid in particular cases. The point, perhaps, is that the notion of repeated sampling with replacement cannot be attacked by use of the notions of unbiasedness and minimum variance without engaging in obvious stupidities. One might take the view that the simple example is misleading. I believe, however, that it illuminates a very deep difficulty. Consider a labeled population of size 6, $(\theta_1, \theta_2, \theta_3, \theta_4, \theta_5, \theta_6)$. Then, evidentially, the two samples $(\theta_2, \theta_3, \theta_2, \theta_4)$ and $(\theta_2, \theta_3, \theta_4)$, for instance, must tell us the same about the unknown vector $(\theta_1, \theta_2, \theta_3, \theta_4, \theta_5, \theta_6)$, and any theory which suggests that they tell us different estimates of $\sum_{i=1}^{6} \theta_i$ must be regarded with very deep suspicion. The reader may then sympathize, perhaps, with my having deep doubts about some of the conventional and oft-quoted theorems of sampling theory. These are surely correct mathematical statements, but I need more than this. I find myself tending to the view that the only logically defensible approach to sampling a population of units of unequal sizes is by stratification, with equiprobable sampling without replacement strata. I realize, of course, that this statement may be regarded as evidence that I know nothing about sampling. What do these small examples tell us about a reasonable form for theory of inference in sampling a finite population? To me the suggestion is that ideas of parametric inference of conventional mathematical statistics cannot be brought to bear. In terms of likelihood theory, to which I shall subsequently refer, the situation is not regular. There is the implication from some of the theory, that if I draw a sample of one, I should use θ_1/p_1, say, whereas if I draw a sample of two and get θ_1 twice, I should use $2\theta_1/p_1^2$, or so it seems. I reiterate my opinion that the well-formed formula of sampling, which are often rather pretty, must be examined on a premathematical level.

As further "grist" for my mill, I turn to the matter of making assertions about θ_1 and θ_2 on the basis of a single observation obtained with prechosen probabilities p and $1 - p$. Can we make any assertion of the form: $g(X) \geq \theta_1 + \theta_2$ with known probability that it is correct? I think the answer is in the negative. Suppose we require $g(X)$ to be unbiased (though why we should do so is quite unclear to me). Then we must use θ_1/p_1 or θ_2/p_2 as the estimator. If we assert $\hat{\theta} \geq \theta_1 + \theta_2$, we shall assert $\theta_1 p_2/p_1 \geq \theta_2$ with probability p_1, and $\theta p_1/p_2 \geq \theta_1$ with probability p_2. One or the other of the assertions will be true and we are guaranteed correctness with probability equal to the minimum of p_1 and p_2, which is necessarily less than or equal to $\frac{1}{2}$. There seems, in fact, to be a loss with the use of unequal probabilities from the viewpoint of frequency statistics and so-called "confidence" assertions.

The relevance and impact of the ideas which come to the front in the simple cases discussed above is most obvious. I am quite sure that the basic logic of sampling from finite populations must be independent of population size and sample size. It may be that extreme cases are misleading, but I do not think so.

6. THE MATTER OF LABELING

It is obvious that labeling enters, and it is interesting that this aspect has engendered a considerable amount of controversy and, I believe, confusion.

I believe, in the first place, that one cannot be sure, in the frequency framework, that one has drawn a random sample from a finite population unless the members of the population are uniquely labeled. This aspect arises in connection with the "new estimation theory for sample surveys" of Hartley and Rao (1968). They state that a theory of sampling which requires labeling "would exclude many important situations. For example, the important areas of acceptance sampling of finite lots of machine parts would be excluded since here the attachment of labels is often impractical". My reaction to this is given by my previous assertion. The question is not whether a label is attached to each part, but whether the process of sampling attaches implicitly or explicitly a label to each member of the population. The whole theory of finite sampling of finite populations is based on the assumption that a repeatable process of selection of units is available, this process having prechosen frequency probability properties. The assumption that a process of drawing a sample has particular probability properties in the absence of a listing, implicit or explicit, of the units of the population seems to me to have no better status than a purely Bayesian assumption. It seems to me that there are three possibilities with regard to labels:

(i) there are no labels, and we are unable to state whether a sample obtained by repeated sampling consists of distinct units or consists of repetitions of some units;
(ii) there are no labels, but we do know that our sample consists of distinct population units;
(iii) every member of the population possesses a label, which labeling we know, and we know the labels of the sampled units.

The sampling theory for these three cases must necessarily be different right from the start, and to fail to separate them seems a basic error, at least of exposition. I do not see how sampling with unequal prechosen probabilities is possible except in case (iii), but this leads to problems.

I shall discuss, in view of the previous statements, only the case in which units are labeled. I then find myself in a dilemma with regards to some of the writing, and in particular, that of Hartley and Rao (1968, p. 527):

> We consider it therefore of interest to develop an estimation theory in which estimators are allowed to depend on the labels only if these can be regarded as informative concomitant variables.

I am unable to understand this statement. What does it mean? The word "information" has been used very loosely in statistics. Suppose I am sampling townships in Iowa. Then I know the label of every township in the population. Do these labels contain information on the acreage of grass in each township? When is one variable "informative" with respect to another variable? I think this can be answered only by assuming that the couple (variable 1, variable 2) is a random variable of *known* distribution in some population that is relevant, a rather rare circumstance. What does concomitant mean? I am asked in every class I teach, and in answering I have problems, which are philosophical, of course.

I note that the Hartley–Rao paper stimulated criticism by Godambe (1970) and I hazard the opinion that underlying the argumentation is the matter of distinguishing the cases vis-à-vis labeling that I have just separated, at least partially.

The theory of equal probability finite sampling is, I believe, based on the assumption that the units are labeled, but the full labels are destroyed before analysis, so that there are no labels to give "information," whatever that is. I think this is true for the expositions of the basic theory in all sampling books I have seen. I also hazard the view that the "unified theory of estimation" is based on the assumption that labels are ignored as soon as the sample has been observed, with retention, however, of the distinctness of the units in the sample (or of stratum index).

I am uncertain about the force of the Hartley–Rao proof of UMVUness of the sample mean. The representation by means of a multinomial is interesting. The proof uses intimately the fact that the multinomial is complete, but the parameter space is taken to consist of all mathematically possible occupanicies of the multinomial classes, so that, for instance, a possible parameter point has the whole population in one cell. I had thought the proof erroneous at one time because I envisaged the parameter space to be all possible occupancies limited to one or zero members in any multinomial class. The other parameter space simply did not occur to me. I think the parameter space appropriate to a real problem should not be taken to be that of Hartley and Rao. I have doubts then that their proof that the sample mean is UMVU has force with regard to real problems. On the other hand, I incline to the view that if there has been equiprobable sampling without replacement and labels are destroyed and hence not available, then the sample mean is UMVU. This does not, however, give me any comfort, except, perhaps, in the case of a large number of strata of equal size.

I am mystified along the lines of my earlier remarks about the case of a population of size 2. It seems to me that given a sample, say (α_1, x_{α_1}), (α_2, x_{α_2}), $\ldots$, (α_n, x_{α_n}), the only estimators which can be considered are functions of the sample. I find the examples of Joshi (1969) and Royall (1968) obscure as I discussed before (Kempthorne, 1969). It would seem that we should consider *only* functions definable from the sample. So in the case of a population of size 3 and samples of size 2, we can have three samples,

$$X_1 = \{(1, \theta_1), (2, \theta_2)\}, \qquad X_2 = \{(1, \theta_1), (3, \theta_3)\},$$
$$X_3 = \{(2, \theta_2), (3, \theta_3)\},$$

arising, say, with probabilities p_{12}, p_{13}, and p_{23}, respectively. Then the only estimators of $\theta_1 + \theta_2 + \theta_3$ that *should* be considered are $g_{12}(X_1)$, $g_{13}(X_2)$, and $g_{23}(X_3)$. Here we could use, for instance,

$$g_{12}(X_1) = \frac{\theta_1 + \theta_2}{2p_{12}} \qquad g_{13}(X_2) = \frac{\theta_1 + \theta_3}{2p_{13}} \qquad g_{23}(X_3) = \frac{\theta_2 + \theta_3}{2p_{23}}.$$

with $p_{12} > 0$, $p_{13} > 0$, $p_{23} > 0$, conditions necessary for unbiasedness, or, for instance,

$$g_{12}(x_1) = \frac{\theta_1}{P_1} + \frac{\theta_2}{P_2} \qquad g_{23}(x_2) = \frac{\theta_1}{P_1} + \frac{\theta_3}{P_3} \qquad g_{23}(X_3) = \frac{\theta_2}{P_2} + \frac{\theta_3}{P_3},$$

where $P_1 = p_{12} + p_{13}$, $P_2 = p_{12} + p_{23}$, $P_3 = p_{13} + p_{23}$. The former is the Midzuno estimator, and the latter the Horvitz–Thompson estimator. Is one of these uniformly better than the other? I imagine that they are both admissible, but this does not help in a real applied problem. It is correct, I believe, to state that both estimators use the labels. Do these estimators "functionally depend" on the labels? I think so. What can we say, then, about estimators that do not "functionally depend" on the labels? Presumably in this case, we observe what is previously given, but we are to use only the unit value numbers in our estimator. Suppose we observe (5, 7), say. If we do not use an estimator that "depends functionally" on the labels, we must, it would seem, discard the name of the sample which gave it, and hence, it would seem also, the probability that this sample was drawn. This is, of course, entirely in disagreement with conventional sampling theory.

Another aspect of the Hartley–Rao paper with which I feel discomfort, as I did with the paper by C. R. Rao (1971), is that with a multinomial of unknown number of classes the problem cannot be tackled with ordinary Fisherian likelihood theory, as I myself failed once to realize (Kempthorne, 1966). Also, the fact that the parameters in the multinomial representation are restricted to positive integers seems to destroy the possibility of differentiating the likelihood, but perhaps this is a minor difficulty. I do not see, then, any possibility of applying usefully classical likelihood theory, but the attempt of Hartley and Rao (1971) is interesting. If one does not use their reduction to a multinomial case, the likelihood function does not involve, except trivially, the unobserved unit values. I think that trying to embed the finite population problem in the conventional parametric model situation, and then calling on the notions of the available theory for that case cannot be sustained.

7. ADMISSIBILITY

I would be remiss if I did not express my views about this. The problem of whether $g(X)$ is admissible with a particular loss function for a parameter θ which indexes the family of distributions from which X arises is obviously a well-formed problem which has engendered much mathematical research. In the finite population case, the behavior of a function of the sample statistic, no matter how it is defined, except that it be defined by the sample, depends on the population parameter, the whole set of unit values. This is, I think, quite different from a parametric case in which every observation gives some "information" on the finite set of parameters, in the sense that the probability of any observation is a function of the parameters. I am of the opinion, then, that essentially all the work on admissibility is of no relevance to the real problem, except for the triviality that one gains nothing by using a unit value more than once if one knows that it occurs more than once. In order not to give a false impression I should also state that I do not find admissibility ideas of conventional mathematical statistics as having bearing on applied statistics, except to remove from consideration "stupid" procedures.

8. PIVOTALITY

It is clear, I believe, that the standard practice of interpretation of sampling data is that the quantity $(\hat{\theta} - \theta)/S\hat{E}(\hat{\theta})$ is distributed in a known way independent of the population structure. I remarked on my disbelief of this, *as a general proposition*, in 1969 (Kempthorne, 1969). But it is rare for this to be discussed. Estimates and standard errors are, it seems, sufficient. I was interested to note that Kerridge (1969) described the problem in elementary and vivid terms. Suppose, he says, we have a finite population of 1000 mice weighing 1 ounce each and one elephant weighing 100 tons. If we take a sample of 10, then with 99% probability we get an average weight of 1 ounce and 1% probability an average weight of 1 ton. He says, in a masterful understatement: "Since the true answer is a hundredth of a ton, neither answer is very helpful." Furthermore, the function which is hoped to have approximately the pivotality of t of the normal distribution does not have properties even approximately those needed. It seems that conventional sampling theory assumes approximate pivotality, and the absence of discussion of this is a grave defect. In the case of equal probability sampling, with "reasonable" populations, and with "largish" samples, I think this may hold. But we should note that this is an informal Bayesian idea—not a formal Bayesian one, of course, for which pivotality is entirely irrelevant.

9. PRIORS

What is one to do about the Bayesian approach which seems to be invading the literature of sampling? I do not know. I note that Hartley and Rao (1968) consider also a formal Bayesian approach, so perhaps my doubts about the whole process need to be reconsidered. But I see statements of the type "if a reasonable prior exists," then do something or other. I can only ask what it means for a prior to be "reasonable". If one is a Bayesian, one's prior is one's prior and cannot but be "reasonable". I note that Godambe (1969) said: "However, in some situations he can be fairly *certain* that his prior distribution belongs to a certain class C of prior distributions". I will only make the comment that I think this statement cannot withstand philosophical analysis. I envisage students of statistics trying to understand it and failing hopelessly, as I do. How about exchangeable priors? Can this notion be sustained in the finite sampling context? I surmise that it can be sustained, in the belief arena, only if labels of units are destroyed. I think that the problem is that labels are always informative in a Bayesian context, or in almost all circumstances one can easily get additional data on the units that are "informative", in Bayesian terms. The problem is, however, how to use all this "information" and in this regard every Bayesian is "his own man". The conclusions are bound to be personalistic, and one can get agreement between statisticians only if they can agree on a prior. To achieve this is a data analysis problem.

10. CONCLUSION

For decades the finite population inference problem has bothered me. It seems to me to be intrinsically different from the finite model randomization inference problem for the comparative experiment. I have read a number of the "standard" texts and have some disappointment. The field has produced many interesting and clever ideas on methods of sampling, choice of estimators and estimators of variance of estimators. But I am of the opinion that fundamental philosophical questions of inference have not been addressed at all. The criticisms of conventional sampling by the Bayesians are to some extent justified, although it would appear that some of these have no suggestions for even the simplest sampling problem because it is too difficult. The reader will have noted that my essay is littered with "I think," "I believe," "I surmise," and so forth. I am merely a student who is trying to understand the philosophy of finite sampling in the context of frequentist statistics. I have very deep sympathy with students, who should, I think, be asking some of the questions I pose. I would have liked to attempt a rational discussion of the superpopulation ideas and the notions of analytical surveys, on both of which deep questioning is needed, I think. I fear that the theory of sampling *inference* is very obscure once one goes past nearly unbiased estimators of means and unbiased estimators of their variances. I incline to the view, perhaps quite erroneously, that sampling with varying probabilities is fraught with deep difficulties at the inference level. I wonder if the whole area of sampling can be saved only by an appeal of the type: the population I am looking at is like such and such a population for which the pivotality I need holds. If this is the basis for inference, I would like the assertion to be stated and sustained, and not merely deemed obvious and not necessary of statement. But perhaps to do so, would open up a Pandora's box! Perhaps also, the formal finite population problem is insoluble without such an input. The finite population problem is, I am sure, an extremely difficult one, much harder than the famous four-color problem, for example. H. O. Hartley has obviously contributed much, and my debt to him in preparing this essay is obvious. He has put forward ideas and suggestions which I consider to merit very deep consideration. This is far more than most of us can hope to achieve. I wish again to state my great respect for all of his life work. I end this essay with my belief that he will continue to make highly original and stimulating contributions to a problem area for which, perhaps, there are no final solutions.

REFERENCES

Cox, D. R. and Hinkley, D. V. (1974). *Theoretical Statistics*, Chapman and Hall, London.

Godambe, V. P. (1969). Discussion on Professor Ericson's paper. *J. Roy. Statist. Soc. Ser. B* **31** 239.

Godambe, V. P. (1970). Foundations of survey-sampling. *Amer. Statist.* **24** (1) 33–38.

Hartley, H. O. (1963). In Dr. Bayes' consulting room. *Amer. Statist.* **17** (1) 22–24.

Hartley, H. O. and Rao, J. N. K. (1968). A new estimation theory for sample surveys. *Biometrika* **55** 547–557.

Hartley, H. O. and Rao, J. N. K. (1971). Foundations of survey sampling (A Don Quixote tragedy). *Amer. Statist.* **25** (1) 21–27.

Joshi, V. M. (1969). Admissibility of estimates of the mean of a finite population. In *New Developments in Survey Sampling.* (N. L. Johnson and H. Smith, Jr., eds.), pp. 188–212, Wiley, New York.

Kempthorne, O. (1966). Some aspects of experimental inference. *J. Amer. Statist. Assoc.* **31** 11–34.

Kempthorne, O. (1969). Some remarks on statistical inference in finite sampling. In *New Developments in Survey Sampling.* (N. L. Johnson and H. Smith, Jr., eds.), pp. 671–692, Wiley, New York.

Kempthorne, O. and Folks, J. L. (1971). *Probability, Statistics, and Data Analysis.* Iowa State Univ. Press, Ames.

Kerridge, D. F. (1969). Discussion on Professor Ericson's paper. *J. Roy. Statist. Soc. Ser B* **31** 239.

Lehmann, E. L. (1959). *Testing Statistical Hypotheses.* Wiley, New York.

Northrop, F. S. C. (1948). *The Logic of the Sciences and Humanities.* Macmillan, New York.

Rao, C. R. (1971). Some aspects of statistical inference in problems of sampling from finite populations. In *Foundations of Statistical Inference.* (V. P. Godambe and D. A. Sprott, eds.), Holt, New York.

Royall, R. M. (1968). An old approach to finite population sampling theory. *J. Amer. Statist. Assoc.* **63** 1269–1279.

Savage, L. J. (1954). *The Foundations of Statistics.* Wiley, New York.

Savage, I. R. (1968). *Statistics: Uncertainty and Behavior.* Houghton, Boston, Massachusetts.

Wolfowitz, J. (1969). Reflections on the future of mathematical statistics. In *Essays in Probability and Statistics.* (R. C. Bose, I. M. Chakravarti, P. C. Mahalanobis, C. R. Rao, K. J. C. Smith, eds.), Univ. of North Carolina Press, Chapel Hill, pp. 739–750.

Predictive Estimation and Internal Congruency

M. R. Sampford

UNIVERSITY OF LIVERPOOL

An estimator, based on a sample, of the total Y of a character y over the units of a finite population can be regarded as the sum of the known total over the sampled units and a predictor of the total of y over the unsampled units. Where no underlying (e.g. regression) model is assumed, this predictor can be held to embody estimates of one or more totals, over the whole population, of characters other than y. It seems logically desirable that these estimators should be structurally identical with the estimator of Y: if this is so, that estimator is said to be *internally congruent*. The Horvitz–Thompson estimator is seen to be the unique internally congruent estimator in a class of linear estimators. An extension to permit the predictor to have the nature of a ratio estimator leads to a new estimator, which appears to perform very favourably in comparison with two well-known estimators.

1. INTRODUCTION

Smith (1976) has recently commented on the apparently unsatisfactory nature of the Horvitz–Thompson estimator regarded as a "predictive estimator." In this paper I maintain that it is not merely satisfactory, but in some respects uniquely so. Consideration of model-free predictive estimation suggests a simple but appealing property, *internal congruency*, the use of which generates a new estimator for with-replacement sampling.

Key words and phrases Finite population, Horvitz–Thompson estimator, internal congruency, predictive estimation, sampling WPPS.

2. PREDICTIVE ESTIMATORS

We consider a population of N units, arbitrarily labelled 1, 2, ..., N. A particular character takes the value y_i on unit i. We wish to estimate the population total of the y_i, $Y = \sum y_i$ (where $\sum$ here and throughout denotes, except as qualified, summation over all units in the population) from observations of the y_i belonging to the units of a sample $\mathscr{S}$ (obtained by any sampling scheme whatever). Commonly, estimators have been derived to satisfy criteria such as (approximate) unbiasedness and relative smallness of variance without direct consideration of the fact that the sampling procedure divides the population into a completely known (in respect of the y_i), and a completely unknown, part. If we write, say,

$$Y = \sum_{i \subset \mathscr{S}} y_i + \sum_{i \not\subset \mathscr{S}} y_i = n\bar{y} + Y'_{\mathscr{S}} \tag{1}$$

where n, $\bar{y}$ are the actual sample size and the sample mean, then the first term of (1) is known. Any estimator $\tilde{Y}$ of Y can be written in the form

$$\tilde{Y} = n\bar{y} + \tilde{Y}'_{\mathscr{S}}; \tag{2}$$

clearly $\tilde{Y}'_{\mathscr{S}}$ must be regarded as a predictor of $Y'_{\mathscr{S}}$. An intuitive idea of the "reasonableness" of $\tilde{Y}$ may then be obtained by the plausibility or otherwise of $\tilde{Y}'_{\mathscr{S}}$ as such a predictor. Fortunately, many of the best-known estimators prove reasonable in this respect. For example, if the sample is simple random, the raised sample mean can be written as

$$\hat{Y} = N\bar{y} = n\bar{y} + (N - n)\bar{y}; \tag{3}$$

$(N - n)\bar{y}$ is an obviously intuitive predictor for the total of the unsampled residuum, and b is the "regression coefficient" estimated from the sample. On tion of each y_i in that residuum by the sample mean $\bar{y}$). Similarly, if x_i is the (known) value for unit i of an ancillary character x, the customary ratio estimator can be written as

$$\begin{aligned}\tilde{Y}_{\text{rat}} &= X \cdot \sum_{i \subset \mathscr{S}} y_i / \sum_{i \subset \mathscr{S}} x_i = \left(\sum_{i \subset \mathscr{S}} x_i + \sum_{i \not\subset \mathscr{S}} x_i \right) \sum_{i \subset \mathscr{S}} y_i / \sum_{i \subset \mathscr{S}} x_i \\ &= n\bar{y} + \sum_{i \not\subset \mathscr{S}} x_i(\bar{y}/\bar{x}). \end{aligned} \tag{4}$$

If y_i/x_i is expected to be approximately constant (which is presumably so if ratio estimation is to be used), it is reasonable to estimate y_i by $x_i\,\bar{y}/\bar{x}$, and so $Y'_{\mathscr{S}}$ by the second term of (4), provided the sample is simple random. Again, the customary regression estimator for simple random sampling takes the form

$$\tilde{Y}_{\text{regr}} = N[\bar{y} + (\bar{X} - \bar{x})b] = n\bar{y} + (N - n)[\bar{y} + (\bar{X}'_{\mathscr{S}} - \bar{x})b], \tag{5}$$

where $\bar{X}$, $\bar{X}'_{\mathscr{S}}$ are means of x over the whole population and the unsampled residuum, and b is the "regression coefficient" estimated from the sample. On the other hand, the sometimes advocated product estimator yields

$$\tilde{Y}_{\text{prod}} = N(\bar{x}\bar{y}/\bar{X}) = n\bar{y} + (N - n)(\bar{x}\bar{y}/\bar{X}'_{\mathscr{S}})A, \tag{6}$$

where

$$A = \left(\frac{N}{\bar{X}} - \frac{n}{\bar{x}}\right)\frac{\bar{X}'_{\mathscr{S}}}{(N-n)}; \tag{7}$$

for large samples A will be approximately unity, but (6) has a more intuitive form if A is replaced by 1.

However, the situation apparently becomes less satisfactory as soon as we turn to unequal probability sampling. If a sample of fixed size n has been selected in such a way that the probability of inclusion of unit i in the sample was $\pi_i = nx_i/X$, then the Horvitz–Thompson estimator becomes

$$\hat{Y}_{\text{HT}} = \sum_{i \subset \mathscr{S}} y_i/\pi_i = \sum_{i \subset \mathscr{S}} y_i + \sum_{i \subset \mathscr{S}} y_i[(1/\pi_i) - 1], \tag{8}$$

where the second term equals

$$\begin{aligned} \sum_{i \subset \mathscr{S}} y_i\left(\frac{X}{nx_i} - 1\right) &= \sum_{i \subset \mathscr{S}} y_i(X - nx_i)/nx_i \\ &= \sum_{i \not\subset \mathscr{S}} x_i\left[\frac{1}{n}\sum_{i \subset \mathscr{S}}\left(\frac{y_i}{x_i}\right)\left(\frac{X - nx_i}{X - n\bar{x}}\right)\right] \qquad (9) \\ &= \sum_{i \not\subset \mathscr{S}} x_i\tilde{R} \qquad (10) \end{aligned}$$

say, where $\tilde{R}$ is obviously an estimator of sorts of the population ratio Y/X, but a decidedly nonintuitive one: to predict y_i by $x_i\tilde{R}$ is by no means an obvious action, as Smith (1976) has pointed out.

3. MODEL-FREE PREDICTION

Most predictions are based either on distributional forms or on assumed models (as in the regression estimator previously considered). However, suppose we are not prepared to assume either a distribution or a (possibly arbitrary) model: what, then, can we say about a quantity such as $Y'_{\mathscr{S}}$ in (1)? Strictly speaking, the answer is nothing. What we *can* do is to use the data from the sample to *estimate* the *expectation* of $Y'_{\mathscr{S}}$. This procedure may seem a rather roundabout one, but it appears to be what we are in fact doing in formula (3): there is no real justification for predicting the individual y_i in the residuum by $\bar{y}$ other than our knowledge that the expectation of their mean is indeed $\bar{Y}$, the population mean, which is estimated by $\bar{y}$. Now if r_i denotes the customary inclusion indicator for unit i, taking values 1 or 0 according as unit i is or is not included in the sample (so that $E(r_i) = \pi_i$), then for any sample $\mathscr{S}$

$$Y = \sum y_i = \sum r_i y_i + \sum (1 - r_i) y_i = \sum_{i \subset \mathscr{S}} y_i + \sum (1 - r_i) y_i,$$

so that

$$Y'_{\mathscr{S}} = \sum (1 - r_i) y_i \tag{11}$$

and

$$E(Y'_{\mathscr{S}}) = \sum (1 - \pi_i) y_i = \sum z_i = Z,$$

where

$$z_i = (1 - \pi_i) y_i = [1 - (n x_i / X)] y_i \tag{12}$$

is a value observable on any sampled unit, just as is y_i. It is therefore logical to require that if we are to estimate Y by

$$\tilde{Y} = n\bar{y} + \tilde{Z}, \tag{13}$$

where $\tilde{Z}$ is an estimator of Z, then $\tilde{Y}$, $\tilde{Z}$ should be estimates of identical form, but using the observed y_i and z_i respectively. Now consider a general linear estimator $\sum r_i \alpha_i y_i$ (where α_i is allowed to be a function of $\mathscr{S}$). Then we require

$$\sum r_i \alpha_i y_i = n\bar{y} + \sum r_i \alpha_i z_i = \sum r_i y_i + \sum r_i \alpha_i (1 - \pi_i) y_i,$$

whence equating coefficients of y_i for all $r_i \neq 0$ we obtain

$$\alpha_i = 1 + \alpha_i (1 - \pi_i), \tag{14}$$

or

$$\alpha_i = 1/\pi_i. \tag{15}$$

Thus the α_i do not in fact depend on $\mathscr{S}$, and yield the estimator

$$\tilde{Y} = \sum r_i y_i / \pi_i = \hat{Y}_{\mathrm{HT}}, \tag{16}$$

the Horvitz–Thompson estimator.

The idea that any estimator involved in the right-hand side of a "predictive representation" of an estimator should be of identical form to the estimator itself is a simple and appealing one. I would have liked to call such estimators "internally consistent", but in view of the specialized meaning of "consistent" I propose instead to call them *internally congruent*.

We may therefore claim that the Horvitz–Thompson estimator, so far from being unreasonable, is the *unique internally congruent* (IC) *linear estimator*. In fact, the apparently implausible form of $\tilde{R}$ in (9) and (10) as an estimator of Y/X merely reflects the fact that Y/X is an inappropriate ratio for prediction in the residuum of the population if units have been selected with unequal probabilities; a more relevant ratio would be

$$\sum (1 - \pi_i)y_i / \sum (1 - \pi_i)x_i, \tag{17}$$

which (as will be seen) is quite reasonably estimated by $\tilde{R}$.

It is perhaps worth pointing out that the derivation of the Horvitz–Thompson estimator in (16) did not involve any *direct* appeal to unbiasedness. However, it is clear that any linear internally congruent estimator must be unbiased, for if the bias is $\sum \beta_i y_i$, expectation of (13) yields

$$Y + \sum \beta_i y_i = \sum \pi_i y_i + Z + \sum \beta_i z_i, \tag{18}$$

or

$$\sum \beta_i y_i = \sum (1 - \pi_i)\beta_i y_i,$$

whence the β_i must be zero. It might appear, therefore, that the uniqueness of the H–T estimator as a linear IC estimator is a simple consequence of its uniqueness as an unbiased linear estimator. However, the latter condition holds only among estimators with fixed α_i, a condition not implied by the internal congruency argument. (For example,

$$\tilde{Y}_{\text{PS}} = n\bar{y} / \binom{N-1}{n-1} P(\mathscr{S})$$

is well known to be linear unbiased, but is not internally congruent.)

4. AN INTERNALLY CONGRUENT RATIO-TYPE ESTIMATOR

More generally, if the y_i/x_i are still expected to be approximately constant, but the π_i are not necessarily proportional to the x_i, we might then consider estimating the numerator and the denominator of (17), and using this ratio, in conjunction with $X'_{\mathscr{S}} = \sum_{i \not\subset \mathscr{S}} x_i$, to obtain the predictor for the unsampled residuum. Since both numerator and denominator are population totals, an internally congruent estimator would need to satisfy

$$\sum r_i \alpha_i y_i = \sum r_i y_i + X'_{\mathscr{S}} \sum r_i(1 - \pi_i)\alpha_i y_i / \left(\sum r_i(1 - \pi_i)\alpha_i x_i\right), \tag{19}$$

where again the α_i may depend on $\mathscr{S}$. Equating coefficients of y_i, we have

$$\alpha_i = 1 + X'_{\mathscr{S}}(1 - \pi_i)\alpha_i / \sum r_i(1 - \pi_i)\alpha_i x_i \qquad i \subset \mathscr{S}. \tag{20}$$

To solve these equations, consider first the system of equations

$$\alpha_i = 1 + X'_{\mathscr{S}}(1 - \pi_i)\alpha_i / C, \tag{21}$$

where C is any constant. These have the trivial unique solution

$$\alpha_i = C/[C - X'_{\mathscr{S}}(1 - \pi_i)], \tag{22}$$

all the α_i being monotonically decreasing with C for

$$C > C_{\min} = X'_{\mathscr{S}}\left(1 - \min_{i \subset \mathscr{S}}\{\pi_i\}\right).$$

Now $\sum r_i(1 - \pi_i)\alpha_i x_i = \xi$, for example, is a linear function of the α_i with positive coefficients, and so also decreases monotonically, from infinity to $\sum r_i(1 - \pi_i)x_i$ as C increases from $C_{\min}$ to infinity. There is thus just one value C_0 of C for which $\xi = C$, and the corresponding values of α_i will be unique solutions of (20). C_0 must satisfy the equation

$$\xi = \sum r_i(1 - \pi_i)x_i C_0/[C_0 - X'_{\mathscr{S}}(1 - \pi_i)] = C_0,$$

or

$$\sum r_i(1 - \pi_i)x_i/[C_0 - X'_{\mathscr{S}}(1 - \pi_i)] = 1. \tag{23}$$

Substituting $C = C_0$ in (22), and writing $K = C_0/X'_{\mathscr{S}}$, the required values of α_i are found to be

$$\alpha_i = K/[K - (1 - \pi_i)], \tag{24}$$

where K satisfies

$$\sum r_i(1 - \pi_i)x_i/[K - (1 - \pi_i)] = X'_{\mathscr{S}} = X - \sum r_i x_i$$

or

$$\sum r_i x_i K/[K - (1 - \pi_i)] = X, \tag{25}$$

which is equivalent to the intuitively reasonable condition

$$\sum r_i x_i \alpha_i = X. \tag{26}$$

Since the left-hand side (LHS) of equation (25) decreases monotonically from infinity to $\sum r_i x_i$ ($< X$) as K increases from $C_{\min}/X'_{\mathscr{S}}$ to infinity, a numerical solution is easily obtained by, for example, the Newton–Raphson method if no explicit solution exists. Four special cases are worth considering.

(i) *Selection with equal probabilities without replacement* Here all the π_i, and hence all the α_i for a given sample, are equal, so that from (26)

$$\alpha_i = X/\sum r_i x_i,$$

and the estimator becomes

$$\tilde{Y} = \sum r_i y_i X/\sum r_i x_i = \tilde{Y}_{\text{rat}},$$

the customary ratio estimator.

(ii) *Selection with probability proportional to x without replacement*
Here $\pi_i = nx_i/X$, the unique solution of (25) is easily seen to be $K = 1$, since then

$$\text{LHS}(25) = \sum (r_i x_i/\pi_i) = (X/n) \sum r_i = X,$$

and $\alpha_i = 1/\pi_i$, leading again to the Horvitz–Thompson estimator. Hence we can generalize the result of the previous section, and say that *when* $\pi_i \propto x_i$, *the unique internally congruent estimator in the class* (19) is $\hat{Y}_{\text{HT}}$.

Comparison of this derivation with that leading to (16) indicates that the denominator in the second term on the RHS of (25) must equal $X'_{\mathscr{S}}$ when $\alpha_i = 1/\pi_i$, a result that is easily verified, while the numerator becomes

$$\sum r_i y_i[(1/\pi_i) - 1].$$

Comparison with (8) and (10) then shows that

$$\tilde{R} = \frac{\sum r_i \alpha_i (1 - \pi_i) y_i}{\sum r_i \alpha_i (1 - \pi_i) x_i},$$

which is a plausible estimator of (17), since with this model $E[\sum r_i \alpha_i (1 - \pi_i) z_i] = \sum (1 - \pi_i) z_i$ for any character z.

(iii) *Selection with probability proportional to x with replacement* This is the traditional Hansen–Hurwitz method, in which a fixed number n of drawings are made, the probability of selection of unit i at any drawing being $p_i = x_i/X$. Here $1 - \pi_i = (1 - p_i)^n$, so that equation (25) becomes

$$\sum r_i p_i [K/(K - (1 - p_i)^n)] = 1, \tag{27}$$

(where r_i is still the simple inclusion indicator, taking the value 1 when the unit i appears one or more times in the sample). There is no explicit solution, so that the internally congruent estimator is more difficult to compute than the inadmissible estimator $\sum n_i y_i/np_i$, but not necessarily more so than the well-known admissible unbiased estimator

$$\hat{Y}_{\text{AU}} = \sum E(n_i | \mathscr{S}) y_i/np_i, \tag{28}$$

where n_i is the number of occurrences of unit i in the sample. It is of interest to compare the internally congruent estimator (which is not here unbiased) with $\hat{Y}_{\text{AU}}$.

Example Suppose $N = 12$, $n = 5$, and the sample contains units 7, 8, 11 and 12, where the corresponding p_i $(= x_i/X)$ are $p_7 = 0.09$, $p_8 = 0.10$, $p_{11} = 0.13$, $p_{12} = 0.14$. Equation (27) then becomes, say,

$$\varphi(K) = K\left\{\frac{0.09}{K - 0.91^5} + \frac{0.10}{K - 0.90^5} + \frac{0.13}{K - 0.87^5} + \frac{0.14}{K - 0.86^5}\right\} - 1 = 0,$$

where K must clearly be greater than $(0.91)^5 = 0.624$. Taking $K_1 = 0.75$ as a first approximation, successive Newton–Raphson approximations as shown in Table 1 yield

$$K = 1.005812.$$

The α_i and the corresponding coefficients for $\tilde{Y}_{\text{AU}}$ are also shown in Table 1. For this particular sample the coefficients are very similar, but greater discrepancies can occur, particularly for small p_i. For example, if unit 7 in the sample is replaced by unit 1, with $p_1 = 0.02$, the coefficients for the three larger units are still very similar, but those for y_1 are 11.36 (IC), 10.51 (AU). Some sampling investigations into the properties of the two estimators are reported in the next section.

TABLE 1

Successive Approximations to K, and Weights for $\tilde{Y}_{\text{IC}}(\alpha_i)$ and $Y_{\text{AU}}(\gamma_i)$

K_j	i	α_i	γ_i
0.75	1	2.634534	2.657005
0.85	2	2.421764	2.434783
0.949	3	1.982321	1.973244
0.998	4	1.878670	1.863354
1.0057			
1.005812			

(iv) *Sample selection probability proportional to sample x total* It has been well known for many years that if

$$P(\mathscr{S}) = \sum_{i \subset \mathscr{S}} x_i \Big/ \binom{N-1}{n-1} X, \tag{29}$$

then the usual ratio estimator is unbiased (a result published independently by Sen, Lahiri, Midzumo, and Hájek). Here

$$\pi_i = p_i + (1 - p_i)(n - 1)/(N - 1) \qquad 1 - \pi_i = (1 - p_i)(N - n)/(N - 1), \tag{30}$$

(where $p_i = x_i/X$), so that clearly the α_i in an internally congruent estimator will not generally all be equal, so that the IC estimator will not equal the ratio estimator. Absorbing the factor $(N - 1)/(N - n)$ into K, we obtain

$$\alpha_i = K/[K - (1 - p_i)], \tag{31}$$

where K is the solution of

$$\sum r_i p_i K/[K - (1 - p_i)] = 1. \tag{32}$$

Again, no explicit solution exists, so that the IC estimator is more inconvenient to calculate than the ratio estimator, and not unbiased. It is interesting to compare the two estimators, however.

Example Suppose as before $N = 12$, $n = 5$, and the sample selected consists of units 1, 7, 8, 11, 12, with p_i as described previously. Then equation (32) has solution $K = 1.704669$, and the α_i from equation (31) yield the estimator $2.3523y_1 + 2.1451y_7 + 2.1185y_8 + 2.0423y_{11} + 2.0182y_{12}$,

$$\sum r_i x_i / X = \sum r_i p_i = 0.48$$

so

$$\hat{Y}_{\text{rat}} = \sum r_i y_i / \sum r_i p_i = 2.0833 \sum r_i y_i .$$

Again, the weights are not dissimilar, that for y_1 (small p_i) showing the only large discrepancy. A sampling investigation is reported in the next section.

5. SOME SAMPLING INVESTIGATIONS

Sampling investigations of the internally congruent estimators of Sections 4(iii) and (iv) have been made using three small populations, details of which are shown in Table 2. Population A ($N = 12$) is a 1 in 3 systematic sample of a population used elsewhere by me, in which the y_i are approximately proportional to the x, population B is identical with A, except that all the y_i

TABLE 2

Test Populations for Comparison of $\tilde{Y}_{\text{IC}}$ and $\tilde{Y}_{\text{AU}}$

A		B		C	
x	y	x	y	x	y
50	17	50	67	1	10
58	16	58	66	2	26
62	20	62	70	2	17
68	20	68	70	3	19
78	23	78	73	3	34
92	34	92	84	4	61
140	43	140	93	40	98
156	45	156	95	55	187
209	70	209	120	83	244
300	59	300	109	269	613
324	128	324	178		
410	72	410	122		
$Y = 547$		$Y = 1147$		$Y = 1309$	

have been increased by 50, and population C is an artificial one, constructed to include a number of very small x values, and display a measure of nonlinearity in the apparent dependence of y on x.

Table 3 shows means and variances of the internally congruent estimator (Y_{IC}) and the admissible unbiased estimator $\tilde{Y}_{AU}$ for sampling with replacement, calculated over 1000 samples for various values of n. The means for $\tilde{Y}_{IC}$ are also expressed as percentages both of the true Y, and of the corresponding means for $\tilde{Y}_{AU}$. Since the two sets of estimates are calculated from the same samples, and $\tilde{Y}_{AU}$ is an unbiased estimator, the latter percentages

TABLE 3

Calculated Means and Variances of $\tilde{Y}_{IC}$ and $\tilde{Y}_{AU}$ for 1000 Samples of Various Sizes from Populations A, B and C of Table 2

Population	n	$\tilde{Y}_{AU}$ Mean (M_{AU})	$\tilde{Y}_{AU}$ Variance (V_{AU})	$\tilde{Y}_{IC}$ Mean (M_{IC})	$\tilde{Y}_{IC}$ Variance (V_{IC})	100 M_{IC} ÷ Y	100 M_{IC} ÷ M_{AU}	100 V_{AU} ÷ V_{IC}
A	4	548.86	4995.46	546.94	4530.26	99.99	99.65	110.27
	6	544.57	2815.68	544.37	2280.89	99.52	99.96	123.45
	8	551.24	1727.55	548.62	1362.77	100.30	99.52	126.77
B	4	1126.15	6103.01	1112.95	6092.83	97.03	98.83	100.17
	6	1141.71	4020.50	1131.99	3782.51	98.69	99.15	106.29
	8	1157.60	2664.93	1140.40	2655.13	99.42	98.51	100.37
C	4	1287.81	117262	1250.41	132700	95.52	97.10	88.37
	6	1292.70	89679	1251.24	69145	95.58	96.79	129.70

might be expected to give a slightly better picture of the degree of bias of $\tilde{Y}_{IC}$. The efficiencies of $\tilde{Y}_{IC}$ relative to $\tilde{Y}_{AU}$ are also shown. Clearly the bias of $\tilde{Y}_{IC}$ is trivial for both of populations A and B, and it tends also to be more efficient than $\tilde{Y}_{AU}$. Results for population C are peculiar. The bias of $\tilde{Y}_{IC}$ is more severe than for the other populations (a consequence, it seems, of the very small x values), but the efficiency relative to $\tilde{Y}_{AU}$ is surprisingly high for $n = 6$ (as it also was for $n = 3$ and 5, not included in the table), but less than 100% for $n = 4$; this appears to be an accidental property of the population, a second run with $n = 4$ giving very similar results to the first.

Table 4 shows exact results for population A only for $\tilde{Y}_{IC}$ and $\tilde{Y}_{rat}$ for sample selection probabilities proportional to total x. It appears that for this population at least, the bias of the internally congruent estimator is trivial, and it is nontrivially more efficient than the ratio estimator.

TABLE 4

Expectations and Variances for the Internally Congruent Estimator and the Ratio Estimator, Sampling with $P(\mathscr{S})$ Proportional to $\sum_{i \in s} x_i$.[a]

	$\tilde{Y}_{\mathrm{rat}}$		$\tilde{Y}_{\mathrm{IC}}$			
n	Exp. $(E_{\mathrm{rat}} = Y)$	Variance (V_{rat})	Exp. (E_{IC})	Variance (V_{IC})	$\dfrac{100\, E_{\mathrm{IC}}}{Y}$	$\dfrac{100\, V_{\mathrm{rat}}}{V_{\mathrm{IC}}}$
4	547	8452.36	552.73	6915.32	101.05	122.23
8	547	2416.97	549.13	2240.00	100.38	107.90

[a] Population A only: based on evaluation of all possible samples.

6. CONCLUSIONS

The concept of internal congruency arose from consideration of the HT estimator as a predictive estimator [Smith (1976): contributions by me to the discussion, and author's reply]. In this context its effectiveness might appear to be due to an algebraic fluke. I find it reassuring that it also gives apparently quite acceptable estimators in other circumstances. I do not suggest that these estimators are likely to be of much practical value, which is why the sampling investigations of the previous section have been limited to showing that the biases and variances of IC estimators are of an acceptable order. Determination of variances looks dishearteningly difficult. Also, of course, the method applies only when one has no model in mind for the dependence of the y_i on the x_i; the use of a realistic and appropriate prediction function should produce better results. However, when one is not prepared to do more than guess that the y_i are "roughly proportional" to the x_i, IC estimators may have a use. In particular, that derived here for with-replacement sampling might prove at least as convenient in practice as $\tilde{Y}_{\mathrm{AU}}$, for large samples with appreciable duplication of units.

REFERENCE

SMITH, T. M. F. (1976). The foundations of survey sampling: a review. *J. Roy. Statist. Soc. Ser. A* **139**, 183–204.

Survey Statistics in Social Program Evaluation

Edward C. Bryant

WESTAT, INC.

Substantial resources are applied annually to the evaluation of government programs, primarily Federal Government programs in health, welfare, education, housing, and manpower. The survey statistician feels right at home in the massive data-collection efforts which accompany some of these evaluations. But the traditionally trained statistician feels uncomfortable in working with the conceptual design and the analysis of evaluations which, with few exceptions, do not employ principles of randomization in the assignment of effect to cause. This paper identifies some of the problems from the viewpoint of the statistician, provides an example of a data-adjustment problem, and makes a plea for greater participation by statisticians in the process of social program evaluation.

1. INTRODUCTION

A social program is an organized effort on the part of the government (we will concentrate on the Federal Government) to impact the well-being of segments of the population, usually the poor, the handicapped, the educationally disadvantaged, etc. As a general rule, such programs originate in the Congress through authorizing legislation and the provision of budgeted funds. Frequently, a small percentage of the funding is earmarked for evaluation, presumably to determine whether the program is doing its intended job

Key words and phrases Evaluation, surveys, social programs, comparison groups, control groups, adjustment of data, longitudinal comparisons.

and hence is worthy of further funding. Historically, however, one can observe that Federal programs rarely are halted as a result of an unfavorable evaluation. Perhaps the most that one can hope for in the way of bureaucratic response to evaluations is a modification of the program to increase its effectiveness.

2. THE SURVEY ROLE IN EVALUATION

Social programs deal with people, and their objectives are multidimensional. For example, a housing program may be designed to upgrade the quality of housing for a specific target group, but may also have objectives relating to racial integration, involvement of the poor in community affairs, and placing the disadvantaged near job opportunities. Thus, a program may meet its stated objectives but may still be judged unsatisfactory in terms of its societal performance, that is, in its impact on other facets of human relationships. Both program effectiveness and societal performance may be measured in a variety of ways—in fact, it is unusual for a single measure to portray adequately any facet of program impact. Generally, then, administrative records and other compilations of data over the universe of interest provide unsatisfactory data for program evaluation. Impacts on families and persons are simply too complex to capture in such aggregative statistics.

In most cases, sample surveys are the only feasible method for collecting complex data on backgrounds (demographic and socioeconomic variables) and on the multidimensional outcome variables needed for program evaluation. In particular, there is no other way to collect data on knowledge, attitudes, perceptions, and satisfactions. Thus, survey methodology is at the heart of social-program evaluation.

Experienced social scientists have developed an ability to summarize and interpret data with regard to program objectives. Some of this ability rests on application of conventional statistical methods, but much of it is based more on an understanding of the social mechanisms which are presumed to have an impact on program outcomes. Thus, the social scientist is more likely to look at survey data as a verification of social policy theories than as an independent and objective vehicle for evaluation. Given that social experimentation is not feasible, this attitude on the part of social scientists may not be unreasonable. It comes close to an exaggerated application of Bayesian inference. When it is carried to the point of ignoring data that do not support conventional wisdom, however, it is clearly being carried too far. In that event, one may as well save the cost of data collection.

The attitude of the social scientist toward the use of data in the evaluation process has an impact on the role of the survey statistician. Emphasis on the use of data to confirm theories may lead to an isolation of the survey statistician from the heart of the evaluation process. He tends to become an order taker where, schematically, an order is placed with him for a given number of interviews on specified subjects. He cannot operate effectively in this role.

Similarly, the statistician has a responsibility to the social scientist to be responsive to his needs and to have a willingness to work within the limitations imposed by availability of existing data and the feasibility of survey–data collection.

The concept of evaluation as a team effort is fundamental to its success, and the survey statistician must be involved with all facets of the evaluation project. Specifically, he must assist in the identification of program objectives, largely through the mechanism of asking questions. He must be heavily involved in the research design which relates program objectives to measurable outcomes and must impart to the evaluation team his knowledge of what is feasible and what its expected sampling error may be under alternative designs. He should also be involved with the analytic design, contributing his knowledge of sampling and nonsampling errors and his experience in the statistical adjustment of survey data. His contributions to the design of data-collection instruments, data-collection procedures, and the processing and summarization of data are so obvious that they do not need comment.

The point being made is that the proper role of the survey statistician in social program evaluation is an active rather than a passive role. In order for him to fulfill this role adequately he may have to learn something about the subject matter, and this should be planned as part of the resource allocation for the project. Anything less than active and informed participation in the team evaluation effort will tend to weaken the total evaluation and may contribute further to the generally poor record of effectiveness of social program evaluation.

3. THE EVALUATION SETTING

Topics that are a continuing source of argument in evaluation circles are: On what basis should a program be evaluated? And how does one accomplish the measurement? Seldom, if ever, is there a single criterion by which a program can be judged a success or a failure. Also, there are various levels at which the evaluation can be accomplished. At one level one can ask what was the total impact of the program. Did it contribute to a reduction in social unrest? To a lessening of racial tensions? To the social and cultural enrichment of the nation? And so on. If it did contribute, how much did it contribute, and what was its unique contribution among the host of overlapping programs that benefit similar subgroups? What was the net cost of the benefit to society? Could the same benefits have been achieved at a lower cost? These are extremely difficult questions to answer, and methodologies for answering them are just beginning to appear in the literature of "public policy evaluation."

The concept that a program should be evaluated on the basis of participant outcomes rather than on inputs is fundamental. The principal evaluation question may be stated as follows:

> To what extent are the participants better off (e.g., economically, socially, in terms of health) as a result of participation than they would have been if there had been no such program?

Without a time dimension the question may be meaningless. It seems obvious, for example, that the position of a jobless person is better economically with a public service job provided under the Comprehensive Employment and Training Act (CETA) than he or she would be without the job assuming, for the moment, that the job pays more than could be received from public assistance. But what of the participant's status one month after termination? One year after termination? Three years after termination? Do benefits, if any, persist over time? And, if so, how long? One might be surprised to find lasting benefits from public service employment, but maybe not—an opportunity to gain paid work experience may be highly valuable to the participant. Certainly one would expect training benefits received under CETA to persist, but how long do they persist?

A further complication for the evaluation is that some programs are designed for participation during an extended period of time and the impact, if any, needs to be measured as a cumulative effect over time. Title I programs in education are a case in point.

It is apparent, then, that the collection of evaluative data may involve longitudinal surveys; that is, surveys which follow the same sample of persons over an extended period of time. The survey statistician will immediately recognize the problems posed by longitudinal studies. They include cumulative nonresponse due to inability to locate respondents, possibly increased resistance to being interviewed, increased costs of interviewing a sample which becomes more widely dispersed, and the possible danger that the interviewing itself will affect the behavior and attitudes of the respondents. This is to say nothing of the problems in analysis created by missing observations in a longitudinal sequence of interviews.

By far the most difficult problem facing the designer and the analyst is the problem of attribution. That is, to what extent can observed outcomes be attributed to participation in the program? Conceptually, the problem is simple if viewed as a problem in experimental design. The biggest stumbling block is that, with few exceptions, public policy will not permit the withholding of social services on a randomized basis.

In most studies where experimental designs have been involved, the contemplated program had not been mounted at the time of the experimentation. The experiments were conducted to assist in the design of the program and to judge its merits a priori. It appears to be more nearly feasible to randomize treatments (including control status) to eligible participants under experimental programs than under programs which have become fully funded and operational.

Even in evaluations using experimental designs, the statistician must be concerned with restrictions on generality of experimental results due to the Hawthorne effect, the need to obtain voluntary cooperation, the artificiality of the treatment in a differently organized system, and the possible impact of repeated interviewing on behavior. In spite of these shortcomings, it seems evident that policy stemming from outcomes of social experimentation is more likely to be "on target" than policy based on intuition and reaction to political pressure. One can hope that there will be more such experimentation in the future.

The problem of overlapping social programs is present in nearly all social program evaluations, whether experimental or nonexperimental. A family may be eligible for housing assistance, welfare, food stamps, free medical care, and manpower training. In addition, the children may benefit from Title I of the Elementary and Secondary Education Act of 1965. How, then, can one quantify the unique contribution to the well-being of the family of any one of these programs? If it were possible to experiment with a batch of programs simultaneously, assigning and withholding combinations of program participation, it might be possible to arrive at unequivocal conclusions. This is clearly impossible under the sociopolitical structure of the United States. Perhaps the most that can be done is to record the participation of persons or families in other programs and, by statistical analysis, attempt to determine to what extent outcome measures might have been contaminated by the substitution of services provided under other programs.

To the statistically unsophisticated, a way of measuring program effect is to

(1) select a sample of participants,
(2) select a sample of nonparticipants, generally by matching on demographic and other characteristics, and
(3) measure program effect as the difference between the participant and nonparticipant groups.

In the literature one frequently finds the nonparticipant group referred to as a "control" group. It is not, of course.

The principal thing which prevents it from being a control group is that eligibility may reflect a characteristic that is unrelated to other things on which matching can be done. Motivation is one such characteristic.

Whatever motivates a person to participate in a program or the environmental factors that influence him or her to participate may be the very things that are highly correlated with outcomes. If, at the same time, these factors are not measurable (and they probably are not) no amount of adjustment by covariance or other means will make the participant and nonparticipant groups equal.

A second factor is equally obvious. Generally, for example, a member of the labor force is eligible for employment and training assistance if he or she is unemployed or underemployed. (There are definitions of these states.) In the natural course of events, and without program participation, many would obtain jobs or upgrade their current jobs so that a year later they would, on the average, have a much improved economic status. This phenomenon is referred to as "regression toward the mean" and has been well known for generations. Unfortunately, it is not always recognized and, as a result, a training program may get credit for gains that should not be attributed to it. With random assignment of eligibles (assuming this were possible) one could create a true control group and, with the reservations expressed above, determine the effect of the training program.

The search for alternatives to experimental controls has been long and unproductive. Perhaps more than anything else, it has discouraged statisticians from becoming involved in program evaluation. It also raises the question as to whether evaluations that cannot produce methodologically supportable conclusions are worthwhile.

In spite of one's reservations about the validity of nonexperimental evaluations, the evaluations will be continued. Funding and program management decisions must be made and, to the extent that the statistician, because of his training in the logic of inference, can contribute to the evaluative decisions he should do so. Coleman [1] has made this point forcefully. The statistician's responsibility obviously includes sufficient warnings against drawing inferences about cause and effect from nonexperimental data. However, he or she also has a responsibility to make a positive contribution to inference in the swampy field of nonexperimental data. How can this be done?

Evaluations frequently are a team effort, the team being comprised of subject-matter specialists and one or more statisticians. Usually the leader of the team (the evaluator) is one of the subject-matter specialists and not the statistician. The customary secondary role of the statistician may be a consequence of the negativism inherent in traditional tests of hypothesis and the possibility that the statistician may perform his cautionary function too well.

The evaluation generally proceeds in logical steps. First, one compares outcome (or status) after participation with outcome (or status) prior to participation. Hopefully, this comparison will be available for both a participant group and a comparison group,† although sometimes only participant data are available. Second, one seeks to explain part or all of the observed differences by differences between participant and comparison groups. The unexplained difference is tentatively labeled as program effect. The subject-matter specialists must then be relied upon to offer alternative explanations of observed differences and to assess the reasonableness of those not subject to statistical analysis. The statistician, of course, has the responsibility of examining the alternatives by analysis of such data as are available. His analysis may have the effect of explaining away part of the observed difference or may have the effect of enhancing the apparent differences.

In any case, the team effort by statistician and subject-matter specialist (sometimes the same person) may dispose of alternative hypotheses, either by analysis or on the basis of subjective judgments concerning their reasonableness. If so, it may be possible for the evaluation team to feel reasonably sure of a program effect which it is willing to report to decision-makers. It is no wonder that the traditional statistician feels uncomfortable in this kind of activity, but program decisions must be made and the statistician has an obligation to contribute to those decisions if he has the capability to do so.

The relationship between smoking and lung cancer provides an excellent example of basing a major decision on statistical observations of a nonexperimental nature. Extensive investigation of alternative hypotheses, both on statistical and biomedical grounds, led the Surgeon General's office to believe that the alternative hypotheses provided unreasonable explanations of the association between smoking and lung cancer and that the public interest would be served by taking action as though the cause-effect relationship had been proved. It is easy to see how the same kind of reasoning can be transferred to social program evaluation.

It is obvious that there are massive problems for the statistician in social program evaluation. It is equally apparent that there are not always satisfactory solutions. For instance, in the smoking and lung cancer example, many prestigious statisticians felt that the cause–effect case had not been made and one can expect conclusions from social program evaluations to be even more controversial.

The remainder of this paper concerns itself with a specific problem in the statistical adjustment of nonexperimental evaluation data. The example is given to illustrate the kinds of problems which face the survey statistician in program evaluation.

† In much of the literature, such groups are referred to as "control" groups. "Comparison group," being a less ambitious term, is preferred here.

4. THE USE OF COMPARISON GROUPS

Among the most common techniques proposed to serve as a substitute for controlled experimentation is the selection of a comparison group comprised of persons who are not participants in the social program but who have characteristics similar to those of participants. Their gain in whatever measure of progress is chosen serves as the benchmark against which the progress of the participants is to be compared. Since both groups will have been subjected to similar aging, economic conditions, and environmental factors during the exposure period, it is argued that any greater increase in the well-being of the participants than of the comparison group is therefore due to participation in the program.

The most obvious error in the logic of attribution through use of such comparison groups is that the participants may differ in fundamental ways from the comparison group, and many of those ways may not be measurable. Attitudes and motivations may be among the most important of these. A table and a dog both may have four legs, be 16 in. tall, weigh 20 lb, be brown, and neither may have an arrest record. The point is that matching on obvious, and measurable, characteristics may not guarantee comparability at all with respect to critical characteristics.

Generally, however, it is possible to find matching characteristics which, at least on the surface, are associated with the outcome measure of interest. It is not difficult, for example, to show that educational outcomes are related to income of parents, to educational attainment of parents, to geographic region, to race, to number (and birth order) of siblings in the home, and to a variety of other characteristics [2]. One is led to suspect that there may be other, but unmeasured, characteristics of pupils which are strongly correlated with educational outcomes; the possibility that they are also correlated with the criteria used for selection of participants in educational programs is (or should be) a constant source of worry to evaluators who attempt to use comparison groups in education–program evaluations. If such a correlation exists, then of course no amount of increase in sample sizes will correct for the bias between comparison and participant groups.

The problem is only partly alleviated by "before and after" studies to compare gains of the participant group with gains of the comparison group. One needs only to replace the word "outcome" in the previous paragraph by the words "gain in outcome" to see the problem. An obvious case in which the measurement and comparison of gains aggravates the problem is if the participants in an education program were chosen in the first place on the basis of their low scores on an achievement test. Here the correlation with gains can be expected to be high because of regression toward the mean. Children with the lowest scores on a test can be expected to improve their relative position, on the average, in a retest with no additional learning.

The statistician is frequently faced with the problems of how to select comparison groups and how to adjust treatment and comparison groups on the basis of auxiliary variables in order to make them more comparable. Several alternatives may be available for selection of the comparison group and increasing its comparability.

5. MATCHING

One approach is case-by-case matching. That is, having selected a program participant, one searches the frame (if such exists) of candidates for the comparison group and finds the nearest match with respect to the set of auxiliary variables available. As Cochran [3] points out, the procedure is tedious and expensive. Perhaps a more serious criticism is that a convenient sample frame for the comparison group is generally not available. For example, sampling from a list of participants in an employment and training program will yield treatment subjects, but how does one find matches? By a general household sample in the same county? Perhaps so, but the impact on evaluation costs may be great.

Stratum (or group) matching is a way of organizing the case-by-case match. One forms a conceptual multiway classification of persons according to the levels or categories of stratification variables presumed to be important and classifies the candidates for the comparison group into the cells. When a participant is selected, the cell in which he/she would fall is identified and a random selection is made to identify the comparison-group partner. In case the participant falls in an empty cell, a match is usually made from an adjacent cell. The same problem as previously noted arises if no comparison-group sampling frame is available.

Another technique is to classify both the participant sample and the comparison-group sample into the multicell framework and to use all of the comparison-group persons. One-for-one matching is not required nor attempted.

The method of selecting the comparison group has an impact on the subsequent analysis. If the matching is imperfect with respect to the matching variables (this is almost always the case) or if additional variables not used in the matching are available, one may wish to make covariance adjustments to increase the comparability of treatment and comparison groups. Cochran [3] has pointed out that one is not likely to have much favorable impact on the precision of comparisons by such adjustments but, to the extent that they reduce bias, such adjustments should be considered.

6. CLASSIFICATION VERSUS REGRESSION

Let us consider first an instance in which both the treatment group and the comparison group can be classified through auxiliary variables into a number of mutually exclusive and comprehensive cells. One can collapse cells so that there are no empty cells in either distribution. Generally, the members of the participant group will have been drawn with different probabilities than the members of the comparison group. This design matches one being considered by Westat in the area of employment and training in which the participant group is drawn from a two-stage clustered sample of participants and the comparison group is drawn from a subset of the Census Current Population Survey. Let $w_{\mathrm{p}i}$ be the sampling weight for the ith member of the participant group, $i = 1, 2, \ldots, n_1$, and $w_{\mathrm{c}i}$ the sampling weight for the ith member of the comparison group, $i = 1, 2, \ldots, n_2$. Then, an estimate of the average gain per person (unadjusted) is

$$\bar{x}_{\mathrm{p}} - \bar{x}_{\mathrm{c}} = \sum_{i=1}^{n_1} w_{\mathrm{p}i} x_{\mathrm{p}i} - \sum_{i=1}^{n_2} w_{\mathrm{c}i} x_{\mathrm{c}i}, \tag{1}$$

where x_{ji} denotes the outcome measure. The weights may contain nonresponse and ratio adjustments and are scaled to sum to unity.

We consider, first, a simple method of adjustment for differences between participant and comparison group characteristics. We suppose that both sample groups have been classified into the cells of a multiway classification, with such collapsing of cells as is necessary to avoid empty cells. The cell means can be denoted by

$$\bar{x}_{jg} = \sum_{i=1}^{n_{jg}} w_{jgi} x_{jgi}, \tag{2}$$

where $j = \mathrm{p}, \mathrm{c}$; $g = 1, 2, \ldots, n_g$; $i = 1, 2, \ldots, n_{jg}$. It will be seen that the result is n_g pairs of means. An estimate of the gain, adjusted for the characteristics used in the multiway classification, is

$$\hat{G} = \sum_{g=1}^{n_g} z_g(\bar{x}_{pg} - \bar{x}_{cg}), \tag{3}$$

where the weights z_g are such that $z_g \geq 0$ for all g, and $\sum_g z_g = 1$, and such that the average gain is extrapolated to the universe of interest (possibly the total members of the target population). Sometimes z_g can be represented by the proportionate weights of the participant group, i.e. $z_g = \sum_{i=1}^{n_g} w_{\mathrm{p}i} / \sum_{i=1}^{n_1} w_{pi}$, on the grounds that the participant group is a probability sample of the target population.

The variance of the estimate of gain can be expressed by

$$\text{var } \hat{G} = \sum_{g=1}^{n_g} z_g^2(\text{var } \bar{x}_{pg} + \text{var } \bar{x}_{cg} - 2 \text{ cov } \bar{x}_{pg}\bar{x}_{cg}). \tag{4}$$

Some cautions are in order in the use of this method. Firstly, if the distributions of the cell weights (z_{pg} and z_{cg}) are materially different in the comparison group and in the participant group, there may be real doubt whether the comparison group matches the participant group and under those circumstances one will want to qualify the conclusions.† Secondly, if cell numbers are small and a lot of collapsing is necessary to avoid empty cells, the variances of the cell means in (4) may become large, causing the variance of estimated gains to be so large as to be useless.

Apart from this latter factor, the method has a great deal to recommend it: It is simple, and it is model-free—no assumptions are made concerning functional relationships. It lends itself easily to the examination of gains over important subsets of the target population—the young, the handicapped, females, etc. One simply adds over the cells in which one is particularly interested. Finally, assuming it was planned for in the sample designs, the variance of the gain is generally easy to compute.

Two factors may prevent the widespread use of this methodology. One is the small-sample problem which prevents one from forming as many comparison cells as desired, and the second is the desire to have a model which, in some sense, explains the relationships among the variables mathematically.

An alternative approach which is widely used is the covariance model:

$$y_j = \alpha + \beta_1 x_{1j} + f(x_{ij}) + \varepsilon_{ij}$$
$$i = 2, \ldots, v; \quad j = 1, 2, \ldots, \quad n_1, n_1 + 1, \ldots, n_1 + n_2, \tag{5}$$

where $x_{1j} = 1$ if the jth observation is a member of the participant group and zero otherwise. There are n_1 members of the participant group and n_2 members of the comparison group. The function $f(x_{ij})$ is generally a linear function of the $k - 1$ socioeconomic and demographic variables with coefficients β_i, $i = 2, \ldots, v$. When the model is fitted to the data, the value of b_1 (the least-squares estimate of β_1) is interpreted as the average gain in the outcome measure, y, "due to" participation in the program.

A variant is to omit the treatment variable, x_{1j}, and to compute the "adjusted treatment means" of both participant and comparison groups.

† This qualification is beyond the usual qualifications concerning the possibility of associations between measured outcomes and unmeasured variables in the selection process.

The adjusted means are:

$$\bar{y}'_p = \bar{y}_p - f(\bar{x}_{pi} - \bar{x}_i), \tag{6}$$

$$\bar{y}'_c = \bar{y}_c - f(\bar{x}_{ci} - \bar{x}_i), \tag{7}$$

where the subscripts p and c indicate participant and comparison groups, respectively. Adjusted gain between participant and comparison groups is then measured by

$$\hat{G} = \bar{y}'_p - \bar{y}'_c. \tag{8}$$

For our subsequent discussion, the "adjusted means" approach is more useful than the combined functional relationship shown in (5).

The common-regression assumption is critical to the validity of the covariance method, i.e., the assumption that the association between the outcome variable and the linear function of x variables is the same (with the same coefficients) for the comparison group as for the participant group. Another assumption is that the mean square residuals from regression are equal in the participant and comparison groups. There are methods for testing these assumptions which will not be given here.

What are the advantages and disadvantages of using the multiple classification (clustering) scheme (3) compared to the covariance approach (8)? Some cases can be distinguished:

Case 1 Perfect match on the x variables between participant and comparison groups If there is a one-for-one match between participant and comparison groups and if there are unit weights,† then the clustered estimate of gain is simply the difference between the unadjusted means of particpant and comparison groups. Since the second terms of (6) and (7) disappear, the covariance-adjusted estimate of gain from (8) is also the difference between the unadjusted means. It is likely that, under these circumstances, one would compute an estimate of gain by the sum (possibly weighted) of the differences between members of the exactly matched pairs, which is a limiting case of the clustering approach. The clustered approach shows clearly those portions of the x domain where gains are substantially higher or lower.

A possible argument in favor of using regression adjustment is that a regression model may preserve degrees of freedom for the estimation of variance if the number of parameters in the model is fewer than the number of clusters in the clustering approach. On intuitive grounds, however, it seems unlikely that in practice one can find models which fit so well that the added degrees of freedom for error will compensate for the increased sum of squares of deviations from regression as compared to the sum of squares of deviations from cluster means.

† The added complications due to differing sample weights are discussed subsequently.

Case 2 Inexact matches and common regression Problems arise in practice because seldom, if ever, can one achieve exact matches. One of the clear advantages of the clustering approach is that the differences in distributions of the x variables become immediately obvious and thus serve to warn the evaluator concerning the comparability of the participant and comparison groups. In the covariance approach these differences tend to become masked by the model. Good practice dictates that one fit the model to both participant and comparison groups and compare the coefficients. One might also fit a common regression equation to the combined data and tabulate the residuals from regression for both participant and comparison groups by the cells of a multiway classification. If the residuals are predominantly† of one sign in a cell (or cluster of cells) it indicates a failure of the regression model to fit the data in that subdomain of the adjustment variables. If the signs of the residuals are predominantly different for participant and comparison groups, there is further evidence that a different model may apply to the two groups. This may have serious consequences in the contribution of bias to the estimation of gains.

Consider a specific cell (or cluster) in a multidimensional classification. In this cell, the average residuals from common linear regression for the participant group may be denoted by

$$\bar{d}_p^{(k)} = \bar{y}_p^{(k)} - \sum_{i=1}^{v} \beta_i x_{ij}^{(p \in k)} - \bar{e}_p^{(k)} + \bar{B}_p^{(k)}, \tag{9}$$

and for the comparison group

$$\bar{d}_c^{(k)} = \bar{y}_c^{(k)} - \sum_{i=1}^{v} \beta_i x_{ij}^{(c \in k)} - \bar{e}_c^{(k)} + \bar{B}_c^{(k)}, \tag{10}$$

where $\bar{B}$ indicates a bias due to inappropriateness of the model, the superscript (k) denotes the kth cell (or cluster), and $p \in k$ and $c \in k$ indicate the x's which lie in the kth cell. In this model it is to be understood that some of the x's may represent transformations of adjustment variables, e.g., square roots, products, etc. It is evident that if the model fits the actual data equally well (or poorly) in the participant and comparison groups so that $\bar{B}_t = \bar{B}_c$, a comparison of the average residuals will be unaffected by the poor fit *provided* that the x variables are similarly distributed in the cell. In this case, then, adjustment by linear regression will produce results similar to the clustering approach and might be preferred if the number of observations per cell is small.

However, the x's are frequently not equally distributed over the cells for participant and comparison groups. Under these circumstances, the differen-

† Methods of determining significance of differences in signs will not be discussed here.

tial weighting of the bias terms can accumulate substantial bias in the comparison of participant and comparison groups over subsets of the x variables of interest, i.e. blacks, low-income participants, youths, etc. The point to make is that separate models may be needed to adjust outcomes data in order to make inferences about population subgroups and all of the usual steps to validate aggregate models must be employed on submodels before inferences are made from them. Unfortunately, evaluation literature is full of examples where this has not been done.

Case 3 No common regression In this case, there is a serious problem with interpretation. If the functional relationship between response variable y and background variables x is different for the participant group and the comparison group, one must either question the validity of the comparison group or conclude that participation itself was responsible for the differential effect. In the former case, one might want to abandon the comparison group, except for whatever analytic perspective it might provide. In the latter case, the difference in the model should be combined with estimates of gains due to the program. This can be done either by adjusting the means of the participant group by using the comparison-group model or by the clustering approach previously described. Freedom of the method from limiting assumptions is likely to direct one toward analysis by the clustering approach provided that there are sufficient observations. In particular, this may be important in cases of nonlinear regression.

A problem closely related to Case 3 is presented when a part of the x domain contains comparison-group observations, but no participant observations. Under these circumstances one is well advised to discard that portion of the target group from consideration and to draw conclusions only about those portions of the x domain in which there are observations in both participant and comparison groups. For example, in the plan to use the CPS as a comparison group for manpower participants mentioned earlier, high-income wage earners will be eliminated before clusters are formed. Otherwise one is trapped into estimating how well a high-income wage earner would have fared in an employment and training program for which he or she does not qualify—an activity which is both pointless and misleading.

Other cases can be envisioned, but the ones identified above are likely to be most frequently encountered.

7. VARIABLE SAMPLING WEIGHTS

The analysis of survey data is complicated by assignment of variable weights which tend to optimize the design for the estimation of population totals or ratios. A simple cross-tabulation of sampling frequencies may be totally misleading unless the sampling weights are taken into account. For this reason, it is customary to cross-tabulate the projected counts, i.e. those weighted up by the sampling weights. However, chi-square and other tests of association cannot be applied directly to such projected data.

One approach which is sometimes employed is to scale the weights so that their sum equals the number of sample observations, to retabulate the reweighted data, rounding to integers, and to apply the tests of association to the reweighted data. SPSS and certain other computing packages are well suited to this treatment.

When regression analysis is applied to weighted survey data, one must remember that the weights are to be applied to the squares and cross products of deviations from means and not to the data themselves. In using a canned program one must be sure that the program operates in this way. A secondary problem is then the subdivision of the total weighted sum of squares into components suitable for testing for common regression, etc. Here, again, an approximation is provided by scaling the weights so that they total the number of observations in the sample. The degrees of freedom appearing in the analysis-of-variance table then sum to the number of sample observations.

The validity of tests using these approximate methods appears, in general, to be unknown.

8. SUMMARY

This paper has not attempted to catalog the problems associated with estimating program effects from survey data nor has it cataloged the statistical methods which might be used in the attempt. Hopefully, it may reinforce awareness of the existence of such problems on the part of evaluation practitioners. Equally hopefully, it may stimulate statisticians to become interested in the application of statistical methods to an area of investigation in which experimental conditions do not apply or, at best, apply only approximately.

REFERENCES

[1] COLEMAN, J. S. (1975). Problems of conceptualization and measurement in studying policy impacts. In *Public Policy Evaluation*, **II.** (Dolbeare, K. M., ed.), Sage Publ., Beverly Hills.

[2] BRYANT, E. C., GLASER, E., HANSEN, M. H., and KIRSCH, A. (1974). *Associations Between Educational Outcomes and Background Variables: A review of selected literature*, National Assessment of Educational Progress.

[3] COCHRAN, W. G. (1953). Matching of analytical studies. *Amer. J. Public Health* **43** 684–691.

Variance Estimation for a Specified Multiple-Frame Survey Design

Morris H. Hansen *Benjamin J. Tepping*

WESTAT, INC.

This paper is concerned with the estimation of the variance of estimates based on a multiple-frame sample design in which alphabetic clusters of elements have been sampled from different lists, and where elements selected from different lists may correspond to the same element of the population of interest. A nesting procedure is used to assure that clusters selected from any one of the lists are also in the sample for any other list in which an equal or greater sampling rate has been specified.

1. BACKGROUND

Hartley has discussed the problem of sampling from multiple overlapping frames (Hartley, 1962, 1974), i.e. the situations in which the same elementary unit is included in more than one frame or list from which selections are to be made for the sample. We are concerned in this article with variance estimation for a multiple-frame sample design in which alphabetic clusters of elements have been sampled from different lists, and where individual elements within clusters from different lists may correspond to the same element of the population of interest.

The particular problem which gave rise to this investigation was that of obtaining estimates relating to registered nurses for individual states, groups

This report is based on a study by Westat, Inc. under contract #231-75-0815 submitted to the Division of Nursing of the Health Resources Administration, Department of Health, Education, and Welfare.

Key words and phrases Variance estimation, multiple frames, alphabetic clusters, list sampling, nesting, generalized variances, composite estimators.

of states, or the United States as a whole. The sample design and considerations that led to it resulted from work by Walter M. Perkins and one of the authors, and are not the subject of this paper, but a brief summary description is needed as background.

Each state maintains a list of nurses registered for practice in that state. That list includes nurses who are residents in other states who may or may not be registered in those other states, and excludes nurses resident in the state who are registered for practice in other states but not in this one. Thus, if one is interested in estimating the number and characteristics of nurses resident or working in a given state or set of States, it is necessary to select the sample from registration lists for all states; the same nurse may appear in more than one of these lists.

Nested samples of alphabetic segments have been adopted as sampling units for the survey as a means of controlling the probabilities of selection of duplicate registrations. Also, since estimates of change over time are an important consideration, it is desirable that the sampling unit be so defined as to be identifiable at different points of time and with high positive correlations over time. It is judged that alphabetic segments of the nurse registration list constitute such sampling units.

An alphabetic segment of a list, which is first arranged alphabetically by the name of the nurse, is defined as consisting of all names at or after a given permutation of letters up to, but not including, another permutation of letters (alphabetically later). Thus, the entire alphabet is divided into alphabetic segments with each segment defined in terms of a cluster of names that are alphabetically adjacent. For example, one segment might start with all licensed nurses having "Condon" as a last name, and proceed alphabetically up to, but not including, any nurse with a last name of "Cooper." To accommodate the various state sampling fractions, nested samples of alphabetic segments of various sizes are used, so that the alphabetic segments included in the sample from a state with a given sampling fraction will include all the alphabetic segments in the sample for a state with a smaller sampling fraction. For example, one state may have in sample the segment as just defined, while for another state with a smaller sample a subsample of that segment will be included, perhaps defined as beginning with "Conte" and going up to, but not including, "Cook." The sample for any given state will be defined as all nurses currently registered whose names fall within the specified alphabetic segments for that state.

Although nurses vary in how many different states they are licensed, the probability that any given nurse will fall in sample in a given year will be equal to the highest sampling rate among the states in which she is licensed. This result is achieved automatically by the nesting of the sample alphabetic segments.

Nesting assures that all names that are in sample in any given state will also be in sample in any other state that has a sampling rate equal to or greater than the sampling rate for the given state—each successively higher sampling rate adds more names by enlarging the alphabetic bounds of the segments. Nesting can be accomplished either:

(1) by selecting a larger number of alphabetic segments for a higher sampling rate, with the smaller number of alphabetic segments a subgroup of the larger number; or

(2) by having the same number of alphabetic segments but, for a lower sampling rate, subsampling ranges of names lying wholly within the segments used by the state sampling at a higher rate.

The latter procedure was adopted for the sample of nurses.

To the extent that a sampling procedure makes selections that are independent from state to state, certain problems are caused by multiple registrations. The outstanding advantage of nested alphabetic segments is that, in theory at least, the complications of multiple registration are avoided. However, some nurses will be registered under different names in different states, and such duplicate registrations will not be controlled by the nested segments. Such duplicates can be resolved by questionnaire information in the data-collection stage of the survey.

The sizes of the segments should, of course, be determined on the basis of statistical efficiency as well as by the use of nesting, but sizes will necessarily vary within a given list at a given time and over time, as well as between lists.

Sampling rates are necessarily taken to vary among the state lists since estimates would be desired for individual states, although not necessarily with the same precision for each state.

A state's total sample will consist of five panels, each panel being a systematic probability selection of alphabetic segments. One panel will be identified as "first-year-in-sample;" a second panel as "second-year;" a third panel as "third-year;" a fourth panel as "fourth-year;" and a fifth panel as "fifth-year." In the next survey year the first four panels will be carried over from the preceding year, and a new panel will be selected as "first-year-in-sample."

2. ESTIMATION FROM SURVEY DATA

For the purpose of deriving expressions for the approximate variances, we consider that the sample design may be described in the following way. A systematic sample of alphabetic segments is selected from the nurse registration file in each state with variable sampling fractions in the states. The segments are nested in such a way that a nurse who falls in the sample in one state will be in the sample in all states with equal or higher sampling fractions in which she is registered. The nurse is then in the sample with a probability equal to the highest sampling fraction of the states in which registered. The states are sequenced in ascending order by size (size based on registered nurses or population). A name that occurs in a selected alphabetic segment in more than one state is retained in the sample only in the earliest listed state in which it appears. Let N_h be the total registration in state h, and n_{gh} the number of names that were in sample alphabetic segments in state h but were retained only in state g. Then the weight (inflation factor) to be applied for any name retained in the sample in state h is w_h, where

$$\begin{aligned} w_1 &= N_1/n_{11} \\ w_2 &= (N_2 - w_1 n_{12})/n_{22} \\ w_3 &= (N_3 - w_1 n_{13} - w_2 n_{23})/n_{33} \\ &\vdots \\ w_h &= \left(N_h - \sum_{g<h} w_g n_{gh}\right) \Big/ n_{hh}. \end{aligned} \tag{1}$$

Note that $\sum_{h=1}^{51} (N_h - \sum_{g<h} w_h n_{gh}) = \sum n_{hh} w_h$ will be equal to the estimated total unduplicated number of registered nurses in the U.S.

3. VARIANCE ESTIMATES UNDER SOME SIMPLIFYING ASSUMPTIONS

For the present, we make the following simplifying assumptions:

(a) The sample of retained names selected in each state is a simple random sample of the nurses that would be associated with the state in a 100 percent sample.

(b) The samples of retained names in different states are independent.

(c) The weights w_h may be treated as constants, i.e., the w_h are subject to negligible variation from sample to sample.

We anticipate, based on some prior studies, that these assumptions will lead to modest understatements of the variance. In particular, modest variance increases arising from alphabetic segment sampling (affecting assumptions a and *b*) are illustrated in Bryant *et al.* (1970). Consideration will be given in Section 5 to measuring the effect of these assumptions.

Totals, Proportions, and Ratios

If x_{hi} denotes any value associated with the ith retained name in state h, an estimate of the total value for the United States (or any state or set of states or other specified area) may be written

$$\hat{x} = \sum_{h}^{M} w_h \sum_{i}^{n_h} x_{hi}, \tag{2}$$

where M is the number of states (all states, not just those for which the estimate is made) and n_h is the retained sample size in state h. Under the assumptions listed, the variance of $\hat{x}$ conditional on the observed sample sizes n_h may be estimated by

$$\hat{\sigma}_{\hat{x}}^2 = \sum_h^M w_h^2[n_h/(n_h - 1)] \sum_i^{n_h} (x_{hi} - \bar{x}_h)^2, \tag{3}$$

where

$$\bar{x}_h = (1/n_h) \sum_i^{n_h} x_{hi}. \tag{4}$$

If $\hat{y}$ is another statistic of the same form as $\hat{x}$, then the variance of $r = \hat{y}/\hat{x}$ may be estimated by

$$\hat{\sigma}_r^2 = r^2[\hat{\sigma}_{\hat{x}}^2/\hat{x}^2 + \hat{\sigma}_{\hat{y}}^2/\hat{y}^2 - 2\hat{\sigma}_{\hat{x}\hat{y}}/\hat{x}\hat{y}], \tag{5}$$

where

$$\hat{\sigma}_{\hat{x}\hat{y}} \equiv \sum_h w_h^2[n_h/(n_h - 1)] \sum_i (x_{hi} - \bar{x}_h)(y_{hi} - \bar{y}_h). \tag{6}$$

In the special case in which the variable x_{hi} takes on the value 1 when the ith unit has a specified characteristic and the value 0 otherwise, $\hat{x}$ is an estimate of the total number possessing the specified characteristic, and the estimated variance given by (3) may be written in the form

$$\hat{\sigma}_{\hat{x}}^2 = \sum_h w_h^2[n_h^2/(n_h - 1)]p_{hx}q_{hx}, \tag{7}$$

where $p_{hx} = x_h/n_h$ denotes the estimated proportion possessing the specified characteristic for the retained sample in state h, and $q_{hx} = 1 - p_{hx}$. If $\hat{y}$ is a similar estimate of the total number possessing a different characteristic, with p_{hy} and q_{hy} denoting the corresponding proportions, and if p_{hxy} denotes the estimated proportion possessing both characteristics for the retained sample in state h, then (6) may be written, with $m_h = n_h^2/(n_h - 1)$,

$$\hat{\sigma}_{\hat{x}\hat{y}} = \sum_h w_h^2 m_h(p_{hxy} - p_{hx}p_g), \tag{8}$$

so that the estimated variance of the ratio $r = \hat{y}/\hat{x}$ given by (5) becomes

$$\begin{aligned}\hat{\sigma}_r^2 &= \left(\frac{\hat{y}}{\hat{x}}\right)^2 \left[\frac{\sum w_h^2 m_h p_{hx} q_{hx}}{\hat{x}^2} + \frac{\sum w_h^2 m_h p_{hy} q_{hy}}{\hat{y}^2} - \frac{2\sum w_h^2 m_h(p_{hxy} - p_{hx}p_{hy})}{\hat{x}\hat{y}}\right] \\ &= \frac{p_y^2}{p_x^2(\sum_h w_h n_h)^2} \sum_h w_h^2 m_h \left[\frac{p_{hx}q_{hx}}{p_x^2} + \frac{p_{hy}q_{hy}}{p_y^2} - \frac{2(p_{hxq} - p_{hx}p_{hq})}{p_x p_y}\right],\end{aligned} \tag{9}$$

where p_x and p_y denote the overall estimates of the proportions, i.e.,

$$p_x = \hat{x} \Big/ \sum_h w_h n_h, \qquad p_y = \hat{y} \Big/ \sum_h w_h n_h .$$

A case of particular interest is that in which y is a subset of x, i.e., in which all units which have the characteristic measured by y also have the characteristic measured by x. In this case, r is an estimate p of the proportion of those having the characteristic x that also have the characteristic y. Since then $p_{hxy} = p_{hy}$, we have

$$\hat{\sigma}_p^2 = \frac{p^2}{p_x^2(\sum_h w_h n_h)^2} \sum_h w_h^2 m_h \left[p_{hx} q_{hx} + \frac{p_{hy} q_{hy}}{p^2} - \frac{2 p_{hy} q_{hx}}{p} \right] \tag{10}$$

Changes Over Time

Formulas (5) and (9) can, in particular, be applied to obtain the estimated variance of the estimated change over time. For example in (5), $\hat{x}$ may be an estimate of a value in a given year and $\hat{y}$ an estimate of the same value in a later year (the next year, or five years later), so that $r = \hat{y}/\hat{x}$ estimates the ratio of the values in the two years. Similarly, in (9) p_x and p_y may represent estimates of some proportion in two different years, so that (9) gives the variance of the ratio of the two proportions. The estimated variance of the difference of the two proportions can also be obtained, in the form

$$\begin{aligned} \hat{\sigma}_{p_y - p_x}^2 &= \hat{\sigma}_{p_x}^2 + \hat{\sigma}_{p_y}^2 - 2\hat{\sigma}_{p_x p_y} \\ &= \frac{1}{(\sum w_h n_h)^2} \sum_h w_h^2 m_h \left[p_{hx} q_{hx} + p_{hy} q_{hy} - 2(p_{hxy} - p_{hx} p_{hy}) \right] \end{aligned} \tag{11}$$

4. GENERALIZED ESTIMATES OF VARIANCE—TO PROVIDE ROUGH BUT SIMPLY COMPUTED APPROXIMATIONS

We give here simply computed generalized estimates of variances as a means of reducing the work of variance estimation, but still based on the assumptions a, b, and c described in Section 3. However, these generalized variances can also be adjusted approximately for the effects of removing these assumptions after evaluating their effects by procedures given in Section 5.

The expressions given in (3), (5), (7), and (9)–(11) are available for the estimation of variance once the particular statistic is specified, and should be used to estimate the variances of a number of the most important statistics.

Ordinarily useful approximations can be obtained by much simpler computations. Thus, variances of many other statistics can be estimated roughly by first estimating the variance as if the sample were a simple random sample and then adjusting that variance by the "design effect." The design effect is defined as the ratio of the true variance to the value that the variance would have if the sample were a simple random sample, and can therefore be estimated for those statistics whose true variances are directly estimated. For example, suppose that for each of a number of estimated totals the variance is estimated by equation (7). If p_x denotes the estimated proportion $\hat{x}/\sum w_h n_h$, $q_x = 1 - p_x$, and $n = \sum n_h$, then the estimated variance of the total for a simple random sample (SRS) of size n would be

$$\hat{\sigma}^2_{\hat{x}(\text{SRS})} \doteq \left(\sum_h w_h n_h\right)^2 (p_x q_x/n) \tag{12}$$

so that the design effect for that estimated total would be estimated by dividing (7) by (12), yielding

$$d(\text{estimated total}) = \frac{(\sum_h n_h) \sum_h w_h^2 m_h p_{hx} q_{hx}}{(\sum_h w_h n_h)^2 p_x q_x} \tag{13a}$$

For an estimated total for which the variance was not directly estimated by (7), the variance could be roughly estimated by the expression (12) multiplied by the design effect (13a) for a statistic which is judged to be similar in its sampling characteristics. For this purpose, the statistics whose variances are directly estimated may be grouped by the similarity of the design effect, and new statistics associated with one or another of the groups. A common grouping to make would be the estimate for a specified state, the United States, or other area.

The design effect for an estimated proportion $p = y/x$ will be different and ordinarily smaller than for an estimated total. By a similar procedure we obtain

$$d(\text{proportion}) = \frac{n^2 p}{q p_x^2 (\sum_h w_h n_h)^2} \times \sum_h w_h^2 m_h \left[p_{hx} q_{hx} + \frac{p_{hy} q_{hy}}{p^2} - \frac{2 p_{hy} q_{hx}}{p} \right]. \tag{13b}$$

As in the case of an estimated total, statistics whose variances are estimated directly by using (10) may be grouped by the similarity of the design effect and additional statistics associated with one or another of the groups. Then the variance for all statistics in the group may be estimated approximately by

$$\hat{\sigma}_p^2 = pq/n_e,$$

where n_e may be termed the "equivalent simple random sample size," defined by

$$n_e = n/d(\text{proportion})$$

5. EVALUATION OF THE ABOVE APPROXIMATIONS BASED ON MORE EXACT VARIANCE ESTIMATES

As previously noted certain simplifying assumptions were made in deriving the expressions for the sampling variances. The effect of these assumptions on the estimates of variance can be evaluated as desired for specified statistics, and approximate additional average allowances made in the design effects adjustments discussed above, as may be found desirable.

To measure the effect of assumption (a), the variance of $\hat{x}$ can be estimated in a way that replaces that assumption by the much weaker assumption

(a′) The systematic selection of segments is equivalent to defining strata of successive alphabetic segments from each of which pairs of segments were independently selected for the sample.

One simple way to estimate the variance is as follows. For state h, let

$$x_{h(1)} = w_h \sum_{(1)} x_{hi},$$

where the summation extends over the retained names in a half-sample that consists of a random one of the two alphabetic segments in each stratum, and let

$$x_{h(2)} = w_h \sum_{(2)} x_{hi},$$

where the summation extends over the retained names in the other half-sample of the stratum. Then the variance of $\hat{x}$ may be estimated by

$$\hat{\sigma}_{\hat{x}}^2 = \sum_h (x_{h(1)} - x_{h(2)})^2. \tag{14}$$

Similarly, the covariance of two estimated totals $\hat{x}$ and $\hat{y}$ can be estimated by

$$\hat{\sigma}_{\hat{x}\hat{y}} = \sum_h (x_{h(1)} - x_{h(2)})(y_{h(1)} - y_{h(2)}), \tag{15}$$

where $y_{h(1)}$ and $y_{h(2)}$ are defined in the same way as $x_{h(1)}$ and $x_{h(2)}$, respectively. This permits the computation of the variance of the ratio, using the relationship (5).

It is also possible to estimate the variance in such a way that not only is assumption (a) replaced by the weaker assumption (a′), but also assumption (c) is eliminated. This is accomplished by calculating weights $w_{h(1)}$ and $w_{h(2)}$

for the two half-samples defined above for each state h and changing the definitions of $x_{h(1)}$ and $x_{h(2)}$, so that

$$x_{h(1)} = w_{h(1)} \sum_{(1)} x_{hi} \qquad x_{h(2)} = w_{h(2)} \sum_{(2)} x_{hi}.$$

The variance of $\hat{x}$ is then estimated by

$$\hat{\sigma}_{\hat{x}}^2 = \tfrac{1}{4} \sum_h (x_{h(1)} - x_{h(2)})^2 \tag{16}$$

and the covariance by

$$\hat{\sigma}_{\hat{x}\hat{y}} = \tfrac{1}{4} \sum_h (x_{h(1)} - x_{h(2)})(y_{h(1)} - y_{h(2)}). \tag{17}$$

Further, we can also estimate the variance in such a way as to eliminate both assumptions (b) and (c), and replacing (a) by (a′). To do this, for state h let $x_{hk(1)}$ denote the sum of the retained elements in one of the two alphabetic segments selected in stratum k (note that it is the *same* or the corresponding nested alphabetic segment in every state) and let $x_{hk(2)}$ be the sum of the retained elements in the other alphabetic segment selected in stratum k. We now define a half-sample as consisting of the retained elements in a random one of the two segments in each stratum, for all states combined. Note that if there are K strata, the total number of half-samples is $H = 2^K$. From any one of these H half-samples, estimates of totals, proportions, and other ratios can be constructed by computing a set of weights for that half-sample, say, $\hat{w}_{hj}$ for state h and the jth half-sample, and then constructing the estimate as for the whole sample. If m of these half-samples are selected, and if x_j denotes the estimate based on half-sample j, the variance of $\hat{x}$ is estimated by

$$\hat{\sigma}_{\mathrm{HS},\,\hat{x}}^2 = \frac{1}{m} \sum (\hat{x}_j - \hat{x})^2. \tag{18}$$

Other statistics may be substituted for $\hat{x}_j$ and $\hat{x}$ in (18).

Of course, maximum precision in the estimate of variance is attained by taking $m = H$. However, by making use of a technique called "balanced half-sample replication," it is possible to attain the same precision with a much smaller value of m, namely, the smallest multiple of 4 which is equal to or greater than K. This has been proved for the variance of an estimated total and other linear estimates, and has been verified empirically for some complex nonlinear estimates. [See McCarthy (1966, 1969), Frankel (1971), and Kish and Frankel (1974).] If K is so large as to make this choice of m too costly, a subset of the balanced set of half-samples will provide variance estimates of acceptable precision; there is also a method of using balanced half-samples from groups of strata which assures (for linear estimates at least) nearly maximum precision attainable with the number of half-samples chosen.

In this particular survey, if 40 segments are drawn from each state nested across states, $M = 20$, which is a multiple of 4, there should be no problem in making the estimate using a full set of 20 balanced half-samples and the total sample.

It would be feasible to avoid even assumption (a′) by actually drawing the sample of segments as a stratified sample with two segments per stratum. However, with rotation of the sample over time, this is somewhat less convenient, operationally. Moreover, we believe the impact of assumption (a′) is in fact negligible.

6. COMPOSITE ESTIMATORS

The estimators previously considered are based on data from the current survey only, and take no account of the information that may be provided by previous surveys in view of the fact that a rotating sample is used so that successive surveys contain many of the same sample segments. An estimator that does take account of such information is a so-called "composite estimator," which can improve the precision of the estimates considerably if the correlations over time are sufficiently high.

We consider here a form of a composite estimator which is a weighted mean of the simple estimate of a total considered earlier in this article and another estimate which is the composite estimate for the earlier time period plus an estimate of the change from the earlier to the next period of time based on the alphabetic segments that are common to the samples for the two periods of time. Differences or ratios of composite estimates can be formed for two different statistics for the same period of time, or for the same statistics at two different periods of time. The variance of the differences between composite estimates for two periods of time depends upon the correlation between the two periods of time for identical alphabetic segments, and this correlation will be a function of correlations among panel estimates over the period of rotation of the panel.

To define the composite estimator with which we are concerned more specifically, we define:

x_t'' the composite estimate of a total for time period t

x_{t-1}'' the composite estimator of the total for the preceding time period $t - 1$

$x_{c,t}'$ the simple estimate of the total for time period t based on only those alphabetic segments that are common to the surveys for time periods t and $t - 1$

$x_{c,t-1}'$ the simple estimate of the total for time period $t - 1$ based on only those alphabetic segments that are common to the surveys for time periods t and $t - 1$

x'_t the simple estimate of the total for time period t based on the whole sample of the survey for time period t.

Thus, the composite estimate for time period t is the weighted mean of the quantities x'_t and $x''_{t-1} + (x'_{c,t} - x'_{c,t-1})$, i.e.,

$$x''_t = K(x''_{t-1} + x'_{c,t} - x'_{c,t-1}) + (1 - K)x'_t, \tag{19}$$

where K is an arbitrary constant between 0 and 1. The optimum value of K depends upon the correlation between different time periods for the same alphabetic segments. Similar composite estimates can also be developed for means, proportions, ratios, and changes over time, with possibly different values of K as the optimum choices. A compromise value of K often can be chosen that is reasonably near optimum and that gives substantial variance reductions in estimates of both level and change for any particular characteristic or for a set of characteristics. As experience with the survey accumulates, information will be available on the appropriate values of K, and more importantly, on the degree to which the use of composite estimators may increase the precision of selected survey estimates.

REFERENCES

Bryant, E. C., Caldwell, N. W., Hansen, M. H., and Palmour, V. E. (1970). *A study of the use of sampling in surveys of scientific and technical personnel.* Nat. Tech. Infor. Serv., Washington, D.C.

Frankel, M. R. (1971). *Inference from survey samples: an empirical investigation.* Inst. Soc. Res., Univ. of Michigan Press, Ann Arbor.

Hartley, H. O. (1962). Multiple frame surveys. *Proc. Social Statist. Sect.*, Amer. Statist. Assoc., 203–206.

Hartley, H. O. (1974). Multiple frame methodology and selected application. *Sankhyā Ser. C* **36** Pt. 3, 99–118.

Kish, L., and Frankel, M. R. (1974). Inference from complex samples. *J. Roy. Statist. Soc. Ser. B* **36** 1–37.

McCarthy, P. J. (1966). *Replication: an approach to the analysis of data from complex surveys.* Nat. Cent. Health Statis., Ser. 2, No. 14, Washington, D.C.

McCarthy, P. J. (1969). Pseudo-replication: half-samples. *Rev. Int. Statist. Inst.* **37** 239–264.

Sampling Designs Involving Unequal Probabilities of Selection and Robust Estimation of a Finite Population Total

J. N. K. Rao

CARLETON UNIVERSITY, OTTAWA

Methods of unequal probability sampling without replacement are broadly classified into two categories:

(I) IPPS (inclusion probability proportional to size) schemes together with the Horvitz–Thompson (H–T) estimator of the total;

(II) non-IPPS schemes using estimators other than the H–T estimator.

A review of some recent work (since 1968) on the methods in I and II is provided. Relative efficiency (for large samples) of the usual estimator in PPS sampling with replacement to the ratio estimator (in simple random sampling) is investigated from a robustness point of view. Some remarks are offered on Royall's criticism of "conventional" sampling theory for ratio estimation, in the context of superpopulation models.

1. INTRODUCTION

Professor Hartley's major contributions to survey sampling include his work on unequal probability sampling without replacement:

(a) asymptotic theory for PPS systematic sampling with initial randomization of the population units [31, 32];

This work is supported by a research grant from the National Research Council of Canada. The author is thankful to a referee for constructive suggestions.

Key words and phrases Unequal probability sampling, ratio estimation, superpopulation models, robust estimation, relative efficiency of estimators, finite population total.

(b) a "random group" method [45];
(c) theory for PPS systematic sampling in which the units are listed in a "particular" order [30].

This work has generated a spurt of activity during the past ten to fifteen years, and Brewer and Hanif [12, 13] and Vos [65] have given a thorough review of the methods prior to 1968–1969. In Section 2, we provide a brief account of some recent developments (since 1967–1968) emphasizing their relationship (where appropriate) to the previously mentioned contributions of Professor Hartley. We confine ourselves to single stage selection of a fixed number n of units from an unstratified population of N units.

Brewer [10], Royall [53, 54] among others have discussed the relevance of superpopulation models for finite population sampling. Professor Hartley's view on these models are given in [29, 30, 33, and 34]. In Section 3 we offer some remarks on Royall's criticism of "conventional" sampling theory for ratio estimation, in the context of superpopulation models. Finally, the large-sample efficiency of the usual estimator, $\hat{Y}_{PPS}$, in PPS with replacement (PPSWR) sampling relative to the ratio estimator $\hat{Y}_R$, in simple random sampling (SRS) is investigated from a robustness point of view in Section 4.

2. UNEQUAL PROBABILITY SAMPLING WITHOUT REPLACEMENT

Methods of unequal probability sampling without replacement may be broadly classified into two categories:

(I) IPPS (inclusion probability proportional to size) schemes together with the well-known Horvitz–Thompson (H–T) estimator of the total $Y = y_1 + \cdots + y_N$, where y_i is the value of a character of interest for the ith unit, $i = 1, \ldots, N$;

(II) non-IPPS schemes using estimators other than the H–T estimator.

The variance of any estimator of Y in either category becomes zero when $y_i \propto x_i$, where $i = 1, \ldots, N$, and x_i is a known "size measure" attached to the ith unit. The following considerations were also taken into account in developing these methods:

(1) The variance of the estimator should be smaller than that of $\hat{Y}_{PPS}$;
(2) the unbiased variance estimator should be nonnegative and stable;
(3) the computations should not be cumbersome.

2.1 Category I Methods

The H–T estimator of Y is given by

$$\hat{Y}_{HT} = \sum_{i \in s} y_i/\pi_i \tag{1}$$

with variance

$$V(\hat{Y}_{\mathrm{HT}}) = \sum\sum_{i<j} W_{ij}(\tilde{z}_i - \tilde{z}_j)^2, \tag{2}$$

where π_i is the probability of inclusion of the ith unit in a sample s, $\tilde{z}_i = y_i/\pi_i$, $W_{ij} = \pi_i\pi_j - \pi_{ij}$, $i < j = 1, \ldots, N$, and π_{ij} is the joint probability of selecting the units i and j in s. For any IPPS scheme $\pi_i = np_i$, where $p_i = x_i/X$, $X = x_1 + \cdots + x_N$ and $\max(np_i) < 1$. The Sen–Yates–Grundy unbiased variance estimator

$$v(\hat{Y}_{\mathrm{HT}}) = \sum\sum_{i<j\in s} \frac{W_{ij}}{\pi_{ij}}(\tilde{z}_i - \tilde{z}_j)^2 \tag{3}$$

is often preferred over the H–T variance estimator

$$\tilde{v}(\hat{Y}_{\mathrm{HT}}) = \sum_{i\in s}(1 - \pi_i)\tilde{z}_i^2 + 2\sum\sum_{i<j\in s} \frac{W_{ij}}{\pi_{ij}}\tilde{z}_i\tilde{z}_j \tag{4}$$

since the former is always nonnegative for most of the well-known IPPS methods whereas $\tilde{v}(\hat{Y}_{\mathrm{HT}})$ might often take negative values. Moreover, empirical evidence indicates that $v(\hat{Y}_{\mathrm{HT}})$ is relatively more stable in practice (Rao and Singh [47]). Vijayan [64] has obtained the necessary form of the nonnegative unbiased variance estimators (NNUVE's) for $\hat{Y}_{\mathrm{HT}}$, demonstrated that $v(\hat{Y}_{\mathrm{HT}})$ belongs to this class and that it is the only possible NNUVE for the important case of $n = 2$ (see also Lanke [39]).

PPS systematic sampling (with or without initial randomization) is easy to execute:

(i) Let $j = 1, \ldots, N$ denote the order of the units (without loss of generality) and let $S_j = \sum_{i=1}^{j} x_i$, $S_0 = 0$ denote the progressive totals of the sizes x_i in that order;

(ii) select a random number d, where $d \leq X/n$; then the sample consists of those n units whose index j satisfies $S_{j-1} \leq d + Xk/n \leq S_j$ for some integer k between 0 and $n - 1$.

This scheme satisfies the IPPS property, i.e., $\pi_i = np_i$, but the *exact* evaluation of the π_{ij} for the randomized case is cumbersome. Hidiroglou and Gray [35], however, have developed a computer algorithm for the π_{ij} using a modification of Connor's [19] exact formula. Their empirical investigation indicates that the asymptotic formula for π_{ij} to $O(N^{-4})$, assuming max π_i is $O(N^{-1})$ (see [31]), is fairly accurate for $N \geq 10$ and $n = 2$. Recently, Asok and Sukhatme [3] have shown that $V(\hat{Y}_{\mathrm{HT}})$ under randomized PPS systematic sampling is uniformly smaller than that under the IPPS scheme of Hanurav [28] when the variance is considered to $O(N)$, i.e., the π_{ij} to $O(N^{-3})$.

The new design of the Canadian Labour Force Survey (LFS) employed randomized PPS systematic sampling for selecting nonselfrepresenting units

(NSRU's); NSRU is a combination of rural areas and small urban centers (Platek and Singh [42]). Gray [26] has demonstrated the flexibility provided by this method for both rotation of NSRU's and increasing the sample size.

Sampford's [57] method for general n appears to be the most promising in I (for $n = 2$ it reduces to Rao's [43] which is equivalent to the methods of Brewer [9] and Durbin [23] in the sense of having the same π_{ij}):

(i) Make the first selection with probabilities p_i, and all subsequent ones with probabilities proportional to $p_i/(1 - np_i)$;

(ii) if all n selections lead to different units, accept the sample; otherwise, reject as soon as a unit appears twice and repeat the whole process until n different units are selected.

This method ensures that $\pi_i = np_i$, $\pi_{ij} > 0$ for all $i < j$, $v(\hat{Y}_{\text{HT}})$ is nonnegative, and the exact evaluation of the π_{ij} is not cumbersome for "smallish" n (say $n \leq 10$). Moreover, Asok and Sukhatme [1] provide simple asymptotic formulae for the π_{ij} to $O(N^{-3})$ and $O(N^{-4})$, hence $v(\hat{Y}_{\text{HT}})$ to $O(N)$ and $O(1)$.

Vijayan [63] proposed an IPPS method, but it has no computational advantage at least for "smallish" n, and $v(\hat{Y}_{\text{HT}})$ can take negative values for $n > 2$. Das and Mohanty [20] developed another IPPS method which ensures $\pi_{ij} > 0$ for all $i < j$ and facilitates the calculation of the π_{ij}, but it is not known if $v(\hat{Y}_{\text{HT}})$ is nonnegative. Dodds and Fryer [22] derived, for $n = 2$, three families of IPPS methods and showed that considerable variation in efficiency may exist within a family. They also suggested some criteria for choosing a particular member of a family; all three families include Brewer's [9] method. Brewer [11] extended his own method [9] to $n > 2$, but the computation of the π_{ij} is formidable. Fuller [24] developed an IPPS method for general n which, in addition, ensures that the probability of selecting the ith unit at each draw is equal to p_i (this property is particularly suited to rotating samples), but the computation of the π_{ij} is somewhat cumbersome. Asok and Sukhatme [2] used preliminary random stratification (similar to the initial step in the "random group" method [45]) in conjunction with Brewer's [9] procedure for $n = 2$ to generate an IPPS method for $n = 2^m$, where m is any positive integer. The exact evaluation of the π_{ij} is cumbersome, but the authors provide simple asymptotic formulae correct to $O(N^{-4})$.

Chromy [16] proposed to find "optimal" π_{ij} for $n = 2$, assuming the often-used superpopulation model

$$y_i = \beta x_i + e_i \qquad \varepsilon(e_i \mid x_i) = 0 \qquad \varepsilon(e_i^2 \mid x_i) = \delta x_i^g \qquad \delta > 0 \quad g \geq 0$$
$$i = 1, \ldots, N, \quad (5)$$

where the errors e_i are independent and ε denotes the superpopulation expectation. Since the average variance $\varepsilon V(\hat{Y}_{\mathrm{HT}})$ is independent of the π_{ij}, Chromy minimized the superpopulation variance of $V(\hat{Y}_{\mathrm{HT}})$ (which depends on the π_{ij}) subject to $\sum_{j \neq i} \pi_{ij} = \pi_i = 2p_i$ and $0 < \pi_{ij} \leq \pi_i \pi_j$. (Constraints of the form $c\pi_i \pi_j \leq \pi_{ij} \leq \pi_i \pi_j$ with $c \leq \frac{1}{2}$, which are expected to yield a more stable variance estimator, were also investigated.) The optimal π_{ij}, however, depend on g; for $g = 1$ they are identical to Brewer's [9] π_{ij} and the method reduces to Jessen's [36] when $g = 2$. A sensitivity analysis indicated that the π_{ij} with $g = 1$ control $\varepsilon V[v(\hat{Y}_{\mathrm{HT}})]$ effectively. Fuller [24] assumed $g = 2$ and normally distributed errors and minimized a linear approximation to $\varepsilon V[v(\hat{Y}_{\mathrm{HT}})]$ subject to $\sum_{j \neq i} \pi_{ij} = 2p_i$. The resulting solution is identical to Brewer's.

2.2 Category II Methods

Four methods in category II have received considerable attention:

(a) PPS without replacement (PPSWOR) sampling together with an estimator depending on the order of selection of the units (Des Raj [21]);

(b) PPSWOR sampling as in (a) but with the estimator unordered (Murthy [40]);

(c) sampling with probabilities proportional to aggregate size (PPAS) together with the ratio estimator $\hat{Y}_{\mathrm{R}} = X(y_s/x_s)$, where $y_s = \sum_{i \in s} y_i$ and $x_s = \sum_{i \in s} x_i$ (Lahiri [38]);

(d) the random group method [45]: the population is divided at random into n groups with N_i units in the ith group G_i and one unit is then selected independently from each G_i with respective selection probabilities p_j/P_i, where $P_i = \sum_{t \in G_i} p_t$.

Des Raj's unbiased estimator of Y is

$$\hat{Y}_{\mathrm{D}} = \frac{1}{n} \sum_{1}^{n} t_i = \bar{t}, \tag{6}$$

where

$$\begin{aligned} t_i &= y'_1 + \cdots + y'_{i-1} + z'_i(1 - p'_1 - \cdots - p'_{i-1}) \qquad i = 2, \ldots, n \\ &= z'_1 \qquad \text{if} \quad i = 1, \end{aligned} \tag{7}$$

$(y'_i, p'_i) = (y_j, p_j)$ if the jth population unit is selected at the ith draw and $z'_i = y'_i/p'_i$, $i = 1, \ldots, n$. This estimator leads to a simple nonnegative unbiased variance estimator for any $n \geq 2$:

$$v(\hat{Y}_{\mathrm{D}}) = [n(n-1)]^{-1} \sum (t_i - \bar{t})^2. \tag{8}$$

However, the unordered estimator in (b):

$$\hat{Y}_{\mathrm{M}} = \sum_{j \in s} \frac{p(s \mid j)}{p(s)} y_j \tag{9}$$

is more efficient than $\hat{Y}_{\mathrm{D}}$, where $p(s)$ is the probability of selecting an unordered sample s and $p(s \mid j)$ is the conditional probability of selecting s given that the jth unit is drawn first. Murthy's unbiased variance estimator $v(\hat{Y}_{\mathrm{M}})$ is also nonnegative, but the computations become cumbersome as n increases:

$$v(\hat{Y}_{\mathrm{M}}) = \sum_{i<j \in s}\sum p_i p_j \cdot \{p(s)p(s \mid i, j) - p(s \mid i)p(s \mid j)\} p^{-2}(s)(z_i - z_j)^2, \tag{10}$$

where $z_j = y_j/p_j$ and $p(s \mid i, j)$ is the conditional probability of selecting s given that units i and j are selected in the first two draws. Bayless [6] developed a computer program to calculate $p(s \mid i, j)$, $p(s \mid i)$, and $p(s)$ for PPSWOR.

The ratio estimator $\hat{Y}_{\mathrm{R}}$ is unbiased under PPAS sampling: $p(s) = x_s/(XM_1)$, $M_1 = \binom{N-1}{n-1}$. Recently, Rao and Vijayan [49] derived the necessary form of NNUVE's for $\hat{Y}_{\mathrm{R}}$ and proposed two new unbiased variance estimators belonging to this class:

$$v(\hat{Y}_{\mathrm{R}}) = \sum_{i<j \in s}\sum \frac{a_{ij}}{\pi_{ij}} x_i x_j (z_i - z_j)^2, \tag{11}$$

$$\tilde{v}(\hat{Y}_{\mathrm{R}}) = \frac{X}{x_s}\left(\frac{N-1}{n-1} - \frac{X}{x_s}\right) \sum_{i<j \in s}\sum x_i x_j (z_i - z_j)^2, \tag{12}$$

where $a_{ij} = 1 - (X/M_1) \sum_{s \ni i, j} x_s^{-1}$,

$$\pi_{ij} = \frac{(n-1)(N-n)}{(N-1)(N-2)} \frac{(x_i + x_j)}{X} + \frac{(n-1)(n-2)}{(N-1)(N-2)}. \tag{13}$$

For $n = 2$ we have $v(\hat{Y}_{\mathrm{R}}) = \tilde{v}(\hat{Y}_{\mathrm{R}})$, but $\tilde{v}(\hat{Y}_{\mathrm{R}})$ is computationally simpler for $n > 2$. Empirical results (based on a variety of real populations) indicated that $v(\hat{Y}_{\mathrm{R}})$ is slightly better than $\tilde{v}(\hat{Y}_{\mathrm{R}})$ as regards its stability and probability of taking a negative value; both variance estimators, however, outperformed those proposed earlier in the literature.

For the random group method, the unbiased estimator of Y proposed by Rao–Hartley–Cochran [45] is

$$\hat{Y}_{\mathrm{RHC}} = \sum_{i \in s} P_i y_i / p_i, \tag{14}$$

where the suffix i denotes the unit selected from group i. The variance of

$\hat{Y}_{\text{RHC}}$ for the optimal choice $N_1 = \cdots = N_n = R$ (assuming $N = nR$) is given by

$$V(\hat{Y}_{\text{RHC}}) = [(N - n)/(N - 1)]V(\hat{Y}_{\text{PPS}}), \tag{15}$$

where $V(\hat{Y}_{\text{PPS}})$ is the variance of $\hat{Y}_{\text{PPS}}$ for PPSWR:

$$V(\hat{Y}_{\text{PPS}}) = \frac{1}{n} \sum p_i(z_i - Y)^2. \tag{16}$$

A simple nonnegative unbiased variance estimator for any $n \geq 2$ is

$$v(\hat{Y}_{\text{RHC}}) = \frac{1}{n-1}\left(1 - \frac{n}{N}\right) \sum_{i \in s} P_i(z_i - \hat{Y}_{\text{RHC}})^2. \tag{17}$$

Many extensions and improvements to the random group method have appeared recently, a few of which are:

(1) Sampling over two occasions ([5, 15, 25]);
(2) PPAS sampling within the random groups ([14]);
(3) controlled sampling for reducing the probability of getting non-preferred samples to the minimum possible extent [4];
(4) efficiency comparison with $\hat{Y}_{\text{PPS}}$ for the same cost [60];
(5) replicated sampling [59];
(6) more efficient estimators than $\hat{Y}_{\text{RHC}}$ [8];
(7) multistage sampling [14, 44].

The new design of the Canadian LFS employed the random group method for selecting self-representing units (SRU's)—SRU's are cities whose population exceeds 15,000 persons or whose unique characteristics demand their establishment as SRU's—in view of its simplicity and the following operational advantages [42]:

(1) Special studies or surveys can be conducted in any one (or more) groups since each random group by itself provides an unbiased estimator of Y;

(2) since only one unit (PSU) is selected from each group, the well-known Keyfitz method of revising selection probabilities can be applied to those random groups which would develop noticeable changes in size measures x in the future, without affecting the selections in the remaining groups (the Canadian LFS is a continuing large-scale multipurpose survey).

Rao and Bayless [46] and Bayless and Rao [7] have given extensive empirical results for $n = 2$ and $n = 3, 4$, respectively, which indicate that Murthy's method might be preferable over other methods (in categories I and II) when a stable estimator as well as a stable variance estimator are required. (Cochran [18] provides an excellent summary of these results). The variance estimator $v(\hat{Y}_{RHC})$ is slightly more stable than $v(\hat{Y}_M)$, but $\hat{Y}_{RHC}$ is less efficient than $\hat{Y}_M$. The new variance estimators $v(\hat{Y}_R)$ and $\tilde{v}(\hat{Y}_R)$ of $\hat{Y}_R$ compare favourably to $v(\hat{Y}_M)$ as regards their stability [49], but $\hat{Y}_R$ is often less efficient than $\hat{Y}_M$. However, from the viewpoint of robustness $\hat{Y}_R$ might perform better than $\hat{Y}_M$ and the other estimators based on the individual ratios z_i; see Sections 2.3 and 4.

2.3 Approach to Normality

Following Hájek [27], Rosén [51, 52] has developed asymptotic theories for several methods in I and II, in particular established asymptotic normality of the estimators of Y under certain regularity conditions. Bayless, in his unpublished thesis [6], provided empirical results on the relative approaches to normality of seven estimators using the skewness and kurtosis coefficients

$$\gamma_1 = E(\hat{Y} - Y)^3/\{V(\hat{Y})\}^{3/2}, \qquad \gamma_2 = E(\hat{Y} - Y)^4/V^2(\hat{Y}),$$

respectively. In view of the recent interest in the asymptotic normality, it might be of interest to summarize the results of this limited study for $n = 2$, 3, and 4. The methods of (1) Sampford, (2) Des Raj, (3) Murthy, (4) Rao–Hartley–Cochran, (5) Lahiri, the estimators $\hat{Y}_{PPS}$ for PPSWR and $(N/n)y_s$ for SRS were chosen for this study; all estimators, excepting the SRS estimator and $\hat{Y}_R$ in (5), depend on the individual ratios z_i. Employing the 20 natural populations described in [46] and [7], the coefficients γ_1 and γ_2 for the methods (1)–(3) and (5) were obtained by computing $p(s)$ for all possible s. For instance, the formula $E(\hat{Y} - Y)^3 = \sum p(s)(\hat{Y}_s - Y)^3$ was used to calculate γ_1. For the method (4) and PPSWR, however, the formulae for γ_1 and γ_2 were first derived explicitly in terms of the quantities $A_{lm} = \sum y_i^l/p_i^m$ (see [6] for details).

With the exception of one population, all the remaining populations seem to obey (5) reasonably well (i.e. the variability of the ratios z_i is fairly small). The exceptional population, however, has one outlier ratio (= 25) relative to the remaining ratios (between 1 and 4). The following major conclusions emerged from Bayless' study:

(a) Approach to normality of the SRS estimator is faster relative to the six unequal probability sampling estimators, especially for those populations having a larger coefficient of variation of x (this conclusion is in agreement with the results of Stenlund and Westlund [61] for the SRS and Des Raj estimators).

(b) Differences in approach to normality of the six estimators involving unequal probability sampling are small for all those "well-behaved" populations satisfying the model (5).

(c) For the exceptional population, the estimators based on the individual ratios z_i lead to very large γ_1 and γ_2 compared to Lahiri's ratio estimator.

Thus the γ_1 and γ_2 coefficients for $\hat{Y}_R$ under PPAS sampling are robust to outlier ratios.

3. VARIANCE ESTIMATORS FOR $\hat{Y}_R$ IN SRS

In simple random sampling the ratio estimator $\hat{Y}_R$ is in general biased, but the bias is negligible for large n. Two "conventional" variance estimators for $\hat{Y}_R$ are well known to samplers:

$$v_1 = \frac{N^2}{n}\left(1 - \frac{n}{N}\right)\frac{1}{n-1}\sum_{i \in s}(y_i - \hat{R}x_i)^2, \tag{18}$$

$$v_2 = \left(\frac{\bar{X}}{\bar{x}_s}\right)^2 v_1, \tag{19}$$

where $\hat{R} = y_s/x_s$, $\bar{X} = X/N$, and $\bar{x}_s = x_s/n$ (Cochran [17], p. 163 and Sukhatme and Sukhatme [62], p. 144 suggest v_1 whereas Kish [37], p. 204 and Murthy [40], p. 373 propose v_2). Royall [54, 55], however, seems to be unaware of v_2 since he gives the impression that v_1 is the only well-known conventional variance estimator for $\hat{Y}_R$. Rao and Rao [50] and Rao and Kuzik [48] have investigated the small-sample properties of v_1 and v_2, by employing the model (5). Their results indicate that the expected bias of v_1 (in absolute value) is larger than that of v_2 for $g \leq 1.5$ and that the expected mean square error (MSE) of v_2 is smaller than that of v_1 for $g > 1$ (the difference in MSE's is small for the case $g = 1$).

Brewer [10] and Royall [54] suggested $\varepsilon(\hat{Y}_R - Y)^2$, under the model (5), as a measure of uncertainty in $\hat{Y}_R$ for a given s. Using $g = 1$ in (5), Royall [54] proposed a "model-unbiased" variance estimator

$$v_w = \hat{\delta}X(X - x_s)/x_s, \tag{20}$$

where $(n-1)\hat{\delta} = \sum_{i \in s}(y_i - \hat{R}x_i)^2/x_i$, i.e. $\varepsilon(v_w) = \varepsilon(\hat{Y}_R - Y)^2$ for every s. Royall argued that his model-based "prediction" approach is preferable to the conventional theory since v_w tends to decrease as x_s increases (which is consistent with intuition and the behavior of $\varepsilon(\hat{Y}_R - Y)^2$ for $g = 1$) whereas v_1 will tend to be large when x_s is large and tend to be small when x_s is small indicating greatest uncertainty when uncertainty, as measured by $\varepsilon(\hat{Y}_R - Y)^2$, is actually least, and smallest uncertainty when in fact uncertainty is greatest.

This observation of Royall (and others) advocating the prediction approach is well taken and probably more attention should be paid to intuitive arguments. However, these facts have been ignored by Royall (and others) in asserting the superiority of the prediction approach:

(1) Table 1 gives the values of $|Z_i| = |\hat{Y}_R - Y|/v_i^{1/2}$, $i = 1, 2$ and $|Z_w| = |\hat{Y}_R - Y|/v_w^{1/2}$ for the extreme samples (LO $= s$ with the smallest x_s and HI $= s$ with the highest x_s) with $n = 4, 8$ selected from several real populations. (The values for $|Z_1|$ and $|Z_w|$ are taken from Royall [54] and the cases where more than one sample was qualified as LO or HI are omitted.) It is clear from this table that v_2 is consistent with Royall's intuitive argument in the sense $|Z_1| < |Z_2|$ for HI samples and $|Z_1| > |Z_2|$ for

TABLE 1

Values of $|Z_1|$, $|Z_2|$, and $|Z_w|$ for the Extreme Samples (LO *and* HI)

Pop.[a] no.	N	n	LO $\lvert Z_1\rvert$	LO $\lvert Z_2\rvert$	LO $\lvert Z_w\rvert$	HI $\lvert Z_1\rvert$	HI $\lvert Z_2\rvert$	HI $\lvert Z_w\rvert$
1	7	4	5.25	2.02	1.22	0.32	0.49	1.05
2	49	4	9.48	1.81	0.65	0.36	1.37	1.50
		8	10.1	2.68	1.12	0.50	1.47	1.65
3	25	4	4.54	1.68	1.57	0.57	1.57	1.74
4	24	4	9.34	1.77	0.57	0.37	1.12	1.14
		8	6.02	1.83	0.50	0.42	0.88	1.10
5	34	4	7.35	1.78	1.52	0.63	2.13	2.72
		8	2.92	0.86	0.79	0.36	1.03	1.02
6	34	4	0.90	0.04	0.18	0.23	0.77	0.84
		8	0.93	0.24	0.21	0.40	1.00	1.28
7	10	4	2.78	2.33	2.23	1.36	1.57	1.67
		8	0.49	0.46	0.41	1.86	1.95	2.21
8	10	4	1.00	0.88	0.85	0.18	0.20	0.21
		8	0.45	0.43	0.38	2.46	2.56	2.80
9	34	4	0.54	0.11	0.12	0.13	0.34	0.41
		8	1.75	0.49	0.46	0.20	0.44	0.61
10	34	4	0.70	0.15	0.13	0.31	0.70	0.77
		8	2.91	0.92	0.73	0.08	0.15	0.18
12	35	4	4.42	1.41	1.34	1.01	2.34	2.55
14	20	4	∞	∞	∞	0.33	0.95	1.52
		8	28.9	3.82	3.67	0.56	1.16	2.44
15	25	4	5.03	1.96	1.46	0.83	1.40	1.46
		8	9.70	5.27	3.69	0.54	0.82	0.92
16	43	4	4.56	1.76	1.73	0.11	0.21	0.22
		8	1.24	0.58	0.57	0.02	0.04	0.05

[a] See Royall [54] for a description of these populations.

LO samples. Also $|Z_2|$ is close to $|Z_w|$ for most of the populations. (Royall notes that the model (5) with $g = 1$ appears to be at least plausible.)

(2) Royall and Eberhardt [55] removed the bias in v_1 under the model (5) with $g = 1$ and obtained a model-unbiased estimator which in fact is approximately equal to v_2 under their own asymptotic set up: $N \gg n$ and n large. (The authors seem to be unaware of this fact.) Moreover, it follows from their investigation that the model-free variance estimator v_2 performs better than v_w when $g \neq 1$ in the model (5), i.e., v_2 is more robust to model deviations in the variance function than v_w. Perhaps the prediction approach is most useful for studying the relative performances of conventional variance estimators which are model-free.

(3) The probability of selecting an extreme sample by SRS is very small for large n, and v_1, v_2 were, in fact, proposed for the case of large n.

4. ROBUST ESTIMATION OF A TOTAL

The model (5) is often used for investigating the relative efficiencies of estimators in unequal probability sampling and the ratio estimator $\hat{Y}_R$ in SRS. However, very little attention has been paid to efficiency robustness to model breakdowns. We consider here one possible model breakdown and investigate the large-sample efficiency robustness for a simple case: $\hat{Y}_{PPS} = \sum_{i \in s} t_i y_i/(np_i)$ in PPS sampling with replacement versus the ratio estimator $\hat{Y}_R$ in SRS with replacement, where t_i = number of times the ith unit is drawn in the sample.

We assume that the population consists of two domains of sizes N_1 and N_2, respectively $(N_1 + N_2 = N)$ and that the units in each domain obey a model of the form (5) but with different slopes:

$$\begin{aligned} y_i &= \beta_1 x_i + e_i \qquad i = 1, \ldots, N_1, \\ y_j &= \beta_2 x_j + e_j \qquad j = 1, \ldots, N_2, \end{aligned} \tag{21}$$

where the errors e_i and e_j have properties identical to those in (5). It is not known which of the domains any particular population unit belongs to and, in fact, N_1 and N_2 are unknown. Let $X_1 = \sum x_i$ and $X_2 = \sum x_j$ so that $X = X_1 + X_2$. Under the model (21), the average variance of $\hat{Y}_{PPS}$ is given by

$$\begin{aligned} \varepsilon V(\hat{Y}_{PPS}) &= \frac{1}{nX}(X_1 X_2^2 + X_2 X_1^2)(\beta_1 - \beta_2)^2 + \frac{\delta}{n}\sum_1^N x_t^{g-1}(X - x_t) \\ &= \frac{1}{n}(A_{PPS} + \delta B_{PPS}) \end{aligned} \tag{22}$$

and the average MSE of $\hat{Y}_R$, for large n, is

$$\varepsilon[\text{MSE}(\hat{Y}_R)] \doteq \frac{N}{nX^2}\left(X_2^2 \sum_1^{N_1} x_i^2 + X_1^2 \sum_1^{N_2} x_j^2\right)(\beta_1 - \beta_2)^2 + \frac{N\delta}{n}\left\{\sum_1^N x_t^g + \frac{(\sum_1^N x_t^g)(\sum_1^N x_t^2)}{X^2} - 2\frac{\sum_1^N x_t^{g+1}}{X}\right\}$$

$$= \frac{1}{n}(A_R + \delta B_R). \tag{23}$$

It may be noted that $V(\hat{Y}_R)$ in PPAS sampling is approximately equal to $\text{MSE}(\hat{Y}_R)$ in SRS with replacement, provided n is large and n/N is negligible.

If $\beta_1 = \beta_2$ (i.e., no model breakdown), we would only have contributions from the error terms δB_{PPS} and δB_R. Since $B_R - B_{\text{PPS}} \doteq N(\sum x_t^g - X \sum x_t^{g-1})$ for large N, it follows that $B_{\text{PPS}} < B_R$ if $g > 1$, $B_{\text{PPS}} > B_R$ if $g < 1$ and $B_{\text{PPS}} \doteq B_R$ if $g = 1$. Suppose now that the first domain consists of a few units with small size measures x_i which have undergone surprise growth by the time of the survey and y_i is the present size (i.e., $x_i < \bar{X}$, $i = 1, \ldots, N_1$, $X_1 \ll X_2$ and $\beta_1 \gg \beta_2$). Then the terms A_{PPS} and A_R could well dominate the corresponding error terms δB_{PPS} and δB_R. Also

$$A_{\text{PPS}} - A_R = \frac{(\beta_1 - \beta_2)^2}{X}\left\{X_2^2\left(X_1 - \frac{\sum_1^{N_1} x_i^2}{\bar{X}}\right) + X_1^2\left(X_2 - \frac{\sum_1^{N_2} x_j^2}{\bar{X}}\right)\right\}. \tag{24}$$

The contribution from the second term in (24) would be small compared to that from the first term since $X_1 \ll X_2$. Noting that $X_1 > \sum x_i^2/\bar{X}$, it then follows that $A_{\text{PPS}} > A_R$. Consequently, the ratio estimator $\hat{Y}_R$ could lead to substantial gains in efficiency over $\hat{Y}_{\text{PPS}}$ even when $g \geq 1$.

To shed further light on the magnitude of (24), we further assume that the x_i ($i = 1, \ldots, N_1$) and the x_j ($j = 1, \ldots, N_2$) are mutually independent gamma variables with parameters $m_1 = \varepsilon(x_i)$ and $m_2 = \varepsilon(x_j)$, respectively, where $m_1 < m_2$. Then

$$\varepsilon(A_{\text{PPS}} - A_R) = (\beta_1 - \beta_2)^2(N_1 N_2 m_1 m_2)(N^* + 2)(N^* + 3) \cdot \{(m_2 - m_1)[(N_2 - N_1) + (N_2^2 m_2 - N_1^2 m_1)] - (N - 3)(N^* + 2)\}, \tag{25}$$

where $N^* = N_1 m_1 + N_2 m_2$. If $N_1 \ll N_2$ and N_2 is moderately large, the leading term in braces in (25), which is $O(N_2^2)$, equals $N_2^2 m_2(m_2 - m_1 - 1)$ which is positive when $m_2 > m_1 + 1$. For example, taking $m_1 = 1$, $m_2 = 5$, $N_1 = 2$, and $N_2 = 20$, we find that the value of the leading term is 6000 compared to the exact value 6080.

We have computed the average efficiency of $\hat{Y}_R$ relative to $\hat{Y}_{\text{PPS}}$ for eight real x populations using the formulae (22) and (23) and $N_1 = 1, 2$ with the smallest x values in the first domain:

$$\frac{\varepsilon V(\hat{Y}_{\mathrm{PPS}})}{\varepsilon \mathrm{MSE}(\hat{Y}_{\mathrm{R}})} \doteq \frac{A_{\mathrm{PPS}} + \delta B_{\mathrm{PPS}}}{A_{\mathrm{R}} + \delta B_{\mathrm{R}}} = \frac{\alpha A'_{\mathrm{PPS}} + B_{\mathrm{PPS}}}{\alpha A'_{\mathrm{R}} + B_{\mathrm{R}}}, \tag{26}$$

where $\alpha = (\beta_1 - \beta_2)^2/\delta$, $A'_{\mathrm{PPS}} = (X_1 X_2^2 + X_2 X_1^2)/X$, and

$$A'_{\mathrm{R}} = N(X_2^2 \sum x_i^2 + X_1^2 \sum x_j^2)/X^2.$$

Tables 2 and 3 gives the average efficiency for $N_1 = 1$ and 2 and selected values of α and g: $\alpha = 0, 25, 50, 100$, and 250; $g = 0, 1, 1.5$, and 2.0. It is seen from these tables that the gain in average efficiency of $\hat{Y}_{\mathrm{R}}$ over $\hat{Y}_{\mathrm{PPS}}$ is very large when $g \leq 1$ and $\alpha \neq 0$ (for $g = 0$ the terms A'_{PPS} and A'_{R} seem to dominate B_{PPS} and B_{R}, respectively); even for $g = 1.5$ the gains could be substantial if α is large. The effect of α for $g = 2.0$ is small, but any loss in average efficiency is moderate, excepting possibly the populations 1 and 8. It appears

TABLE 2

Average Efficiency of $\hat{Y}_{\mathrm{R}}$ Relative to $\hat{Y}_{\mathrm{PPS}}$ for Selected Values of $\alpha = (\beta_1 - \beta_2)^2/\delta$ and g: $N_1 = 1$

	$g = 0$					$g = 1$				
Pop.[a] no. α:	0	25	50	100	250	0	25	50	100	250
1	2.70	34.6	40.6	44.6	47.5	1.02	2.04	3.02	4.86	9.60
2	1.46	1.87	2.13	2.45	2.84	1.00	1.19	1.34	1.59	2.07
3	1.80	1.98	2.16	2.52	3.54	1.00	1.04	1.07	1.14	1.34
4	1.76	31.5	31.5	31.5	31.5	1.00	9.44	14.2	19.5	25.2
5	1.59	8.50	8.50	8.50	8.50	1.00	7.73	8.09	8.29	8.41
6	1.38	2.08	2.62	3.41	4.06	1.00	1.12	1.24	1.45	2.02
7	1.54	4.43	4.43	4.43	4.43	1.00	3.90	4.14	4.28	4.37
8	4.40	6.15	7.63	9.98	14.3	1.00	1.10	1.18	1.36	1.87
	$g = 1.5$					$g = 2$				
1	0.72	0.80	0.88	1.04	1.51	0.52	0.53	0.53	0.54	0.58
2	0.85	0.95	1.04	1.21	1.58	0.74	0.79	0.84	0.92	1.15
3	0.84	0.85	0.86	0.89	0.97	0.73	0.73	0.74	0.75	0.77
4	0.86	1.09	1.33	1.78	3.06	0.76	0.76	0.77	0.78	0.80
5	0.85	1.77	2.49	3.55	5.26	0.74	0.75	0.77	0.80	0.89
6	0.91	0.96	1.00	1.08	1.32	0.85	0.87	0.88	0.91	1.00
7	0.82	1.47	1.92	2.50	3.30	0.69	0.71	0.74	0.80	0.97
8	0.66	0.68	0.69	0.72	0.80	0.51	0.51	0.51	0.51	0.53

[a] See Table 3 for a description of the x populations.

TABLE 3

Average Efficiency of $\hat{Y}_R$ Relative to $\hat{Y}_{PPS}$ for Selected Values of $\alpha = (\beta_1 - \beta_2)^2/\delta$ and g: $N_1 = 2$

Pop.[a]	$g = 0$					$g = 1$				
no. α:	0	25	50	100	250	0	25	50	100	250
1	2.70	4.63	4.63	4.64	4.64	1.02	3.70	4.10	4.35	4.52
2	1.46	2.12	2.43	2.73	3.01	1.00	1.34	1.59	1.93	2.45
3	1.80	2.53	3.11	3.97	5.37	1.00	1.17	1.33	1.62	2.37
4	1.76	14.8	14.8	14.8	14.8	1.00	11.3	12.8	13.7	14.3
5	1.59	8.38	8.38	8.38	8.38	1.00	7.99	8.18	8.28	8.34
6	1.38	2.62	3.40	4.34	5.49	1.00	1.24	1.45	1.84	2.73
7	1.54	4.31	4.31	4.31	4.31	1.00	4.05	4.17	4.24	4.89
8	4.40	7.62	9.96	13.1	17.7	1.00	1.18	1.36	1.70	2.67
	$g = 1.5$					$g = 2$				
1	0.72	1.42	1.91	2.55	3.41	0.52	0.58	0.64	0.75	1.04
2	0.85	1.04	1.20	1.46	1.96	0.74	0.84	0.92	1.08	1.44
3	0.84	0.90	0.97	1.09	1.43	0.73	0.75	0.77	0.82	0.95
4	0.86	1.63	2.32	3.51	6.01	0.76	0.77	0.79	0.82	0.91
5	0.85	2.48	3.54	4.81	6.39	0.74	0.77	0.80	0.86	1.04
6	0.91	1.00	1.08	1.25	1.68	0.85	0.88	0.91	0.97	1.14
7	0.82	1.91	2.48	3.08	3.68	0.69	0.75	0.80	0.91	1.21
8	0.66	0.69	0.72	0.77	0.93	0.51	0.51	0.51	0.52	0.54

[a] Description of the x populations: (1) size of a large US city in 1920 [17], p. 156; (2) number of villages in an administrative circle [62], p. 256; (3)–(5): area in square miles, cultivated area and number of persons for a village in 1951 [41], p. 128; (6) strip length [41], p. 131; (7) geographical area of a village [41], p. 178; (8) number of dwellings in a block [37], p. 625.

from the previous investigation that, from a robustness viewpoint, $\hat{Y}_R$ in SRS (or PPAS sampling) might perform better than $\hat{Y}_{PPS}$ in PPS sampling.

Assuming the model (5), Brewer [10] and Royall [53] derived the "best" model-unbiased estimator $\hat{Y}$ of Y (i.e. $\varepsilon(\hat{Y}) = \varepsilon(Y)$ for every s) and showed that the "optimal" fixed sample design *purposively* selects those n units with the largest x values. In this connection it is interesting to note that Hartley [29] was aware of such model-dependent strategies twenty years ago: "Note that this far only the fact that the e's are a random sample has been used. These theorems hold regardless of how the x's were chosen; they may be systematic, fixed, biased. Suppose, for example, only large farms have been sampled. Then $\bar{Y}$ can be estimated from the computed regression line by evaluating it at $\bar{X}$ *provided* that the e's are random .... In practice, unfortunately, the reason that biases the x's usually biases the e's."

Sarndal [58] obtained a similar result assuming (5) with a gamma distribution for x. He has shown that the average variance of $\hat{Y}_{\mathrm{PPS}}$, with selection probabilities $p_t^{(k)} \propto x_t^k$, decreases as k increases. Sarndal, however, also investigated a model breakdown specified by

$$y_t = \beta x_t^b + e_t \qquad \varepsilon(e_t | x_t) = 0, \quad \varepsilon(e_t^2 | x_t) = \delta x_t^g, \quad b \neq 1, \tag{27}$$

where the errors e_t are independent and b is unknown. Under the model (27), the average bias of $\hat{Y}_{\mathrm{PPS}}$, $B(\hat{Y}_{\mathrm{PPS}}) = E\varepsilon(\hat{Y}_{\mathrm{PPS}} - Y)$, is nonzero when $k \neq 1$ and its absolute value increases with $k \geq 2$. Similar results for $\hat{Y}_{\mathrm{R}}$ were also obtained. We provide here some parallel results for $\hat{Y}_{\mathrm{PPS}}$ and $\hat{Y}_{\mathrm{R}}$ under the model breakdown specified by (21). The average bias of $\hat{Y}_{\mathrm{PPS}}$ is

$$B(\hat{Y}_{\mathrm{PPS}}) = (\beta_1 - \beta_2)\left(\frac{X \sum_1^{N_1} x_i^k}{\sum x_t^k} - X_1\right). \tag{28}$$

If we further assume that the x_i and the x_j are independent gamma variables with parameters m_1 and m_2, respectively $(m_1 < m_2)$, then for $k \geq 2$

$$\begin{aligned} \varepsilon B(\hat{Y}_{\mathrm{PPS}}) &\doteq (\beta_1 - \beta_2)\left\{\frac{\varepsilon(X \sum x_i^k)}{\varepsilon(\sum x_i^k)} - \varepsilon(X_1)\right\} \\ &= (\beta_1 - \beta_2)N_1 m_1\left\{\frac{(k + m_1 - 1)\cdots(m_1 + 1)}{(k + m_2 - 1)\cdots(m_2 + 1)} - 1\right\} \end{aligned} \tag{29}$$

if N_2 is large. It follows from (29) that $\varepsilon B(\hat{Y}_{\mathrm{PPS}})$ for $k \geq 2$ is negative and that its absolute value increases with k; if $k = 0$ we obtain $\varepsilon B(\hat{Y}_{\mathrm{PPS}}) = (N_1 N_2/N)(\beta_1 - \beta_2)(m_2 - m_1) > 0$.

Turning to $\hat{Y}_{\mathrm{R}}$ and noting that $E(\hat{Y}_{\mathrm{R}}) = X(\sum p_t^{(k)} y_t)/(\sum p_t^{(k)} x_t)$ for large n we get for $k \geq 1$,

$$\varepsilon B(\hat{Y}_{\mathrm{R}}) \doteq (\beta_1 - \beta_2)N_1 m_1\left\{\frac{(k + m_1)\cdots(m_1 + 1)}{(k + m_2)\cdots(m_2 + 1)} - 1\right\} \tag{30}$$

which is negative for $k \geq 1$ (for the SRS case $k = 0$, $\varepsilon B(\hat{Y}_{\mathrm{R}}) \doteq 0$). It may be noted that the average bias of $\hat{Y}_{\mathrm{R}}$ with $k \neq 0$ is independent of n so that the average MSE, for large n, might be substantially larger than that of $\hat{Y}_{\mathrm{R}}$ with $k = 0$. (A similar result holds for $\hat{Y}_{\mathrm{PPS}}$ with $k \neq 1$.)

The sampling practitioner, consequently, is caught in a dilemma: The sampling design with a rather large k gives large efficiency gains if the model (5) is correct, but the same large k brings about a heavy bias penalty if the model is wrong. From the robust viewpoint, the prediction approach might suggest a sampling design which gives a zero or a small ε bias under model breakdowns. Royall and Herson [56] considered specifically a situation in which the breakdown of the model (5) consists of latent higher order polynomial terms in x and/or an intercept β_0. They demonstrated that a "balanced" sampling design (which essentially involves the selection of a sample such that certain sample moments of x equal (at least approximately) the corresponding population moments of x) together with $\hat{Y}_{\mathrm{R}}$ provides the desired robustness, namely a zero or small ε-bias. They also showed that the strategy is more efficient than the one involving SRS and $\hat{Y}_{\mathrm{R}}$, *provided* the model deviation is of the polynomial type. Their approach, however, seems to fail under the model breakdown specified by (21). For instance, with $\bar{x}_s = \bar{X}$,

$$B_s(\hat{Y}_{\mathrm{R}}) = \varepsilon(\hat{Y}_{\mathrm{R}} - Y) = \beta_1\left(\frac{\sum_1^{n_1} x_i}{n} - \frac{X_1}{N}\right) + \beta_2\left(\frac{\sum_1^{n_2} x_j}{n} - \frac{X_2}{N}\right), \quad (31)$$

where n_1 = number of sample units in the first domain and $n_2 = n - n_1$. It is clear from (31) that the ε bias cannot be made equal to zero since the domain frames are unknown; even $\varepsilon B_s(\hat{Y}_{\mathrm{R}}) = \beta_1 m_1(n_1/n - N_1/N) + \beta_2 m_2(n_2/n - N_2/N)$ cannot be reduced to zero. It appears to us that if protection against unforeseen model breakdowns of the type (21) is also needed, it might be advisable to employ a strategy $(p(s), \hat{Y})$ for which the p bias $E(\hat{Y}) - Y$ is either zero or small for large n (see Hartley and Sielken [34]). Since there will be many such strategies, one could study efficiency robustness (as shown here) for choosing suitable strategies.

In as much as SRS is an approximation to balanced sampling for large n, the strategy (SRS, $\hat{Y}_{\mathrm{R}}$) might be preferred, in large samples, to (PPAS, $\hat{Y}_{\mathrm{R}}$) since the former provides protection (in the sense of ε bias) against model breakdowns of the polynomial type and performs equally well in terms of efficiency robustness to outliers in the individual ratios y_t/x_t.

REFERENCES

[1] Asok, C. and Sukhatme, B. V. (1974). On Sampford's procedure of unequal probability sampling without replacement. *Proc. Soc. Statist. Sec.*, Amer. Statist. Assoc., 220–225.

[2] Asok, C. and Sukhatme, B. V. (1975). Unequal probability sampling with random stratification. *Proc. Soc. Statist. Sec.*, Amer. Statist. Assoc., 283–288.

[3] Asok, C. and Sukhatme, B. V. (1976). On the efficiency comparison of two πps sampling strategies. Paper presented at the Amer. Statist. Assoc. meetings, Boston, Massachusetts.

[4] Avadhani, M. S. and Sukhatme, B. V. (1968). Simplified procedures for designing controlled simple random sampling. *Austral. J. Statist.* **10** 1–7.

[5] Avadhani, M. S. and Sukhatme, B. V. (1970). A comparison of two sampling procedures with an application to successive sampling. *Appl. Statist.* **19** 251–259.

[6] Bayless, D. L. (1968). *Variance Estimation in Sampling from Finite Populations.* Ph.D. Thesis, Texas A & M University, College Station, Texas.

[7] Bayless, D. L. and Rao, J. N. K. (1970). An empirical study of stabilities of estimators and variance estimators in unequal probability sampling ($n = 3$ or 4). *J. Amer. Statist. Assoc.* **65** 1645–1667.

[8] BERG, S. (1974). A note on the RHC-method. *Scan. Actua. J.* **57** 108–114.
[9] BREWER, K. R. W. (1963). A model of systematic sampling with unequal probabilities. *Austral. J. Statist.* **5** 5–13.
[10] Brewer, K. R. W. (1963). Ratio estimation and finite populations: some results deducible from the assumption of an underlying stochastic process. *Austral. J. Statist.* **5** 93–105.
[11] BREWER, K. R. W. (1974). A simple procedure for sampling πpswor. *Austral. J. Statist.* **17** 166–172.
[12] BREWER, K. R. W. and HANIF, M. (1969). Sampling without replacement with probability of inclusion proportional to size, I: methods using the Horvitz and Thompson estimator. Unpublished manuscript seen by courtesy of the authors.
[13] BREWER, K. R. W. and HANIF, M. (1969). Sampling without replacement with probability of inclusion proportional to size, II: methods using special estimators. Unpublished manuscript seen by courtesy of the authors.
[14] CHIKKAGOUDAR, M. S. (1967). Sampling with preliminary random stratification. *Austral. J. Statist.* **9** 57–70.
[15] CHOTAI, J. (1974). A note on Rao–Hartley–Cochran method for PPS sampling over two occasions. *Sankhyā Ser. C* **36** 173–180.
[16] CHROMY, J. R. (1974). Pairwise probabilities in probability nonreplacement sampling. *Proc. Soc. Statist. Sec.*, Amer. Statist. Assoc., 269–274.
[17] COCHRAN, W. G. (1963). *Sampling Techniques*, 2nd ed. Wiley, New York.
[18] COCHRAN, W. G. (1974). Two recent areas of sample survey research. In *A Survey of Statistical Design and Linear Models.* (J. N. Srivastava, ed.), North-Holland, Amsterdam, pp. 101–115.
[19] CONNOR, W. S. (1966). An exact formula for the probability that two specified sampling units will occur in a sample drawn with unequal probabilities and without replacement. *J. Amer. Statist. Assoc.* **61**, 384–390.
[20] DAS, M. N. and MOHANTY, S. (1973). On *pps* sampling without replacement ensuring selection probabilities exactly proportional to size. *Australian J. Statist.* **15** 87–94.
[21] DES RAJ (1956). Some estimators in sampling with varying probabilities without replacement. *J. Amer. Statist. Assoc.* **60**, 278–284.
[22] DODDS, D. J. and FRYER, J. G. (1971). Some families of selection probabilities for sampling with probability proportional to size. *J. Roy. Statist. Soc. Ser. B* **33** 263–274.
[23] DURBIN, J. (1967). Estimation of sampling errors in multistage surveys. *Applied Statistics* **16** 152–164.
[24] FULLER, W. A. (1971). A procedure for selecting nonreplacement unequal probability samples. Unpublished manuscript seen by courtesy of the author.
[25] GHANGURDE, P. D. and RAO, J. N. K. (1969). Some results on sampling over two occasions. *Sankhyā Ser. A* **31** 463–472.
[26] GRAY, G. B. (1973). On PSU rotation and further remarks about increasing sample size. Unpublished manuscript seen by courtesy of the author.
[27] HÁJEK, J. (1964). Asymptotic theory of rejective sampling with varying probabilities from a finite population. *Ann. Math. Statist.* **35** 1491–1523.
[28] HANURAV, T. V. (1967). Optimum utilization of auxiliary information: πps sampling of two units from a stratum. *J. Roy. Statist. Soc. Ser. B* **29** 374–391.
[29] HARTLEY, H. O. (1956). *Theory of Advanced Design of Surveys*, lecture notes, Dept. of Statistics, Iowa State University, Ames, Iowa.
[30] HARTLEY, H. O. (1966). Systematic sampling with unequal probability and without replacement. *J. Amer. Statist. Assoc.* **61**, 739–748.
[31] HARTLEY, H. O. and RAO, J. N. K. (1962). Sampling with unequal probabilities and without replacement. *Ann. Math. Statist.* **33** 350–374.

[32] HARTLEY, H. O. and CHAKRABARTY, R. P. (1967). Evaluation of approximate formulas in sampling with unequal probabilities and without replacement. *Sankhyā Ser. B* **29** 201–208.

[33] HARTLEY, H. O., RAO, J. N. K., and KIEFER, G. (1969). Variance estimation with one unit per stratum. *J. Amer. Statist. Assoc.* **64** 841–851.

[34] HARTLEY, H. O. and SIELKEN, R. L. (1975). A "super-population viewpoint" for finite population sampling. *Biometrics* **31** 411–422.

[35] HIDIROGLOU, M. A. and GRAY, G. B. (1975). A computer algorithm for joint probabilities of selection. *Survey Methodology* (Statistics Canada) **1** 99–108.

[36] JESSEN, R. J. (1969). Some methods of probability non-replacement sampling. *J. Amer. Statist. Assoc.* **64**, 175–193.

[37] KISH, L. (1965). *Survey Sampling*. Wiley, New York.

[38] LAHIRI, D. B. (1951). A method of sample selection providing unbiased ratio estimates. *Bull. Int. Statist. Inst.* **33** (2) 133–140.

[39] LANKE, J. (1974). On non-negative variance estimators in survey sampling. *Sankhyā Ser. C* **36** 33–42.

[40] MURTHY, M. N. (1957). Ordered and unordered estimators in sampling without replacement. *Sankhyā* **18** 379–390.

[41] MURTHY, M. N. (1967). *Sampling Theory and Methods*. Statist. Pub. Soc., Calcutta.

[42] PLATEK, R. and SINGH, M. P. (1972). Some aspects of redesign of the Canadian Labour Force Survey. *Proc. Soc. Statist. Sec.*, Amer. Statist. Assoc., 397–402.

[43] RAO, J. N. K. (1965). On two simple schemes of unequal probability sampling without replacement. *J. Indian Statist. Assoc.* **3** 173–180.

[44] RAO, J. N. K. (1975). Unbiased variance estimation for multistage designs. *Sankhyā Ser. C* **37** 133–139.

[45] RAO, J. N. K., HARTLEY, H. O., and COCHRAN, W. G. (1962). On a simple procedure of unequal probability sampling without replacement. *J. Roy. Statist. Soc. Ser. B* **24** 482–491.

[46] RAO, J. N. K. and BAYLESS, D. L. (1969). An empirical study of the stabilities of estimators and variance estimators in unequal probability sampling of two units per stratum. *J. Amer. Statist. Assoc.* **64** 540–559.

[47] RAO, J. N. K. and SINGH, M. P. (1973). On the choice of estimator in survey sampling. *Australian J. Statist.* **15** 95–104.

[48] RAO, J. N. K. and KUZIK, R. A. (1974). Sampling errors in ratio estimation. *Sankhyā Ser. C* **36** 43–58.

[49] RAO, J. N. K. and VIJAYAN, K. (1977). On estimating the variance in sampling with probability proportional to aggregate size. *J. Amer. Statist. Assoc. 72*

[50] RAO, P. S. R. S. and RAO, J. N. K. (1971). Small sample results for ratio estimators. *Biometrika* **58** 625–630.

[51] ROSÉN, B. (1972). Asymptotic theory for successive sampling with varying probabilities without replacement, I and II. *Ann. Math. Statist.* **43** 373–397 and 748–776.

[52] ROSÉN, B. (1974). Asymptotic theory for Des Raj's estimator, I and II. *Scand. J. Statist.* **1** 71–83 and 135–144.

[53] ROYALL, R. M. (1970). On finite population sampling theory under certain linear regression models. *Biometrika* **57** 377–387.

[54] ROYALL, R. M. (1971). Linear regression models in finite population sampling theory. In *Foundations of Statistical Inference*. (V. P. Godambe and D. A. Sprott, eds.) 259–274, Holt, New York.

[55] ROYALL, R. M. and EBERHARDT, K. R. (1975). Variance estimates for the ratio estimator. *Sankhyā Ser. C* **37** 43–52.

[56] Royall, R. M. and Herson, J. (1973). Robust estimation in finite populations, 1. *J. Amer. Statist. Assoc.* **68** 880–889.

[57] Sampford, M. R. (1967). On sampling without replacement with unequal probabilities of selection. *Biometrika* **54** 499–513.

[58] Sarndal, C. E. (1974). Continuous survey sampling models. Unpublished manuscript seen by courtesy of the author.

[59] Singh, R. and Singh, B. (1974). On replicated samples drawn with Rao, Hartley, and Cochran's scheme. *Sankhyā Ser. C* **36** 147–150.

[60] Singh, R. and Kishore, L. (1975). On Rao, Hartley, and Cochran's method of sampling. *Sankhyā Ser. C* **37** 88–94.

[61] Stenlund, H. and Westlund, A. (1974). Sampling with equal and unequal probabilities; a comparative Monte Carlo study of some aspects of inference. Unpublished manuscript seen by courtesy of the authors.

[62] Sukhatme, P. V. and Sukhatme, B. V. (1970). *Sampling Theory of Surveys with Applications.* 2nd ed. FAO of the U.N., Rome.

[63] Vijayan, K. (1968). An exact πps scheme—generalization of a method of Hanurav. *J. Roy. Statist. Soc. Ser. B* **30**, 556–566.

[64] Vijayan, K. (1975). On estimating the variance in unequal probability sampling. *J. Amer. Statist. Assoc.* **70** 713–716.

[65] Vos, J. W. E. (1974). Sampling with unequal probabilities. Translation of a paper in *Statistica Neerlandica* **28** 11–49 and 69–107.

Selection Biases in Fixed Panel Surveys

W. H. Williams

BELL LABORATORIES
MURRAY HILL

The fixed panel survey design (and variants of it called rotation sampling) is often used in socioeconomic surveys that are repeated systematically at different points in time. In this paper, the potential effect of selection bias on the basic fixed panel design is studied. One important observation is that bias, unlike variance, is not necessarily minimized by fixed panel surveys.

Selection probabilities may cause difficulty in a fixed panel survey when (a) the original sample is selected, (b) some sample units are inevitably lost from the survey, and (c) a replacement policy is instituted. Each of these can have very serious bias effects.

Finally, the difficulty in achieving a replenishment policy which actually balances the sample is discussed.

1. INTRODUCTION

1.1. The Use of Fixed Panel Surveys

Surveys are often repeated systematically at different points in time in order to follow the changing characteristics of the target population. The specific objectives of such surveys are not always the same, however, with the result that the details of the designs may differ. In one study emphasis may be placed on the development of good estimates at each point in time, while in another emphasis may be on estimates of change through time. Consequently, in the former case, the repeated observations may be made on a

Key words and phrases Selection biases, nonresponse, panel surveys

completely new set of sample units, while in the latter, observations may be made on an identical (matched) set of units. This latter procedure, which is sometimes referred to as a fixed panel, and variants of it are used as the basis for all kinds of socioeconomic studies.

In practice, many sample designs are modifications of fixed panel surveys in that *some* units are replaced each month (say) and some are carried over from the previous months. The reason for this is that the correlation between observations made on the *same* units at different periods can be used to improve the precision of both estimates of changes through time and estimates for particular points in time. The possibility of such overlap leads to questions of which, and how many, units are to be replaced each month. These issues are the subject matter of rotation sampling; see, for example, Blight and Scott [2], Cochran [3], Eckler [4], Kish [6], Hansen *et al.* [5], Patterson [8], and Rao and Graham [9].

In general, it is felt that the more overlap there is in the sample from one observation period to the next, the greater the information on changes through time. Surveys in which change estimates are important frequently have the characteristic of large fractions of sample overlap. In fact, Stephan and McCarthy [10] have argued that comparisons on identical units are free from statistical errors and hence any differences must be due to real changes. As we shall show, this is misleading because it ignores the possibility of systematic selection bias. Similarly, Cochran [3, p. 342] states; "For estimating change, it is best to retain the same sample throughout all occasions." This statement is valid in terms of the variance model discussed in his textbook, but it is highly suspect if bias issues are also included. In fact, in some of the simple bias models we have considered the opposite conclusion emerges. Specifically, for best estimates of change, it is best to *replace the sample completely*. In practice, it appears that in many surveys, bias considerations will completely dominate variance; see, for example, the paper by Bailar [1] and the references contained in it.

In this paper only the basic fixed panel survey is considered. The comparison of fixed panels and complete replacement sampling will be discussed subsequently. In this paper, we show that very large biases can be created by very slight changes in the selection probabilities. These biases are much larger than the precision that one would normally expect to be associated with socioeconomic estimates. Furthermore, if the various selection probabilities have systematic behavior in time, then these same systematic changes may occur in the expectation of the estimates; this is true even though *no change* actually occurs in the population. This suggests that many of the rules of thumb used in designing surveys need to be more carefully considered.

Finally, as in earlier papers [12, 13], the problem is described in terms of employed and unemployed persons. The reason for this is estimation of unemployment is the original source of the problem. Readers should take care in extrapolating to the real problem of unemployment which is very complex.

1.2 Selection and Nonresponse in Fixed Panel Surveys

We consider a specific set of N individuals who have been designated by the sample design to be included in the fixed panel. These individuals could, for example, be in a selected geographical area. This permits us, without loss of generality, but with a substantial gain in simplicity, to consider estimates conditionally on the sample design.

When the survey is actually conducted, the designated N persons may or may not be observed. There are many reasons for this, the simplest of which is that they may not be at home. But this does mean that the selection probabilities for the individuals, which should be equal to 1, are actually less than 1. Clearly these selection probabilities could also be called response probabilities, but to minimize possible confusion with the area of research described by "response errors," we mostly use the description "selection probabilities."

Three different points at which selection probabilities may cause difficulty in a fixed panel survey are: when the original sample is selected, when some units are inevitably lost from the survey, and when a replenishment policy is instituted. We shall discuss each of these briefly.

(a) *Original selection of the sample* The problem of nonresponse in the original sample selection has been a point of concern for many years. There is a large literature on the subject most of which attempts to deal with the effects of differential nonresponse for various parts of the population, for example, employed versus unemployed (see Waksberg [11]). It does not, however, seem to have been previously recognized that selection difficulties at the first observation period can create systematic *changes* in the characteristics of the estimates as the survey progresses.

(b) *Attrition* Most fixed panel surveys experience a loss of sample units. People die, move, refuse to cooperate further, and for many reasons are lost from the survey. Such losses are by no means confined to socio-economic surveys. For example, in an experiment involving buried telephone cable, some experimental units were lost because the location records were misplaced. In general, the way in which these sample units are lost from the survey can cause major systematic biases in the estimates. Even a very slight correlation between attrition and the characteristics under measurement can cause large distortions, which in some cases will get continuously worse as the survey progresses.

(c) *Replenishment policies* As a result of attrition losses, most surveys will have compensating replenishment policies. Peculiar selection probabilities at this stage can also create peculiar changes in the estimates. This can happen quite independently of biases that result from the first two sources.

Unfortunately, some designers of fixed panel surveys have created major difficulties for themselves by the implementation of unwise replenishment policies. Specifically, they have attempted to replace the units that are lost from the survey in some nonrandom way by using a replenishment policy which is also nonrandom. For example, in a telephone survey, it was suspected that the more mobile customers would be lost from the fixed panel. Consequently, the replenishment policy required that new panel members should be selected from *new* telephone customers. The feeling was that this would tend to make up for the loss of the more mobile customers. In some geographical areas, this objective was successful, but only for the mobility characteristic. It turned out that people in this preferred group were very different in other measurable ways from both the population at large and the rest of the sample.

2. A SIMPLE TWO CATEGORY MODEL REPEATED AT TWO OBSERVATION TIMES

2.1. The Model

In order to focus on questions of bias, we shall assume that we have a frame which is to be sampled 100% both at time T_1 and time T_2. This can equivalently be regarded either as a census or as a model which is conditional upon a particular group of persons having been drawn into the sample; the effect is the same in either case and there is no loss in generality. But the assumption does permit us to assess biases without the additional mathematical complexity which would result from the consideration of a hierarchical statistical sampling plan. The formulas are simpler. In any particular study, the unique design characteristics could be superimposed with no conceptual difficulty.

Fof further simplicity, the population is divided into two categories referred to as (E) employed and (U) unemployed.† Next, it is assumed that the population frame contains the same N persons at both T_1 and T_2 and that these persons fall into four categories such that $N = N_{uu} + N_{ue} + N_{eu} + N_{ee}$, where N_{ue} denotes the number of persons unemployed at T_1 and

† This research was originally suggested to us by the "first-month" bias problem in unemployment estimates; see reference [7].

employed at T_2 with analogous interpretations for N_{uu}, N_{eu}, and N_{ee}. Of course, this formulation ignores the significant group of persons who are not in the work force at all, but this does not hinder the ideas developed in this paper. Then the actual ratio of unemployed to employed at T_1 is given by

$$R_1 = (N_{uu} + N_{ue})/(N_{eu} + N_{ee}), \tag{2.1}$$

and at T_2 by

$$R_2 = (N_{uu} + N_{eu})/(N_{ue} + N_{ee}). \tag{2.2}$$

An obvious condition for the unemployed/employed ratio to remain unchanged, i.e., $R_1 = R_2$, is that $N_{ue} = N_{eu}$. This simply means that the number of persons who found employment during the period is equal to the number of persons who lost it.

Notice that (2.1) and (2.2) are U/E ratios and not the fraction or rate of unemployment $U/(U + E)$. The latter is somewhat more common in the real employment–unemployment case, but the ratio U/E is simpler to handle algebraically and is entirely equivalent.

2.2. The Estimates

When the sample is drawn, the persons who are *actually observed* can then be associated with their proper U or E category. This can be done at both T_1 and T_2. Unobserved persons, of course, cannot be classified so that the population can now be described as in Table 2.1.

TABLE 2.1

Classification of the Designated Panel after Interviewing

	Status at time T_2		
Status at time T_1	Observed unemployed	Observed employed	Not observed
Observed unemployed	F_{uu}	F_{ue}	F_{uo}
Observed employed	F_{eu}	F_{ee}	F_{eo}
Not observed	F_{ou}	F_{oe}	F_{oo}

Consideration of Table 2.1 reveals that three estimates of the U/E ratio can be constructed at each of the observation times. The first of these is based on the *total* number of observed persons at each observation time,

$$R_1^t = (F_{uu} + F_{ue} + F_{uo})/(F_{eu} + F_{ee} + F_{eo}), \tag{2.3}$$

at T_1, and

$$R_2^t = (F_{uu} + F_{eu} + F_{ou})/(F_{ue} + F_{ee} + F_{oe}), \tag{2.4}$$

at T_2.

Note that the estimates R_1^t and R_2^t are based on all observed individuals at both T_1 and T_2, regardless of whether they appeared in the survey on both occasions or appeared only at T_1 or T_2.

A second estimator of U/E that is suggested by Table 2.1 is one based on those individuals who appear at both T_1 and T_2. At T_1 its form is

$$R_1^i = (F_{uu} + F_{ue})/(F_{eu} + F_{ee}), \tag{2.5}$$

and at T_2,

$$R_2^i = (F_{uu} + F_{eu})/(F_{ue} + F_{ee}). \tag{2.6}$$

These are the estimates which are usually recommended for obtaining maximum accuracy on estimates of change (see Section 1).

Finally, there is a "singles" estimate which is based *only* on people who appear at that specific observation time. Consequently, the "singles" estimate at T_1 is

$$R_1^s = F_{uo}/F_{eo}, \tag{2.7}$$

and at T_2 is,

$$R_2^s = F_{ou}/F_{oe}. \tag{2.8}$$

In practice the estimates based on "total" and "identical" persons are commonly used. The "singles" estimate is used less often, usually for comparative purposes.

2.3. The Response Probabilities

The basis for examination of fixed panel surveys is an elementary probability model similar to the one used by Williams and Mallows [13]. To do this let P_u be the probability that a response is actually obtained at T_1 from a person who is unemployed at T_1; P_e the probability that a response is actually obtained at T_1 from a person who is employed at time T_1; and P_{uu} the probability of obtaining an observation at T_2 from a person who was unemployed at both T_1 and T_2 and who *was* observed at T_1. Similar interpretations are given to P_{ue}, P_{eu}, and P_{ee}. For example, P_{eu} is the probability that an individual is actually observed at time T_2 given that he was observed at T_1 *and* was employed at T_1 and unemployed at T_2. Finally, Q_{uu}, Q_{ue}, Q_{eu}, and Q_{ee} are probabilities similar to the P's except that the Q's are conditional upon the individual *not* appearing in the sample at T_1. So, for example, Q_{uu} is the probability that an individual is observed at T_2 given that he was *not* observed at T_1 *and* was unemployed at both T_1 and T_2. In summary, P_u and P_e are the original selection probabilities; P_{uu}, P_{ue}, P_{eu}, and P_{ee} are the retention probabilities; and Q_{uu}, Q_{ue}, Q_{eu}, and Q_{ee} are the replenishment probabilities. We shall discuss each of these briefly.

The first stage probabilities, P_u and P_e, can obviously affect the estimates at sample time T_1. The systematic continuing effects which they can have on a fixed panel throughout the duration of the survey are less obvious, however. In fact, as we shall show, if $P_u \neq P_e$, the effects will be felt throughout the duration of the survey and not simply at T_1. This is true for both R^t and R^i, and hence is true even in the case when 100% of the originally *observed* panel is retained throughout the survey.

The second stage probabilities, P_{uu}, P_{ue}, P_{eu}, and P_{ee} are the retention probabilities. In fixed panel surveys, the goal is to retain as many of the original panel as possible. This means that P_{uu}, P_{ue}, P_{eu}, and P_{ee} are all *ideally* equal to 1. In practice, of course, this rarely happens. Some members of the panel are invariably lost to the survey. Consequently in virtually all applications these second stage P's are somewhat less than 1. And, as with the first stage P's, if these probabilities are not all equal, then the effects will be felt systematically throughout the survey.

Finally, the second stage Q's, Q_{uu}, Q_{ue}, Q_{eu}, and Q_{ee} are the replenishment probabilities. Ideally, in a fixed panel survey, these Q probabilities will be zero because there will be no need for sample replenishment. In practice, of course, virtually every continuing survey loses sample units. Sometimes there is no effort to replace these lost units and in this case the Q probabilities are zero. In other cases, and perhaps in most cases, there is an operative replenishment policy. This means that the Q probabilities are nonzero and if they are also related to the classifications and measurements under study, then some systematic biases will appear in the survey.

Using the specified probabilities, it is possible to construct the expectations given in Table 2.2. In the ideal fixed panel survey in which everyone appears in the survey who is supposed to (i.e., first stage P's equal 1), and everyone is retained (i.e., second stage P's equal 1), so that no new sample units are drawn as replacements (i.e., the second stage Q's are equal to zero), the expectations are given in Table 2.3. These are exactly the desired expectations, but as a comparison with Table 2.2 suggests, these ideal expectations probably never apply due to the influence of the response probabilities.

Other special cases can be easily written down using Table 2.2. For example, if we wanted to study the effects of having no replenishment policy, then we would assume that $Q_{uu} = Q_{ue} = Q_{eu} = Q_{ee} = 0$ and the expectations are as shown in Table 2.4.

TABLE 2.2

Table of General Expectations[a]

Status at time T_1	Status at time T_2		
	Unemployed	Employed	Not interviewed
Unemployed	$N_{uu}P_uP_{uu}$	$N_{ue}P_uP_{ue}$	$N_{uu}P_u(1-P_{uu}) + N_{ue}P_u(1-P_{ue})$
Employed	$N_{eu}P_eP_{eu}$	$N_{ee}P_eP_{ee}$	$N_{eu}P_e(1-P_{eu}) + N_{ee}P_e(1-P_{ee})$
Not interviewed	$N_{uu}(1-P_u)Q_{uu} + N_{eu}(1-P_e)Q_{eu}$	$N_{ue}(1-P_u)Q_{ue} + N_{ee}(1-P_e)Q_{ee}$	$N_{uu}(1-P_u)(1-Q_{uu}) + N_{ue}(1-P_u)(1-Q_{ue}) + N_{eu}(1-P_e)(1-Q_{eu}) + N_{ee}(1-P_e)(1-Q_{ee})$

[a] Notation: P_u, P_e, original response probabilities, P_{uu}, P_{ue}, P_{eu}, P_{ee}, retention probabilities, Q_{uu}, Q_{ue}, Q_{eu}, Q_{ee}, replenishment probabilities.

TABLE 2.3

Ideal Expectations for a Fixed Panel

Status at time T_1	Status at time T_2		
	Unemployed	Employed	Not interviewed
Unemployed	N_{uu}	N_{ue}	0
Employed	N_{eu}	N_{ee}	0
Not interviewed	0	0	0

TABLE 2.4

Expected Cell Numbers in a Fixed Panel with Incomplete Original Sample, Attrition, and No Replenishment

Status at time T_1	Status at time T_2		
	Unemployed	Employed	Not interviewed
Unemployed	$N_{uu}P_uP_{uu}$	$N_{ue}P_eP_{ue}$	$N_{uu}P_u(1-P_{uu}) + N_{ue}P_u(1-P_{ue})$
Employed	$N_{eu}P_eP_{eu}$	$N_{ee}P_eP_{ee}$	$N_{eu}P_e(1-P_{eu}) + N_{ee}P_e(1-P_{ee})$
Not interviewed	0	0	$N_{uu}(1-P_u) + N_{ue}(1-P_u) + N_{eu}(1-P_e) + N_{ee}(1-P_e)$

It is unnecessary to write down the expectations for more of such cases; so we shall proceed to discuss the characteristics of the estimators and some illustrative numerical examples.

2.4. Characteristics of the Estimates

2.4.1. The "Total" Estimator

Using Table 2.2 the approximate expectations of the estimates R_1^t, R_2^t, R_1^i, R_2^i, R_1^s, and R_2^s can be written down. The approximation is that of the well-known ratio estimator, [3]. First, the expectations of R_1^t and R_2^t are given by (2.9) and (2.10):

$$E(R_1^t) = \frac{N_{uu}P_uP_{uu} + N_{ue}P_uP_{ue} + N_{uu}P_u(1 - P_{uu}) + N_{ue}P_u(1 - P_{ue})}{N_{eu}P_eP_{eu} + N_{ee}P_eP_{ee} + N_{eu}P_e(1 - P_{eu}) + N_{ee}P_e(1 - P_{ee})}$$

$$= \frac{P_u}{P_e}\frac{N_{uu} + N_{ue}}{N_{eu} + N_{ee}}, \tag{2.9}$$

$$E(R_2^t) = \frac{N_{uu}P_uP_{uu} + N_{eu}P_eP_{eu} + N_{uu}(1 - P_u)Q_{uu} + N_{eu}(1 - P_e)Q_{eu}}{N_{ue}P_uP_{ue} + N_{ee}P_eP_{ee} + N_{ue}(1 - P_u)Q_{ue} + N_{ee}(1 - P_e)Q_{ee}}. \tag{2.10}$$

It can be easily seen that if the response probabilities take on their ideal values, i.e., $P_u = P_e = 1$; $P_{uu} = P_{ue} = P_{eu} = P_{ee} = 1$; and $Q_{uu} = Q_{ue} = Q_{eu} = Q_{ee} = 0$, then the expectations are exactly equal to the desired values. Specifically,

$$E(R_1^t) = (N_{uu} + N_{ue})/(N_{eu} + N_{ee}) \quad \text{and} \quad E(R_2^t) = (N_{uu} + N_{eu})/(N_{ue} + N_{ee}). \tag{2.11}$$

From (2.9) and (2.10) a number of algebraic results can be obtained. First we have two general remarks:

(1) Neither $E(R_1^t)$ nor $E(R_2^t)$ are necessarily equal to the true value and each can be distorted by almost all of the response probabilities. The exception is that $E(R_1^t)$ does not depend upon the Q's and the retention probabilities.

(2) $E(R_1^t)$ and $E(R_2^t)$ are not generally the same *even* when the population ratio does *not* change, i.e., $N_{eu} = N_{ue}$. That is, there is a change in the expectation of the total estimator even when there is no change in the population.

Next it is helpful to separate out the effects of each of the three stages of response probabilities. To do this we consider three different kinds of restrictions on the selection probabilities. Under H_1, P_u and P_e are not restricted but the second stage P's and Q's are, so that we may study the effect of differences in the first stage response probabilities. Under H_2, P_{uu}, P_{ue}, P_{eu}, and P_{ee} are not restricted; so the effect of differences in the second stage retention probabilities are brought out. Finally H_3 is formulated to examine the effect of the replenishment probabilities, the Q's.

$$H_1: \quad P_{uu} = P_{ue} = P_{eu} = P_{ee} = P_2$$

$$Q_{uu} = Q_{ue} = Q_{eu} = Q_{ee} = Q_2$$

$$H_2: \quad P_u = P_e = P_1$$

$$Q_{uu} = Q_{ue} = Q_{eu} = Q_{ee} = Q_2$$

$$H_3: \quad P_u = P_e = P_1$$

$$P_{uu} = P_{ue} = P_{eu} = P_{ee} = P_2$$

With these simplifying assumptions some additional results can be obtained.

(a) *Under* H_1
(1) R_1^t is unbiased iff $P_u = P_e$.
(2) R_2^t is unbiased iff
 (i) $P_u = P_e$ or
 (ii) $P_2 = Q_2$ or
 (iii) $N_{uu}/N_{ue} = N_{eu}/N_{ee}$.

In case (2), it is interesting to observe further that

$$\text{Bias}\,(R_2^t) = \frac{aN_{uu} + bN_{eu}}{aN_{ue} + bN_{ee}} - \frac{N_{uu} + N_{eu}}{N_{ue} + N_{ee}},$$

$$= \frac{(a-b)(N_{uu}N_{ee} - N_{ue}N_{eu})}{(aN_{ue} + bN_{ee})(N_{ue} + N_{ee})},$$

where

$$a = P_u P_2 + (1 - P_u)Q_2, \qquad b = P_e P_2 + (1 - P_e)Q_2.$$

Now a is the unconditional probability that a person who is in the U category at T_1 is interviewed at T_2. Similarly, b is the unconditional probability that a person who is E at T_1 is observed at T_2. Under H_1, $a = b \Leftrightarrow P_u = P_e$ or $P_2 = Q_2$. We shall see that the requirement of equality in the unconditional probabilities at T_2 occurs in a number of other similar places.

(3) If $N_{eu} = N_{ue}$, $E(R_2^t) - E(R_1^t) = 0$, iff $P_u = P_e$.

(4) As a result of (2) and (3), we see that the first round probabilities, P_u and P_e, can reach from T_1 to T_2 and create biases in both R_2^t and in the change estimate $R_2^t - R_1^t$.

(b) *Under* H_2

(1) R_1^t is unbiased always. That is P_{uu}, P_{ue}, P_{eu}, P_{ee} cannot affect the estimate at T_1.

(2) R_2^t is unbiased

$$\text{iff} \qquad a = b = c = d,$$

where

$$a = P_1 P_{uu} + (1 - P_1)Q_2, \qquad b = P_1 P_{eu} + (1 - P_1)Q_2$$

$$c = P_1 P_{ue} + (1 - P_1)Q_2, \qquad d = P_1 P_{ee} + (1 - P_1)Q_2$$

which implies Bias $(R_2^t) = 0$,

$$\text{iff } P_{uu} = P_{ue} = P_{eu} = P_{ee}.$$

Again we see that it is the unconditional second stage probabilities a, b, c, and d which need to be equal in order that the estimate at T_2 be unbiased.

(3) If $N_{eu} = N_{ue}$, $E(R_2^t) - E(R_1^t) = 0$ under exactly the same conditions as Bias $(R_2^t) = 0$. This must be true because R_1^t is unbiased at all times under H_2.

(c) *Under* H_3

(1) The estimator R_1^t is always unbiased under H_3. The Q replacement probabilities cannot influence the "totals" estimator at T_1.

(2) R_2^t is unbiased

$$\text{iff } a = b = c = d,$$

where

$$a = P_1 P_2 + (1 - P_1)Q_{uu}, \qquad b = P_1 P_2 + (1 - P_1)Q_{eu}$$

$$c = P_1 P_2 + (1 - P_1)Q_{ue}, \qquad d = P_1 P_2 + (1 - P_1)Q_{ee}$$

which implies $Q_{uu} = Q_{ue} = Q_{eu} = Q_{ee}$. Notice again that the estimator at T_2 is unbiased if the *un*conditional probabilities a, b, c, and d are equal.

2.4.2. The Identicals or Matched Estimator

As was discussed earlier, it has been argued that estimates based on identical sets of individuals who are followed through time give unambiguous estimates of change through time. Since estimates based on identical or matched sets of individuals are commonly used for the above (false) reason, it is important that we turn now to consideration of the estimates R_1^i and R_2^i.

(1) In general, $E(R_1^i) = E(R_2^i)$ iff $P_u/P_e = P_{eu}/P_{ue}$ when $N_{eu} = N_{ue}$. This disturbing result says that unless the ratio of response probabilities at T_1 is the same at T_2 for those persons who changed employment status, then the expectation of the identical estimates R_1^i and R_2^i will differ even though the population has *not* changed. Next we again consider the three hypotheses H_1, H_2, H_3 as specified in Section 2.4.1.

(a) *Under* H_1

(1) R_1^i is unbiased iff $P_u = P_e$.

(2) R_2^i is unbiased iff $P_u = P_e$ and again we see that a difference between P_u and P_e can bias an estimator at T_2.

(3) Assuming $N_{ue} = N_{eu}$, $E(R_1^i) - E(R_2^i) = 0$ iff $P_u = P_e$. That is, a response probability distortion at T_1 can cause a bias in the estimate of change from T_1 to T_2.

(b) *Under* H_2

(1) R_1^i is unbiased iff $P_{uu} = P_{ue} = P_{eu} = P_{ee}$. Here we have the reverse of the earlier phenomenon in that differences among the second stage retention probabilities can reach back and bias the estimator at T_1. This comes about because the development of a matched set of individuals for T_1 and T_2 depends on the retention probabilities at T_2.

(2) R_2^i, like R_1^i, is unbiased iff $P_{uu} = P_{ue} = P_{eu} = P_{ee}$.

(3) Assuming $N_{ue} = N_{eu}$, it can be shown that under H_2 $E(R_1^i - R_2^i) = 0$ iff $P_{ue} = P_{eu}$. The interesting aspect of this case is that both estimators R_1^i and R_2^i may be biased while the estimate of change is unbiased.

(c) *Under* H_3

R_2^i and R_1^i are always unbiased. The probabilities Q_{uu}, Q_{ue}, Q_{eu}, and Q_{ee} do not affect the estimators R_1^i and R_2^i.

2.4.3. *The Singles Estimator*

The singles estimator R^s is used less than either R^t or R^i. However, since it is used occasionally, it is worthwhile examining briefly.

(a) *Under* H_1

(1) It can be shown that R_1^s and R_2^s are unbiased and $E(R_1^s - R_2^s) = 0$ (when $N_{ue} = N_{eu}$) iff $P_u = P_e$.

(b) *Under* H_2

(1) R_1^s is unbiased iff $P_{uu} = P_{ue} = P_{eu} = P_{ee}$ and we see again that the effect of the second stage retention probabilities can reach from T_2 to T_1 to bias an estimator.

(2) R_2^s is always unbiased under H_2.

(3) Assuming $N_{ue} = N_{eu}$, $E(R_1^s - R_2^s) = 0$ iff $P_{uu} = P_{ue} = P_{eu} = P_{ee}$. This can be seen easily from the consideration of points 1 and 2 simultaneously.

(c) *Under* H_3

(1) R_1^s is always unbiased under H_3. The Q probabilities do not affect R_1^s.

(2) R_2^s is unbiased iff $Q_{uu} = Q_{ue} = Q_{eu} = Q_{ee}$.

(3) If $N_{eu} = N_{ue}$, $E(R_1^s - R_2^s) = 0$ iff $Q_{uu} = Q_{ue} = Q_{eu} = Q_{ee}$.

2.5. Sample Balancing Conditions

We remarked earlier that some sample surveys had been implemented with a replenishment policy (the Q's) which was intended to offset suspected biases resulting from systematic attrition. To this point, it can be shown that R_2^t, the total estimator at T_2 is unbiased iff $a = b = c = d$, where

$$a = P_u P_{uu} + (1 - P_u)Q_{uu}, \qquad b = P_u P_{ue} + (1 - P_u)Q_{ue}$$

$$c = P_e P_{eu} + (1 - P_e)Q_{eu}, \qquad d = P_e P_{ee} + (1 - P_e)Q_{ee}.$$

This is a more general form of conditions which arose earlier in this paper. Specifically, a is the unconditional probability that a UU person appears in the sample at T_2 with similar interpretations for b, c, and d.

In theory, if P_u, P_e, P_{uu}, P_{ue}, P_{eu}, and P_{ee} were all known, then Q's could be found which would unbias the estimate at T_2. (The Q's would have to be scaled to some appropriate expected sample size.) In practice this would have major difficulties. First, it is univariate. Virtually every survey is multivariate in character and unbiasing one variate will not necessarily help any others. In fact, it could make their biases worse. Second, it is difficult to see how to implement the desired Q's because membership in the U, E categories is not known in advance of sampling.

2.6. Numerical Examples

Seven numerical examples are included. Each one is constructed to demonstrate specific characteristics of the various biases. Five of the examples are based on a population assumed to have a $U/(U + E)$ ratio of 0.05. [While the algebra was discussed in terms of the U/E ratio, the numerical examples are in terms of the more familiar $U/(U + E)$.] We shall refer to these as the employment examples. The last two examples have a population $U/(U + E)$ ratio equal to 0.50. This ratio allows comparisons of the results for a population with medium size $U/(U + E)$ with the first population and its relatively small $U/(U + E)$ fraction.

In Example A, the numerical values selected for the parameters are consistent with H_1. Specifically, they indicate the influence of the two first stage P's. Furthermore, the probabilities are selected so that the response rate is relatively high and increases somewhat from T_1 to T_2. Such an increase in overall response rate is consistent with general experience. The high level of the response rate is *not* usual, however. Rates in the neighborhood of 40 to 70% are more common in practice with response rates of 90% being characteristic mostly only of surveys run by the U.S. Census Bureau. In short, the probabilities are constructed so that in advance one might expect that the biases would be relatively innocuous. However, examination of the estimates shows that they are not negligible, even in this case in which there is no prior reason to suspect large biases. R^t and R^i are biased badly enough but R^s is much worse.

EXAMPLE A

I. Population

$N_{uu} = 3500 \qquad N_{ue} = 1500 \qquad N_{eu} = 1500 \qquad N_{ee} = 93{,}500.$

II. Response probabilities

First stage: $P_u = 0.94 \qquad P_e = 0.88$

Retention: $P_{uu} = 0.90 \qquad P_{ue} = 0.90 \qquad P_{eu} = 0.90 \qquad P_{ee} = 0.90$

Replenishment: $Q_{uu} = 0.90 \qquad Q_{ue} = 0.90 \qquad Q_{eu} = 0.90 \qquad Q_{ee} = 0.90$

III. Expected responses

$E(n_1) = 88{,}300 \qquad E(n_2) = 90{,}000$

IV. Expected value of estimators

	True	R^t	R^i	R^s
T_1	0.0500	0.0532	0.0532	0.0532
T_2	0.0500	0.0500	0.0522	0.0333
$T_2 - T_1$	0.0000	−0.0032	−0.0010	−0.0199

V. Expected sample numbers

	Observed status at T_2			
Observed status at T_1	Unemployed	Employed	Not interviewed	Total
Unemployed	2,961	1,269	470	4,700
Employed	1,188	74,052	8,360	83,600
Not interviewed	351	10,179	1,170	11,700
Total	4,500	85,500	10,000	100,000

Example B is consistent with H_2 and indicates the effect of the second stage attrition probabilities. Again the biases are not trivial for R^t and R^i and the singles estimate R^s is very bad.

EXAMPLE B

I. Population

$N_{uu} = 3500$ $N_{ue} = 1500$ $N_{eu} = 1500$ $N_{ee} = 93{,}500$

II. Response probabilities

First stage: $P_u = 0.89$ $P_e = 0.89$

Retention: $P_{uu} = 0.86$ $P_{ue} = 0.93$ $P_{eu} = 0.87$ $P_{ee} = 0.95$

Replenishment: $Q_{uu} = 0.90$ $Q_{ue} = 0.90$ $Q_{eu} = 0.90$ $Q_{ee} = 0.90$

III. Expected responses

$E(n_1) = 89{,}000$ $E(n_2) = 94{,}036$

IV. Expected value of estimators

	True	R^t	R^i	R^s
T_1	0.0500	0.0500	0.0466	0.1089
T_2	0.0500	0.0461	0.0456	0.0500
$T_2 - T_1$	0.0000	−0.0039	−0.0010	−0.0589

V. Expected sample numbers

	Observed status at T_2			
Observed status at T_1	Unemployed	Employed	Not interviewed	Total
Unemployed	2,679	1,242	530	4,450
Employed	1,161	79,054	4,334	84,550
Not interviewed	495	9,405	1,100	11,000
Total	4,335	89,701	5,964	100,000

Example C is consistent with H_3 and suggests what effect the replenishment policy could have in the event of no biasing effects at the other stages. Notice that in this example, the effect of the Q probabilities is not as large as in the two earlier examples. This seems to be true in many cases; however, it is possible to construct examples in which the Q's have a very large effect.

EXAMPLE C

I. Population

$N_{uu} = 3500$ $N_{ue} = 1500$ $N_{eu} = 1500$ $N_{ee} = 93{,}500$

II. Response probabilities

First stage:	$P_u = 0.89$	$P_e = 0.89$		
Retention:	$P_{uu} = 0.90$	$P_{ue} = 0.90$	$P_{eu} = 0.90$	$P_{ee} = 0.90$
Replenishment:	$Q_{uu} = 0.86$	$Q_{ue} = 0.93$	$Q_{eu} = 0.87$	$Q_{ee} = 0.95$

III. Expected responses

$E(n_1) = 89{,}000$ $E(n_2) = 90{,}499$

IV. Expected value of estimators

	True	R^t	R^i	R^s
T_1	0.0500	0.0500	0.0500	0.0500
T_2	0.0500	0.0495	0.0500	0.0456
$T_2 - T_1$	0.0000	−0.0005	0.0000	−0.0044

V. Expected sample numbers

	Observed status at T_2			
Observed status at T_1	Unemployed	Employed	Not interviewed	Total
Unemployed	2,804	1,202	445	4,450
Employed	1,202	74,894	8,455	84,550
Not interviewed	475	9,924	601	11,000
Total	4,480	86,019	9,501	100,000

In Example D, the same population is used as in the first three examples. In this case, however, the probabilities at all three levels are allowed to vary so that we may observe their effects in combination. Notice that the biases are bigger than in the first three cases, and that this happens *in spite of a larger expected response at* T_2.

EXAMPLE D

I. Population

$N_{uu} = 3500$ $N_{ue} = 1500$ $N_{eu} = 1500$ $N_{ee} = 93{,}500$

II. Response probabilities

First stage: $P_u = 0.94$ $P_e = 0.88$

Retention: $P_{uu} = 0.86$ $P_{ue} = 0.93$ $P_{eu} = 0.87$ $P_{ee} = 0.95$

Replenishment: $Q_{uu} = 0.86$ $Q_{ue} = 0.93$ $Q_{eu} = 0.87$ $Q_{ee} = 0.95$

III. Expected responses

$E(n_1) = 88{,}300$ $E(n_2) = 94{,}535$

IV. Expected value of estimators

	True	R^t	R^i	R^s
T_1	0.0500	0.0532	0.0496	0.1154
T_2	0.0500	0.0456	0.0477	0.0304
$T_2 - T_1$	0.0000	−0.0076	−0.0020	−0.0850

V. Expected sample numbers

	Observed status at T_2			
Observed status at T_1	Unemployed	Employed	Not interviewed	Total
Unemployed	2,829	1,311	559	4,700
Employed	1,148	78,166	4,286	83,600
Not interviewed	337	10,743	620	11,700
Total	4,315	90,220	5,465	100,000

In Example E, the response probability for U category persons has been dropped substantially at time T_2. Notice, however, that there is no operational way of distinguishing between Examples D and E because the response rates are virtually identical (and very high) in both cases. Nevertheless, the biases have gone from very bad in Example D to something much worse in Example E. The singles estimate at T_2 is over five times as large as it ought to be. Again, it needs to be stressed that there is no immediately available operational way of detecting this situation.

EXAMPLE E

I. Population

$N_{uu} = 3500$ $N_{ue} = 1500$ $N_{eu} = 1500$ $N_{ee} = 93{,}500$

II. Response probabilities

First stage: $P_u = 0.94$ $P_e = 0.88$

Retention: $P_{uu} = 0.50$ $P_{ue} = 0.93$ $P_{eu} = 0.50$ $P_{ee} = 0.95$

Replenishment: $Q_{uu} = 0.50$ $Q_{ue} = 0.93$ $Q_{eu} = 0.50$ $Q_{ee} = 0.95$

III. Expected responses

$E(n_1) = 88{,}300$ $E(n_2) = 92{,}720$

IV. Expected value of estimators

	True	R^t	R^i	R^s
T_1	0.0500	0.0532	0.0361	0.2675
T_2	0.0500	0.0270	0.0282	0.0178
$T_2 - T_1$	0.0000	−0.0263	−0.0080	−0.2497

V. Expected sample numbers

	Observed status at T_2			
Observed status at T_1	Unemployed	Employed	Not interviewed	Total
Unemployed	1,645	1,311	1,744	4,700
Employed	660	78,166	4,774	83,600
Not interviewed	195	10,743	762	11,700
Total	2,500	90,220	7,280	100,000

In Examples F and G, the P and Q parameters are the same as in Examples D and E, respectively. The population, however, has been changed to one which has an even split in the two categories. In Example F, the biases are very bad. In Example G, they are again much worse. These last two examples are included lest the reader be misled by the belief that large biases would be less likely in populations with U/E fractions more moderate than those in Examples A to D. We refer to F and G as election examples. Notice that in both of these cases all of the evidence points to a swing from candidate U to candidate E, when in fact the only thing that has changed is the response rates for the two categories.

EXAMPLE F

I. Population

$N_{uu} = 40{,}000 \quad N_{ue} = 10{,}000 \quad N_{eu} = 10{,}000 \quad N_{ee} = 40{,}000$

II. Response probabilities

First stage: $P_u = 0.94 \quad P_e = 0.88$

Retention: $P_{uu} = 0.86 \quad P_{ue} = 0.93 \quad P_{eu} = 0.87 \quad P_{ee} = 0.95$

Replenishment: $Q_{uu} = 0.86 \quad Q_{ue} = 0.93 \quad Q_{eu} = 0.87 \quad Q_{ee} = 0.95$

III. Expected responses

$E(n_1) = 91{,}000 \quad E(n_2) = 90{,}400$

IV. Expected value of estimators

	True	R^t	R^i	R^s
T_1	0.5000	0.5165	0.4999	0.6710
T_2	0.5000	0.4768	0.4867	0.3778
$T_2 - T_1$	0.0000	−0.0397	−0.0132	−0.2931

V. Expected sample numbers

	Observed status at T_2			
Observed status at T_1	Unemployed	Employed	Not interviewed	Total
Unemployed	32,336	8,742	5,922	47,000
Employed	7,656	33,440	2,904	44,000
Not interviewed	3,108	5,118	774	9,000
Total	43,100	47,300	9,600	100,000

EXAMPLE G

I. Population

$N_{uu} = 40{,}000$ $N_{ue} = 10{,}000$ $N_{eu} = 10{,}000$ $N_{ee} = 40{,}000$

II. Response probabilities

First stage: $P_u = 0.94$ $P_e = 0.88$

Retention: $P_{uu} = 0.50$ $P_{ue} = 0.93$ $P_{eu} = 0.50$ $P_{ee} = 0.95$

Replenishment: $Q_{uu} = 0.50$ $Q_{ue} = 0.93$ $Q_{eu} = 0.50$ $Q_{ee} = 0.95$

III. Expected responses

$E(n_1) = 91{,}000$ $E(n_2) = 72{,}300$

IV. Expected value of estimators

	True	R^t	R^i	R^s
T_1	0.5000	0.5165	0.4212	0.7595
T_2	0.5000	0.3458	0.3548	0.2602
$T_2 - T_1$	0.0000	−0.1707	−0.0664	−0.4994

V. Expected sample numbers

	Observed status at T_2			
Observed status at T_1	Unemployed	Employed	Not interviewed	Total
Unemployed	18,800	8,742	19,458	47,000
Employed	4,400	33,440	6,160	44,000
Not interviewed	1,800	5,118	2,082	9,000
Total	25,000	47,300	27,700	100,000

3. SAMPLING AT THREE OBSERVATION TIMES

To develop a model for three observation periods, it is first necessary to extend the two-period notation slightly. Specifically, let N_{uuu} be the number of persons who are unemployed at each of the three interview times, T_1, T_2, and T_3. Similar interpretations are given to the other seven possibilities, N_{uue}, N_{ueu}, etc. Using this notation, it can easily be seen that

$$\begin{aligned} R_1 = R_2 &\Leftrightarrow N_{ueu} + N_{uee} = N_{euu} + N_{eue} \\ R_2 = R_3 &\Leftrightarrow N_{uue} + N_{eue} = N_{ueu} + N_{eeu} \\ R_1 = R_3 &\Leftrightarrow N_{uue} + N_{uee} = N_{euu} + N_{eeu}. \end{aligned} \tag{3.1}$$

These conditions are the generalization of the two-stage condition that $R_1 = R_2 \Leftrightarrow N_{ue} = N_{eu}$. The new conditions have both the same proof and the same logical interpretation.

Next, P_{uuu} is the probability of actually obtaining a response from an unemployed person at T_3 given that he was unemployed at T_1, T_2, and T_3 and was interviewed at both T_1 and T_2. An analogous interpretation applies to P_{uue}, P_{ueu}, P_{euu}, etc. In this three-period model, we shall allow only for fixed panel surveys where the sample is selected and retained subject only to inadvertent losses and with no replenishment. One of the reasons for this is that the extension of the conditional P, Q probabilities to three stages requires new notation for each of the four possible sequences of interviewed and not interviewed; and since there are eight possible sequences of U and E, there are 32 new parameters. Consequently, some simplification is in order, and we have chosen to discuss the fixed panel with no replenishment.

Consequently, in this specific case the expectations of the estimators based on the total number of persons available at T_1, T_2, and T_3 are approximately

$$E(R_1^t) = \frac{P_u(N_{uu} + N_{ue})}{P_e(N_{eu} + N_{ee})} = \frac{P_u N_u}{P_e N_e}, \tag{3.2a}$$

$$E(R_2^t) = \frac{P_u P_{uu} N_{uu} + P_e P_{eu} N_{eu}}{P_u P_{ue} N_{ue} + P_e P_{ee} N_{ee}}, \tag{3.2b}$$

and

$$E(R_3^t) = \frac{P_u P_{uu} P_{uuu} N_{uuu} + P_u P_{ue} P_{ueu} N_{ueu} + P_e P_{eu} P_{euu} N_{euu} + P_e P_{ee} P_{eeu} N_{eeu}}{P_u P_{uu} P_{uue} N_{uue} + P_u P_{ue} P_{uee} N_{uee} + P_e P_{eu} P_{eue} N_{eue} + P_e P_{ee} P_{eee} N_{eee}}. \tag{3.3}$$

And for the identicals at each interview period

$$E(R_1^i) = \frac{P_u P_{uu} P_{uuu} N_{uuu} + P_u P_{uu} P_{uue} N_{uue} + P_u P_{ue} P_{ueu} N_{ueu} + P_u P_{ue} P_{uee} N_{uee}}{P_e P_{eu} P_{euu} N_{euu} + P_e P_{eu} P_{eue} N_{eue} + P_e P_{ee} P_{eeu} N_{eeu} + P_e P_{ee} P_{eee} N_{eee}} \tag{3.4}$$

$$E(R_2^i) = \frac{P_u P_{uu} P_{uuu} N_{uuu} + P_u P_{uu} P_{uue} N_{uue} + P_e P_{eu} P_{euu} N_{euu} + P_e P_{eu} P_{eue} N_{eue}}{P_u P_{ue} P_{ueu} N_{ueu} + P_u P_{ue} P_{uee} N_{uee} + P_e P_{ee} P_{eeu} N_{eeu} + P_e P_{ee} P_{eee} N_{eee}} \tag{3.5}$$

$$E(R_3^i) = E(R_3^t). \tag{3.6}$$

Using these expressions, some interesting results can be derived. The first result is that even more than (3.6) is true; in fact, R_3^i and R_3^t are identical estimates. This is clear because all the people in the survey at T_3 are people who have been in the survey since T_1. With no replenishment this will always be true for the latest observation time.

Furthermore, consider the simple case that

$$P_u = P_{uu} = P_{eu} = P_{uuu} = P_{ueu} = P_{euu} = P_{eeu} = P$$

and

$$P_e = P_{ue} = P_{ee} = P_{uue} = P_{uee} = P_{eue} = P_{eee} = \pi \tag{3.7}$$

and assume that (3.1) hold. Then

$$E(R_1^t) = E(R_3^t) \qquad \text{iff} \qquad P = \pi \tag{3.8}$$

or

$$\frac{N_u}{N_e} = \frac{N_{uuu}}{N_{uue}} = \frac{N_{ueu}}{N_{uee}} = \frac{N_{euu}}{N_{eue}} = \frac{N_{eeu}}{N_{eee}}. \tag{3.9}$$

Similar conditions can be found for the other change biases.

It should be pointed out, however, that for many populations condition (3.9) is *not* likely to hold. For example, if the overall unemployment rate is 4% at T_1, it seems likely that the unemployment rate at T_3 among those persons unemployed at T_1 and T_2 will be much higher than 4%.

Condition (3.8) is simple and to the point. It says that the response rate is the same for employed persons as for unemployed persons. Consequently, if (3.9) and the analogous conditions for R_2^t and R_3^t are not true, then the three total estimates, R_1^t, R_2^t, R_3^t, will not be equal unless the probability of actually obtaining a response from an employed person is the same as the probability of actually obtaining a response from an unemployed person, i.e., $\pi = P$.

This result is quite disturbing because it says that even if there is no real change in the population unemployment rate, a *constant* difference $\pi - P$ can cause the estimates to be different at each of the three survey periods. In an extreme case, if $\pi \neq P$ and there is no shifting in employment status from one period to the next, i.e., $N_{ue} = N_{eu} = N_{ueu} = \text{etc.} = 0$, then the unemployment ratio is constant at all observation periods and is equal to N_u/N_e. In this case, it may be seen that

$$E(R_k^t) = \frac{P^k N_u}{\pi^k N_e}, \tag{3.10}$$

so that the estimate either increases or goes to zero depending upon which group is being lost at a faster rate.

It can be shown also that $E(R_1^i)$, $E(R_2^i)$, and $E(R_3^i)$ are not necessarily equal even though they are based on an identical set of individuals. In fact, it can be shown that $E(R_1^i - R_2^i) = 0 \Leftrightarrow$

$$P_u P_{ue} P_{ueu} N_{ueu} + P_u P_{ue} P_{uee} N_{uee} = P_e P_{eu} P_{euu} N_{euu} + P_e P_{eu} P_{eue} N_{eue}, \tag{3.11}$$

which in the simple case of

$$P_u = P_{uu} = P_{eu} = P_{uuu} = P_{ueu} = P_{euu} = P_{eeu} = P$$

$$P_e = P_{ue} = P_{ee} = P_{uue} = P_{uee} = P_{eue} = P_{eee} = \pi$$

reduces to

$$E(R_1^i - R_2^i) = 0 \qquad \Leftrightarrow \qquad \pi = P.$$

The same result will be found in the other cases, i.e.

$$E(R_2^i - R_3^i) = 0 \Leftrightarrow \pi = P$$

and

$$E(R_1^i - R_3^i) = 0 \Leftrightarrow \pi = P.$$

This unfortunately means that any difference in the probabilities of selection which is related to the characteristic being measured, creates a systematic bias in the expectation of the estimator even in the case in which no change occurs in the U/E ratio in the population.

Explicitly, in this simplified case, we have

$$E(R_1^i) = \frac{P^3 N_{uuu} + \pi P^2 N_{uue} + P^2 \pi N_{ueu} + P\pi^2 N_{uee}}{P^2 \pi N_{euu} + \pi^2 P N_{eue} + P\pi^2 N_{eeu} + \pi^3 N_{eee}},$$

$$E(R_2^i) = \frac{P^3 N_{uuu} + P^2 \pi N_{uue} + P^2 \pi N_{euu} + P\pi^2 N_{eue}}{P^2 \pi N_{ueu} + P\pi^2 N_{uee} + P\pi^2 N_{eeu} + \pi^3 N_{eee}},$$

$$E(R_3^i) = \frac{P^3 N_{uuu} + P^2 \pi N_{ueu} + P^2 \pi N_{euu} + P\pi^2 N_{eeu}}{P^2 \pi N_{uue} + P\pi^2 N_{uee} + P\pi^2 N_{eue} + \pi^3 N_{eee}}.$$

Consequently, the systematic response of persons in the fixed panel survey can cause systematic behavior in the expectation of the estimator.

4. SUMMARY DISCUSSION

Systematic behavior in the probabilities of nonresponse can cause unfortunate biases in estimates obtained from fixed panel surveys. In particular, if nonresponse is correlated with the characteristic under measurement then substantial difficulties can arise. Differential nonresponse can occur at any or all of three stages. These are: the original sample selection, the attrition of sample units from one observation period to the next, and finally in any replenishment scheme that may be used. Unfortunately, the magnitude of these biases appears to be potentially very large and affects all of the estimates commonly used in panel surveys. Specifically, it affects those estimators based on: all persons, matched (identical) individuals, and persons who appear in the survey irregularly.

REFERENCES

[1] Bailar, B. A. (1973). A common problem in the analysis of panel data. Unpublished manuscripts. U.S. Census Bureau.

[2] Blight, B. J. N. and Scott, A. J. (1973). A stochastic model for repeated surveys. *J. Roy. Statist. Soc. Ser. B* **35** 61–66.

[3] Cochran, W. G. (1963). *Sampling Techniques.* 2nd ed. Wiley, New York.

[4] Eckler, A. R. (1955). Rotation sampling. *Ann. Math. Statist.* **26** 664–685.

[5] Hansen, M. H., Hurwitz, W. N., and Madow, W. G. (1953). *Sample Survey Methods and Theory.* **I, II**. Wiley, New York.

[6] Kish, L. (1965). *Survey Sampling.* Wiley, New York.

[7] Measuring employment and unemployment. Report of the President's Committee to Appraise Employment and Unemployment Statistics. (1962). The White House, Washington, D.C.

[8] Patterson, H. D. (1950). Sampling on successive occasions with partial replacement of units. *J. Roy. Statist. Soc. Ser. B* **26** 241–255.

[9] Rao, J. N. K. and Graham, J. E. (1964). Rotation designs for sampling on repeated occasions. *J. Amer. Statist. Assoc.* **59** 495–509.

[10] Stephan, F. F. and McCarthy, P. J. (1958). *Sampling Opinions.* Wiley, New York.

[11] Waksberg, J. and Pearl, R. B. (1964). The effects of repeated interview in the current population survey. Paper presented at the 47th National Conference of the American Marketing Association, Dallas, Texas.

[12] Williams, W. H. (1969). The systematic bias effects of incomplete response in rotation samples. *Public Opinion Quart.* **33** 593–602.

[13] Williams, W. H. and Mallows, C. L. (1970). Systematic biases in panel surveys due to differential nonresponse. *J. Amer. Statist. Assoc.* **65** 1338–1349.

Sampling in Two or More Dimensions

Robert S. Cochran

UNIVERSITY OF WYOMING

Populations of interest can often be subdivided into strata using more than one variable as the stratification consideration. This multiple stratification often leads to more strata cells than can be accommodated with a traditional one-way stratification design. Jessen, Goodman and Kish, and Bryant, Hartley, and Jessen have addressed this problem and have proposed procedures for drawing the sample that permits cross-classification restrictions to be met with less sample units than a one-way design. In this paper these procedures are reviewed with instances of competition between the procedures discussed. This leads to some recommendations for choices among them. Some examples of where each procedure has been used are also mentioned.

1. INTRODUCTION

Given a population to study, the beginning research worker first considers a complete enumeration, or census. When the resources are not available to carry out such a 100% count, attention switches to the possibility of sampling from the population. Estimates of population parameters are then made on the basis of the sample information.

The simplest form of sampling is simple random sampling where each unit in the population has the same chance to be selected. Estimates are easily made for population parameters and in some cases ratio, difference, or regression procedures can be used to provide more efficient estimates than with a single variable.

Key Words and Phrases. Two-dimensional sampling, square and cubic lattices, probability lattice, controlled selection, two-way stratification.

Simple random samples do not always provide adequate sample information about subsections of the population. Since such information may be valuable to research workers, steps can be taken to guarantee that the subsections are represented in the sample. The process of dividing the population into nonoverlapping subsections is known as stratifying the population. A sample design that gives a predetermined collection of sample units to the strata created is known as stratified sampling. With stratified sampling the allocation may be proportional to the number of population units in the strata or it may be allocated according to some other criterion depending upon the purpose of the study.

After the idea of stratifying the population on one variable, say section of town, has been suggested, other variables that are possible for stratification are often considered. In this case the type of ownership, age of the building, occupation of the head of household, employment status of the housewife, etc., are all possibilities for stratification variables. Some are more practical than others for implementation, but it is conceivable that at least one other proposed variable for stratification could be utilized.

Jessen (1975) gives an illustration of a study of San Diego where the individual census tracts, blocks or groups of blocks, were classified by both their geographic location in the city and the income class in a two-variable stratification scheme. In a survey of the fishing on Pole Mountain, Wyoming, Bryant (1961) used four different types of strata, location in area, time of day, season of summer, and type of day. Hess and Srikantan (1966) referred to a hospital study with size of hospital (in bed count), region of the state (Michigan) and size of community (SMSA) used as stratification variables. Moore *et al.* (1974) describe the 1970 sample for the National Assessment of Educational Progress (NAEP). In the study, strata were established for regions of the country, states within regions, socioeconomic levels, and size of community levels.

Unfortunately, the proliferation of types of strata rapidly increases the number of strata cells created by the cross-classification of the stratification variables. In Jessen's study there were 12 geographic areas and 12 income classes, or 144 strata cells. In Bryant's study there were 5 locations, 2 times of day, 4 seasons, and 2 types of days for $5 \times 2 \times 4 \times 2 = 80$ strata cells. In the hospital study there were four sizes, four regions, and three sizes of SMSA's for a total of 48 strata cells. In the NAEP the country had 4 regions, each region had 12 to 15 states, there were three socioeconomic levels and three sizes of community levels. In a 12-state region there was the potential for $12 \times 3 \times 3 = 108$ strata cells.

This rapid growth of the number of cells often leads to more cells than the research worker can cover with the budget for his study. In Jessen's, funds were available to sample just 24 of the cells; in Bryant's only 46 cells could be covered. Two different sample sizes were investigated in the hospital study, $n = 50$ and $n = 100$. In the NAEP study each region with at least 108 cells was to have a sample of 27 cells selected. Faced with the prospect of having to sacrifice the balance of having the multiple stratification in studies such as these, several procedures have been proposed. Three of these will be discussed in this paper. These are based on work done by Jessen (1970, 1973, and 1975), by Goodman and Kish (1950) as detailed by Hess and Srikantan (1966) and by Bryant, Hartley, and Jessen (1960). The NAEP study also used the Goodman and Kish procedure.

2. GENERAL CONSIDERATION

The general problem is the same. With two or more stratification criteria there are not enough resources to provide each cell with an adequate sample size, particularly if estimates of variances are desired, or if cell selection is to be with probability proportional to the number of units in the strata cells. The solutions differ and depend upon additional specifications and characteristics of the cross classification arrangements.

A. Jessen

When there are two stratification variables, each with the same number of categories L, and the sample size is to be $n = rL$, where r is usually 1 or 2, the name "square" lattice was used by Jessen (1975). The selection procedures given are for either equal probabilities for each cell, using either a "general" lattice or a "Latin" lattice arrangement depending on the exact dimensions of the problem, or unequal probabilities for each cell using a "probability" lattice scheme. The extension to three dimensions (referred to as "cubic" lattice), where there are L^3 cells and the sample is of $n = r^2L$ cells, is also made. Both the equal-probabilities and the unequal-probabilities cases are covered. The San Diego study was used as an illustration of both the square and cubic probability lattice procedures. In the square lattice example the number of housing tracts for sampling purposes in each cell of the 12×12 table ranged from none to 14. When a third dimension of "ethnicity" was added and the 12 classes of the original dimensions were split to make 24 classes, the result was a possible 24^3 cells, many of which were empty. With a large number of empty cells it was necessary to use a scheme that would reflect this in the sampling plan. It will often not be possible to have equal numbers of classes for each dimension of the problem. This situation of an unequal number of classes was covered in Jessen's 1970 and 1973 papers.

In these papers he discussed "Probability Sampling With Marginal Constraints" and looked further into the properties of these plans. The dimensions of the two-way design are simply R rows and C columns with a measure of size A_{rc} assigned to each cell and an expected sample size of nA_{rc} in each cell. These are combined so that there is an expected sample size of $nA_{r.}$ in each row and $nA_{.c}$ in each column. The expected sample sizes, nA_{rc}, are also considered to be inclusion probabilities for the cells taken over all the possible sample designs. The sample designs are actually established subject to the restraint of maintaining the expected sample sizes in each cell.

Jessen (1970) suggests two methods of construction and assignment of probabilities to the sample designs. Both methods meet the criteria of the expected sample sizes previously mentioned. However, they differ in the number of possible designs considered and in the assignment of probabilities to the designs that may be generated by both methods. Due to an earlier paper dealing with the selection of sampling units with unequal probability in one dimension (Jessen, 1969), Jessen (1970) refers to the methods as Method 2 and Method 3.

The goal of Method 2 is to generate the set of feasible samples that contains as few samples as possible for a specified marginal arrangement of the sample units.

In Method 3 concentration is placed on generating combinations of sample designs. The combination is then chosen as a first stage using a specified probability. From within the selected combination of sample designs a specific design is selected with equal probability for all designs in the combination. The combinations and their probabilities are assembled such that the conditions of expected sample size in the cells are maintained. With Method 3 more feasible samples are generated and there is a positive probability, P_{ij}, for all pairs of cells. Method 2 will be illustrated in detail later in this paper.

Using an artificial population Jessen (1970) ran a comparison of 16 different sampling schemes, half with replacement and half without, half with equal probability and half with unequal, and where there was no stratification, row only, column only, and both row and column stratification. By far the most efficient estimation was the one using both row and column stratification, unequal probabilities (proportional to a variable highly correlated with the variable of interest) and without replacement (using sampling Method 3).

In the 1970 paper an illustration was also given of a three-dimensional application of Method 2. The "probability" lattice schemes of the 1975 paper are adaptations of the Method 2 procedure previously discussed.

B. Goodman and Kish

Goodman and Kish (1950) developed a procedure under the title "Controlled Selection" to handle this same situation. The technique assigns a probability of selection to each of several possible samples so that "preferred" combinations of units are given a higher probability than "non-preferred" combination. Hess and Srikantan (1966) list three advantages that can be anticipated for "Controlled Selection" over traditional one-way stratified sampling.

(1) "Controls may be imposed to secure proper distribution geographically or otherwise and to insure adequate sample size for subgroups that are domains of study."

(2) "to secure moderate reductions in the sampling errors of a multiplicity of characters simultaneously."

(3) "the significant reduction of the sampling error in the global estimates of specified key variable."

One caution pointed out by Hess and Srikantan is that there is no guarantee of a lower sampling variance for a specified estimate than might be obtained by any other sampling procedure. They give an illustration of this case for the aforementioned hospital study and compare the controlled selection procedure with three possible alternatives, and for two sample sizes. The three alternatives were

(a) Stratification by bed size only

(b) Stratification by bed size and regions of the state

(c) Stratification by bed size, regions of the state, and by three SMSA classes.

The Controlled Selection used the same cells as alternative C but treated the allocation of sample units in its own way. The allocation of sample units to the three alternatives was using Neyman optimum allocation (the number to be selected from the *h*th stratum is proportional to both the number in the *h*th stratum, N_h, and the within-stratum variability as measured by the standard deviation, S_h, i.e., $n_h = n[N_h S_h / \sum_h^H N_h S_h]$), subject to the limitations of rounding to integer values. Therefore their comparisons were for sample sizes large enough to have stratification alternatives, and these do not deal with the type of situation that is most advantageous to multiway stratification selection procedures.

Unfortunately they did not concern themselves with any of Jessen's procedures and in his papers he makes only passing reference to Controlled Selection. Jessen (1970) comments "The method (C.S.) is somewhat complicated and its use in applied sampling appears limited." He does state that "Controlled Selection . . . provides samples that seem to meet the cell and marginal constraints." In 1973 he states as his second case of two-way sampling:

"If N_{rc} (cell sizes) are known and $N_{rc} \geq 0$, i.e., cells are unequal, disproportionate and even empty, controlled selection and probability lattices can be used." Later in the same article he indicates that while the situations in which they can be used are similar, they do differ in the sample selection procedure and "may thereby differ in the statistical properties of the resulting samples." He goes on to discuss probability lattice sampling and does not mention Controlled Selection again.

The actual operation of Controlled Selection is very similar to Method 2 of Jessen. A pattern of expected sample sizes is determined for the cells of the cross classification; these are the same as the nA_{rc} of Jessen. Usually they reflect proportionality of the cell with respect to some measure of total size for the population. Based on these expected cell sample sizes, different sample patterns of cell selections are formed. Each pattern is assigned a probability of being chosen. The probabilities for the entire set of patterns are chosen such that the sum of probabilities over the entire set is 1 and such that the sample selection of cells in a pattern, when matched with the probabilities of the pattern of selection, yields the predetermined expected sample size. It is the determination of the sample patterns and their respective probabilities that lead to Jessen's comment about the limited usefulness of Controlled Selection. In any large problem these determinations could be very unwieldy. Hess and Srikantan do admit this and offer suggestions for a computer procedure to determine the sample patterns and their respective probabilities.

The monograph by Moore *et al.* (1974), The National Assessment Approach to Sampling, deals with the many problems involved in the establishment of the expected sample sizes in each cell of a very large national survey but does not go into details on how the patterns and probabilities were obtained.

The National Assessment Controlled Selection permitted the selection of a probability sample from a two-dimensional grid so that restrictions placed on marginal sample totals were satisfied. One dimension dealt with region of the country, a separate stratification in itself, community size, and socioeconomic groups. The other dimension was the states of the region such that each state had at least one sampling unit selected. There were nine sizes of communities by socioeconomic class levels for the first dimension of stratification. Not all nine were present in all regions (in fact, one class was not present in any region). Three regions had 12 states each and the other (West) had 15. (The District of Columbia was counted as a state.)

After a fixed allocation of large cells (expected sample size over 1) the West region had 19 cells to select from a $6 \times 15 = 90$ cell grid that had 36 empty cells. There were 33 different sample patterns produced along with their accompanying probabilities. For the actual study one pattern was selected using the probabilities associated with the patterns.

C. Bryant

The Pole Mountain study (1961) utilized an extension of material presented by the Bryant, Hartley, Jessen paper (1960) for two-way stratification sampling. The sampling and estimation technique was developed by Bryant in his Ph.D. dissertation at Iowa State written under H. O. Hartley. It is discussed in Cochran's book (1963) and mentioned in the papers of Jessen discussed earlier as well as in Kish's book (1965). The basic sampling and estimation procedure has been extended to more than two dimensions for a number of projects besides the Pole Mountain study carried out by the University of Wyoming. Its basic strength in some problems is also its biggest weakness as an operational procedure in others. The sample selection procedure used to choose the cells of the cross classification is designed to adhere to only marginal sample size constraints. Usually it is not possible to adhere to exact levels of cell selection probabilities. Provision is made for handling cells with large expected sample sizes through a preliminary allocation before the random allocation.

Since only marginal constraints are utilized, there is no requirement to enumerate a full set of feasible sampling plans and determine associated probabilities. Instead, only the sample design to be used is constructed utilizing a procedure that guarantees the marginal constraints.

In cross-classifications in which some cells may be empty there is unfortunately no guarantee that a sample unit may not be assigned to that cell. Jessen (1973) comments that the most appropriate case for the use of this technique is when the cell sizes are proportional to the marginal sizes. Without this proportionality there are many problems encountered in the application of this technique.

The Pole Mountain project mentioned earlier utilized the extension of the basic selection technique to determine the sampling plan that was actually used. In projects such as this, where time is represented by at least one dimension and the physical location of sample points is one dimension, the problems of disproportionality are not encountered. The sampling unit is designated as a location during a specified time interval; thus, there usually will not be any missing cells.

3. SPECIFIC EXAMPLES OF SAMPLING PROCEDURES

The sampling plans mentioned in Section 2, Jessen's Lattices, Jessen's Method 2, M2, Goodman and Kish's Controlled Selection, CS, and Bryant,

Hartley and Jessen's procedure, BHJ, will now be discussed as to their applicability in certain types of survey design problems. Also some of the specific sampling procedures will be presented. In the following discussion the design problems will be broadly classified into two groups, equal cell and marginal probabilities and unequal cell and marginal probabilities.

A. Equal Probabilities

For the class of equal probabilities the discussion will include the square and cubic lattices of Jessen and the work of Bryant *et al.* (1960). Equal probability problems can also be handled as special cases under the unequal probability procedures but generally the specific procedures discussed here should be used.

When two variables are used for stratification and each has the same number of levels, say L, there are L^2 cells in the design. Jessen (1975), following the lead of Yates (1960), discusses two techniques for drawing samples of size $n = rL$, where r units are selected from each level of each dimension (variable). The first technique, the random lattice, begins with a selection of r units at random from the L units of the second factor (or columns of the two-way table) that lie within the first level of the first factor (or rows). The second step is to select r units (columns) at random from the second row. This process continues until r units have been selected from one of the columns. The sample from the next row contains only $r - 1$ units. The selection continues until all but the last row have been covered. The columns selected from the last row are determined by the selections through row $L - 1$. Jessen refers to this technique as a "general" lattice. He also presents an alternative scheme known as a "Latin" lattice. When $L/4$ is an integer, $r \times r$ squares are designated along the diagonal of the $L \times L$ square. The rows and columns are then permuted to obtain the final design. When L is odd and r is even, the diagonal can be filled by using a combination of $r \times r$ complete squares and $(r + k) \times (r + k)$ incomplete squares.

To illustrate these procedures consider a 6×6 square with $r = 2$. First, for the "general" lattice design, columns 2 and 4 were selected from row 1. Then in row 2 columns 1 and 4 were selected. Now for row 3, column 4 is no longer available and the choices are columns 2 and 6. The procedure continues until in row 6 only columns 3 and 6 are left. If this design was for a six-week study of a store that is open 6 days a week, during the first week day 2 (Tuesday) and day 4 (Thursday) would be sampled. If there were several stores available, each day one would be selected at random on Tuesday and one on Thursday.

Sample

	Day					
Week	1	2	3	4	5	6
1		×		×		
2	×			×		
3		×				×
4	×				×	
5			×			
6			×			×

With six days of a week as the time dimension and six boat docks, on the first day docks 2 and 4 would be covered by the observer to determine the number of fish caught and other characteristics.

With the "Latin" procedure there are three steps. First set up three 2 × 2 squares along the main diagonal (Step 1); second, permute the rows of the square (Step 2); third, permute the columns of the square (Step 3).

STEP 1

	1	2	3	4	5	6
1	×	×				
2	×	×				
3			×	×		
4			×	×		
5					×	×
6					×	×

STEP 2

	1	2	3	4	5	6
4			×	×		
2	×	×				
5					×	×
3			×	×		
6					×	×
1	×	×				

STEP 3

	4	2	1	3	5	6
4	×			×		
2		×	×			
5					×	×
3	×			×		
6					×	×
1		×	×			

For the same examples as mentioned above this design would call for sampling day 1 and day 4 during the first week, or observing dock 1 and dock 4 on the first day of the project.

The "Latin" procedure is a special case of a technique in which any r "letters" of a classical Latin square experimental design may be selected and then permuted to obtain the specific design. The advantage of the Latin procedure is in the estimation of the variance of the estimate. With either scheme the sample can be divided into r parts and the contrasts among the r simple estimates of the mean (or total) are used to produce the estimate of the variance of the estimate. Jessen (1975) indicates that the Latin procedure results in r degrees of freedom for the estimate of the variance while the general lattice gives $r - 1$. When r is only 2 or 3, which is very common, the extra degree of freedom is quite valuable. The Latin designs involve more work in getting them set up and they are a bit more restrictive in their dimensionality, but they may be worth the effort for a higher quality estimate of the variance.

Jessen (1975) uses an analysis of variance argument to show that generally a two-way design such as this will be better than both a simple random sample of cells and a one-way stratification using either rows or columns.

There are some obvious limitations to the use of these designs. The equal number of levels for the strata classifications is one. However, with a little thought the number of categories of a variable may be shrunk, or stretched, in order to create a square where none exists naturally. Of more concern are the stringent sample size considerations; n has to be a multiple of the number of categories, $n = rL$, and there must be an equal allocation to each row and column.

A procedure that can be used if either or both of the above limitations are a problem in the design is presented by Bryant *et al.* (1960): A two-way grid of size $n \times n$ is laid out (graph paper works well) and the sample of columns is selected one row at a time without replacement until the sample of n cells is completed (one in each row and one in each column). Then a coarser grid that represents the planned marginal sample sizes is placed over the sampled grid. The cells of the coarse grid that have been chosen are then noted. For example, in the 6×6 problem, the sample is of size 12. Thus a 12×12 grid is produced and sampled as will be shown. The heavy grid lines split the sample so that 2 units would be selected from each row and 2 from each column.

Sample

Day

Week	1		2		3		4		5		6		$n_{i\cdot}$
1			×										2
									×				
2							×						2
						×							
3				×									2
										×			
4											×		2
								×					
5												×	2
					×								
6	×												2
		×											
$n_{\cdot j}$	2		2		2		2		2		2		12

One practical problem in the application of this procedure is exemplified by what is found in the lower left-hand corner of the layout. This sample configuration indicates that two units should be drawn from the sixth week and on the first day. If there are two or more population units in that cell, there is no problem. However, if the study is for a certain store and the sample is just of days of the week, then something would have to be done.

This type of dilemma would also present itself if the allocation called for one unit from day 3 and three from day 4. In that case two units would be needed for day 4 in week 2. Other examples of this same sort can be found by examination of the table.

Notice that while the same situation was used as in the lattice design, there is nothing in the sampling structure to prohibit a study over 5 or 7 or any other number of weeks. The only problem would be to decide upon the necessary unequal allocation of units to the days.

When three variables are available for stratification Jessen (1975) has a procedure to use when there are L levels of each variable, $N = L^3$, and the sample is of size $n = r^2L$, where r is an integer and L/r is also an integer. The basic scheme is to construct a set of $r \times r \times r$ Latin cubes along an internal diagonal, then randomly permute the rows, columns, and levels to obtain the design. As before, the major problems with these designs are the same—finding three stratification variables with equal number of levels, the inflexibility of the sample sizes permitted, and the equal numbers required for all levels of all factors.

Though not presented by Bryant *et al.* (1960), the basic selection procedure for this method has been extended to multidimensions for various studies, including the Pole Mountain study mentioned earlier. The Pole Mountain study had four dimensions and a sample size of $n = 46$. Thus a four-dimensional cube of size 46^4 was envisioned. The sample of cells was chosen in four dimensions so that once a row, column, level, or flat was selected, it could not be chosen again. One method would be to start in row 1 and choose three random numbers from 01 to 46 (duplications are allowed at this step); then for row 2 choose a new set of three random numbers from 01 to 46, being careful not to select any duplicates from the first set for the specific position in the set. This process would continue until the entire 46^4 grid is filled. The number of samples to be selected from each four-dimensional cell is found by putting the coarse grid that reflects the desired marginal frequencies over the fine 46^4 grid sample. The actual sample found for the Pole Mountain study called for one observation out of the ten weekday mornings at check station 5 during the third two-week time period of the study.

A weakness of the BHJ method is that more complete balance may be desired than can be guaranteed with the selection of cells at random. For example, during a two-week time period the design can guarantee 10 observations and for a single location the design can guarantee 10 observations, but with the 46^4 structure there is no guarantee that this location will be sampled during this time period. For that guarantee the design, and the resulting estimator, will have to be completely altered. The lack of symmetry would also prohibit the use of multidimensional analogs of the square lattices. The unequal probability procedures that follow offer more guarantees of sample allocation to cells.

B. Unequal Probabilities

When the cells of a cross-classification contain an unequal number of units, it is often desired to sample the cells in a manner that reflects this unevenness. This is especially true in situations where the sampling is done in two stages; first a sample of the cells is drawn and second a sample of the units is drawn from the selected cells. In 1970 Jessen (1973, 1975) discussed "Probability Sampling with Marginal Constraints" and wrote on probability lattice sampling. In these works he discussed the situation where the cells of a two-way or a three-way cross-classification did not contain the same number of population units, and it is desired to select the cells for the sample to reflect this condition. The techniques developed will work for any probability of selecting cells, even equal. Probability proportional to size is often a convenient procedure to follow.

The procedures Jessen dealt with in the 1970 paper include probability lattice sampling from the 1975 paper as a special case; therefore only the details of the 1970 procedures will be reviewed in this presentation.

Each cell of the cross-classification is assigned a probability, A_{rc}, such that the sum of all the probabilities in the table is equal to 1. These probabilities are then multiplied by the planned sample size of n to obtain expected cell and marginal sample sizes. If any of these multiplications result in an expected sample size greater than 1, a sample unit is assigned automatically to that cell and an adjustment to the cell expected sample size is made.

The Bryant *et al.* plan can be used with the marginal sample allocations determined by the relative sizes of the margins, but there is no good control over the cell expected sizes unless the rows and columns are independent in the sense of not interacting. Also, if any cells are missing in the population, there is no way to keep them from possibly being designated for the sample.

If there are no cells with zero population units and the frequencies of the cells are a result of the row and column stratifications' being unrelated, the ease of administering the Bryant *et al.* procedure is worth considering.

Two methods of selection were presented by Jessen (1970). They were two-dimensional analogs of two methods discussed in Jessen (1969) and are referred to simply as "Method 2" and "Method 3." Only "Method 2" with integer marginals will be presented in detail.

Jessen's example will be used to illustrate this technique. The problem is small and the steps are easy to follow. There are two stratification variables; each has three levels which together produce 9 cells with the expected sample size matrix shown in Matrix A.

MATRIX A

Strata variable 1	Strata variable 2			
	1	2	3	$n_{i\cdot}$
1	.8	.5	.7	2.0
2	.7	.8	.5	2.0
3	.1	.7	.8	2.0
$n_{\cdot j}$	2.0	2.0	2.0	6.0

Thus, all feasible samples will have two units in each row and two in each column. There are six possible samples but only three are required to satisfy the expected frequency restriction.

The first step is to designate a sample of cells that satisfies the marginal restrictions. There is an art to doing this so that a minimum number of feasible samples are in a complete set. For this problem the first feasible sample could be as shown in Sample 1.

SAMPLE 1

×		×
×	×	
	×	×

Note that this sample contains the cells of a row that have the highest available probabilities subject to earlier selections. After the statement of a feasible sample, the probability of selecting this sample is determined. The original expected sample size matrix is then rewritten by using the original expected sample sizes wherever ×'s occur and using 1 minus the original expected sample sizes for the other cells. In the present example this operation produces the configuration shown in Matrix B.

MATRIX B

.8	1 − .5 = .5	.7
.7	.8	1 − .5 = .5
1 − .5 = .5	.7	.8

The minimum value from this table, called D_1, $D_1 = 0.5$, is then given to the sampling pattern as its selection probability.

This selection probability, D_1, is then subtracted from 1 to determine the amount of probability remaining to be allocated to subsequent sample plans, called B_1, $B_1 = 0.5$. Also D_1 is subtracted from the original expected sample sizes for those cells in the first sample plan, (1). These then form a new matrix of expected sample sizes, (C). This step yields Matrix C.

MATRIX C

$.8 - .5$ $= .3$	$.5$	$.7 - .5$ $= .2$
$.7 - .5$ $= .2$	$.8 - .5$ $= .3$	$.5$
$.5$	$.7 - .5$ $= .2$	$.8 - .5$ $= .3$

The second feasible sample and its selection probability are now determined following the same procedure as before. Sample 2 is the second sample.

SAMPLE 2

×	×	
	×	×
×		×

Based on this sample plan the expected sample size matrix now becomes that shown in Matrix D.

MATRIX D

$.3$	$.5$	$.5 - .2$ $= .3$
$.5 - .2$ $= .3$	$.3$	$.5$
$.5$	$.5 - .2$ $= .3$	$.3$

The minimum in this table is 0.3, i.e., $D_2 = 0.3$, so this is the probability associated with this sample arrangement. Now subtracting 0.3 from the expected sample sizes before Sample 2 was drawn yields the remaining expected sample sizes shown in Matrix E. The final feasible sample would be that shown in Sample 3, where the remaining 0.2 is the assigned probability of selection of plan (3), ($D_3 = 0.2$).

MATRIX E

$.3 - .3$ $= 0$	$.5 - .3$ $= .2$	$.2$
$.2$	$.3 - .3$ $= 0$	$.5 - .3$ $= .2$
$.5 - .3$ $= .2$	$.2$	$.3 - .3$ $= 0$

SAMPLE 3

	×	×
×		×
×	×	

The specific design used for the survey would then be chosen using the associated probabilities. Estimates of means and totals would also use the design probabilities. Note that cell (1, 1) had an original expected frequency of 0.8. Cell (1, 1) is to be chosen in Sample 1 with a probability of 0.5 and in Sample 2 with a probability of 0.3. These together $(0.5 + 0.3)$ provide the designed expected sample size for this cell.

In the paper Jessen gives a more formal statement of the selection algorithm that could be adapted to a computer program for use in larger problems in which many more stages would be used. He also gives the details for the case of noninteger marginal expected sample sizes.

It is this procedure that he illustrates in the 1975 paper with a sample of 24 drawn from a 12×12 frame for San Diego, California. A number of the cells of the San Diego frame are empty so the Bryant *et al.* procedure would be very difficult, if not impossible, to use.

The other procedure referred to earlier in the paper, Goodman and Kish's Controlled Selection, is very similar to Jessen's techniques referred to above. The idea of expected sample sizes in the cells of the cross classification of two or more strata variables is the starting place for the procedure. Next, possible sample plans are determined that meet the criteria of the design, and a probability is associated with each possible sample plan. These probabilities are again chosen so that the predetermined expected sample sizes for the cells are realized. In the monograph, The National Assessment Approach to Sampling (1974), only the selection patterns and their probabilities are given.

Groves and Hess writing in Probability Sampling of Hospitals and Patients [Hess *et al.* (1975)] present a very elaborate computer algorithm for the implementation of controlled selection. The main difference between their approach, besides its more formal presentation, and Jessen's Method 2 appears to lie with the strategy employed for the selection of the first cell of the sampling plan. They select the cell with the minimum probability associated with a possible cell value (in most cases a cell value would be 0 or 1, present or absent). The final set of plans and their associated probabilities do meet all the requirements of the expected cell and marginal sizes.

Since Jessen's plans and Controlled Selection both appear to be different ways of attaining a set of possible sample plans that have as their goal the achievement of a set of predetermined expected sample sizes, the choice becomes one of ease of operation and implementation rather than of precision of estimates obtained.

There are some concerns about the possible problems of estimating variances with all three procedures, including Bryant's. The general feeling seems to be that to estimate variances with Jessen's plans and with Controlled Selection, the sample should be divided into at least two nonoverlapping subsamples and that the differences among the separate mean, or total, estimates be used in obtaining the estimate of the sampling variance. The Bryant procedure could also be used in the same way although the authors indicate that estimation of variance depends on cells that have two or more observations.

REFERENCES

BRYANT, E. C. (1961). Sampling methods. Seminar paper, Iowa State Univ., July 1961.

BRYANT, E. C., Hartley, H. O., and Jessen, R. J. (1960). Design and estimation in two-way stratification. *J. Amer. Statist. Assoc.* **55**, 105–24.

COCHRAN, W. G. (1963). *Sampling Techniques*, 2nd ed. Wiley, New York.

GOODMAN, J. R. and KISH, L. (1950). Controlled selection—a technique in probability sample. *J. Amer. Statist. Assoc.* **45**, 350–372.

HESS, I., RIEDEL, D. C., and FITZPATRICK, T. B. (1975). *Probability Sampling* of *Hospitals and Patients*, 2nd ed. Health Admin. Press, Chapter VII.

HESS, I. and SRIKANTAN, K. S. (1966). Some aspects of the probability sampling technique of controlled selection. Health Serv. Res., Summer 1966, 8–52.

JESSEN, R. J. (1969). Some methods of probability non-replacement sampling. *J. Amer. Statist. Assoc.* **64**, 175–193.

JESSEN, R. J. (1970). Probability sampling with marginal constraints. *J. Amer. Statist. Assoc.* **65**, 776–796.

JESSEN, R. J. (1973). Some properties of probability lattice sampling. *J. Amer. Statist. Assoc.* **68**, 26–28.

JESSEN, R. J. (1975). Square and cubic lattice sampling. *Biometrics* **31** 449–471.

KISH, L. (1965). *Survey Sampling*. Wiley, New York.

MOORE, R. P., CHROMY, J. R., and ROGERS, W. T. (1974). The National Assessment Approach to Sampling, Nat. Assess. of Educ. Progress, Denver.

YATES, F. (1960). *Sampling Methods for Censuses and Surveys*, 3rd ed. Hafner, New York.

Part II

THE LINEAR MODEL

The Analysis of Linear Models with Unbalanced Data

Ronald R. Hocking *O. P. Hackney* *F. M. Speed*

MISSISSIPPI STATE UNIVERSITY

The purpose of this paper is to describe the hypotheses commonly tested in linear models with unbalanced data, including the case of zero cell frequencies. Historically, the sums of squares for the test statistics have been developed either on heuristic principles or because of computational convenience. Precise statements of the corresponding hypotheses are rarely found in the literature and, in those cases where the hypotheses are stated, they are usually described in terms of the parameters of the non-full rank model which may be difficult to interpret. In this paper, the hypotheses associated with the $R(\ )$ notation for general sets of conditions are described in terms of the means of the observed populations. The discussion is restricted to two-way models, but the concepts are quite general. Examples are included to illustrate the hypotheses tested with missing cells.

1. INTRODUCTION

The purpose of this paper is to discuss the analysis of the classical, fixed effect, linear model for designed experiments when the data are not balanced. In particular, the emphasis here is on the situation in which some of the cell frequencies are zero. For example, in the two-way classification model, given by

$$\begin{gathered} y_{ijk} = \mu + \alpha_i + \beta_j + \gamma_{ij} + e_{ijk} \\ i = 1, \ldots, a, \quad j = 1, \ldots, b, \quad k = 0, 1, \ldots, n_{ij} \end{gathered} \tag{1}$$

the special case, $n_{ij} = 0$, for some (i, j) combinations is of interest.

Key words and phrases Linear models, tests of hypotheses, unbalanced data, missing cells.

In general, consider the linear model written in matrix form as

$$\mathbf{Y} = \mathbf{X}\boldsymbol{\beta} + \mathbf{e}, \tag{2}$$

where $\mathbf{X}$ is the $n \times p$ design matrix of rank $r \leq p$ consisting of 0's and 1's, $\mathbf{e}$ is $N(0, \sigma^2 \mathbf{I}_n)$, and $\boldsymbol{\beta}$ is the p vector of unknown parameters. The analysis of (2) focuses on inferences about $\boldsymbol{\beta}$ with particular emphasis on tests of linear hypotheses.

The analyses of such linear models for unbalanced data, e.g. unequal n_{ij} in (1), is not well understood. The primary source of confusion is that (2) is overparametrized. That is, if $r < p$, there are only r independent parameters. As a consequence, the individual parameters have no meaning. This overparametrization allows for the imposition of "nonestimable" conditions on $\boldsymbol{\beta}$, say,

$$\mathbf{K}\boldsymbol{\beta} = \mathbf{0}. \tag{3}$$

Here $\mathbf{K}$ has rank $p - r$ and the rows of $\mathbf{K}$ are not in the row space of $\mathbf{X}$. The choice of a particular matrix $\mathbf{K}$ gives meaning to the components of $\boldsymbol{\beta}$ and allows for the interpretation of hypotheses about $\boldsymbol{\beta}$. Unfortunately, the choice of $\mathbf{K}$ often makes this interpretation difficult with the result that there is much confusion as to the meaning of certain hypotheses.

Hocking and Speed [7] proposed a return to the basic model in which the parameters are just the cell means, say μ_{ij} in (1). The advantage is that the cell means are clearly the relevant quantities and, further, hypotheses about the cell means are easy to interpret and not confused by the concept of estimability. Their model does allow for linear constraints on the cell means, but these are based on known or assumed relations on these means. The most common constraint is based on the assumption of no interaction, but other constraints do arise.

The cell means model (μ model), which will be used in this paper to provide a mechanism for understanding the analysis of (2) is written as

$$\mathbf{Y} = \mathbf{W}\boldsymbol{\mu} + \mathbf{e} \tag{4}$$

subject to

$$\mathbf{G}\boldsymbol{\mu} = \mathbf{0}. \tag{5}$$

Here $\boldsymbol{\mu}$ is the q vector of cell means, $\mathbf{W}$ is the $n \times q$ matrix of 0's and 1's indicating the number of times a particular cell, or population, has been observed and $\mathbf{G}$ is a matrix which describes known linear relations on $\boldsymbol{\mu}$, if any exist.

For example, the μ model for (1) is given by

$$y_{ijk} = \mu_{ij} + e_{ijk}. \tag{6}$$

That is, (6) simply states that a total of ab populations have been observed with n_{ij} observations on the (i, j)th population having mean, μ_{ij}. In this case, the model imposes no additional constraints on the means.

If the constraints

$$\mu_{ij} - \mu_{i'j} - \mu_{ij'} + \mu_{i'j'} = 0 \tag{7}$$

for $i, i' = 1, \ldots, a, j, j' = 1, \ldots, b$ are adjoined to (6), the two-way classification model, without interaction, is obtained. This model is more commonly written as

$$y_{ijk} = \mu + \alpha_i + \beta_j + e_{ijk}. \tag{8}$$

The constraints (7) thus distinguish model (1) from model (8).

It is of interest to note that (6) applies equally well to the two-fold nested model commonly written as,

$$y_{ijk} = \mu + \alpha_i + \gamma_{ij} + e_{ijk}. \tag{9}$$

The fact that (6) can be used to describe either (1) or (9) would seem to be confusing since they reflect distinctly different experimental situations. The difference becomes evident when hypotheses on the μ_{ij} are considered. Specifically, a hypothesis such as H: $\mu_{.j} = \mu_{.j'}$ is meaningful if the data are cross-classified but not if the data are nested.

The objective of this paper is to describe, in terms of the μ model, the hypotheses tested by some of the standard computational procedures. The fact that this is even necessary seems contrary to logic since statistical theory indicates that the hypothesis should be formulated first followed by the development of an appropriate test statistic. Unfortunately, this has not been the case with the analysis of unbalanced data, as the sums of squares used for testing are frequently dictated by computational convenience rather than by a precise statement of the hypothesis.

The recent paper by Speed *et al.* [12] describes most of the methods used and relates the sums of squares to the $R(\ )$ notation described by Speed and Hocking [11]. In this paper, we shall not discuss specific methods, but simply relate the hypotheses to the appropriate $R(\ )$. The $R(\ )$ notation is reviewed in Section 2.

For simplicity, we restrict the discussion in this paper to the model (6) with or without (7). The concepts are easily extended to cell means with more than two subscripts.

In Section 3, the main effect and interaction hypotheses associated with various procedures are developed for the two-way classification model with interaction. The question of which main effect hypothesis should be considered, or indeed, if a test for main effects is meaningful in the presence of interaction will not be answered. The objective here is just to reveal what hypotheses are being tested. Reference is made to the papers by Elston and Bush [2] and Finney [3] for a discussion of this point.

An important consideration, when one is faced with zero cell frequencies, is the concept of connectedness of the design. This is discussed in Section 3.2 where a simple test for connectedness is developed.

The two-way model without interaction (8) is discussed in Section 4. The appropriate interaction constraints for the zero cell case are derived from Section 3.2 where the rationale for "estimating" data for the missing cells is observed.

Finally, the two-fold nested model (9) is discussed in Section 5.

2. COMPUTATIONAL PROCEDURES

The numerator sums of squares generated by most valid computational procedures can be described in terms of the $R(\ )$ notation. Speed and Hocking [11] observed that there are essentially two $R(\ )$ procedures. These two procedures will be reviewed.

To describe Procedure 2, consider a model of full rank obtained by imposing a set of nonestimable conditions on the model (2). That is, (3) is solved for $p - r$ components of $\boldsymbol{\beta}$ and substituted into (2) to obtain the model

$$\mathbf{Y} = \mathbf{X}^*\boldsymbol{\beta}^* + \mathbf{e}. \tag{10}$$

Writing the r vector $\boldsymbol{\beta}^*$ as $\boldsymbol{\beta}^{*\prime} = (\boldsymbol{\beta}_1^{*\prime}, \boldsymbol{\beta}_2^{*\prime})$ with a conformable partition of $\mathbf{X}^*$, define

$$R(\boldsymbol{\beta}^*) = \hat{\boldsymbol{\beta}}^{*\prime}\mathbf{X}^{*\prime}\mathbf{Y} \tag{11}$$

and

$$R(\boldsymbol{\beta}_2^*) = \tilde{\boldsymbol{\beta}}_2^{*\prime}\mathbf{X}_2^{*\prime}\mathbf{Y}. \tag{12}$$

In (11), $\hat{\boldsymbol{\beta}}^*$ is the least squares estimate of $\boldsymbol{\beta}^*$ in (10) and in (12), $\tilde{\boldsymbol{\beta}}_2^*$ is the least squares estimate of $\boldsymbol{\beta}_2^*$ in the reduced model obtained by setting $\boldsymbol{\beta}_1^* = \mathbf{0}$ in (10). The sum of squares to be considered for the test statistic is then written as

$$R_2(\boldsymbol{\beta}_1^* | \boldsymbol{\beta}_2^*) = R(\boldsymbol{\beta}^*) - R(\boldsymbol{\beta}_2^*). \tag{13}$$

The sum of squares given by (13) will, in general, depend on the choice of nonestimable conditions. For reference, Table 1 lists several sets of conditions which might be applied to the two-way classification model (1).

Several comments are in order on the conditions listed in Table 1.

(i) Condition 6 is not used with the $R(\ )$ notation. Indeed, this condition leads to the μ model, (6).

TABLE 1

Nonestimable Conditions for the Two-Way Classification Model with Interaction[a]

1. $\gamma_{i\cdot} = \gamma_{\cdot j} = 0$
2. $\gamma_{i1} = \gamma_{1j} = 0$
3. $\sum_i n_{ij}\gamma_{ij} = \sum_j n_{ij}\gamma_{ij} = 0$
4. $\sum_i n_{i\cdot}\gamma_{ij} = \sum_j n_{\cdot j}\gamma_{ij} = 0$
5. $\sum_i n_{ij}(\alpha_i + \gamma_{ij}) = \sum_j n_{ij}(\beta_j + \gamma_{ij}) = \sum_i n_{i\cdot}\alpha_i = \sum_j n_{\cdot j}\beta_j = 0$
6. $\mu = \alpha_i = \beta_i = 0$

[a] All conditions hold for $i = 1, \ldots, a$, $j = 1, \ldots, b$.

(ii) In Conditions 1–4, if $n_{ij} = 0$, the stated conditions are augmented by $\gamma_{ij} = 0$. If Condition 5 is used, the condition $\alpha_i + \gamma_{ij} = \beta_j + \gamma_{ij} = 0$ is adjoined if $n_{ij} = 0$.

(iii) Conditions 1–4 represent $a + b - 1$ linearly independent conditions on the γ_{ij}. In the case of connected designs, one additional condition on both the α_i and β_j is necessary to achieve full rank. It will be shown that the conditions on α_i and β_j are arbitrary and have no effect on the hypothesis tested. In Condition 5 the conditions on α_i and β_j must be as specified.

The $R(\ )$ notation described by Searle [9, 10] and labeled Procedure 1 by Speed and Hocking [11] represents a sum of squares by a formula which is similar to (13) with two major exceptions. First, the parameters are those of the original model (2) as opposed to the reparametrized model (10), and second, not all of the parameters of the model need be included. That is, the $R(\ )$ notation, Procedure 1, may be based on a model in which some of the parameters have already been deleted. Thus, for example, even though the original model is (1), the notation

$$R_1(\boldsymbol{\alpha} \mid \mu, \boldsymbol{\beta}) = R(\mu, \boldsymbol{\alpha}, \boldsymbol{\beta}) - R(\mu, \boldsymbol{\beta}) \tag{14}$$

indicates that the computations are performed in terms of model (8). Similarly,

$$R_1(\boldsymbol{\alpha} \mid \mu) = R(\mu, \boldsymbol{\alpha}) - R(\mu) \tag{15}$$

is based on the model

$$y_{ijk} = \mu + \alpha_i + e_{ijk}. \tag{16}$$

It is of interest to note that, with the appropriate set of nonestimable conditions, we can obtain the sums of squares (14) and (15) from (13) using Procedure 2. In particular, if we use Condition 3 from Table 1, then

$$R_1(\boldsymbol{\alpha} \mid \mu, \boldsymbol{\beta}) = R_2(\boldsymbol{\alpha}^* \mid \mu^*, \boldsymbol{\beta}^*, \boldsymbol{\gamma}^*) \tag{17}$$

and, if Condition 5 is used,

$$R_1(\boldsymbol{\alpha} \mid \mu) = R_2(\boldsymbol{\alpha}^{**} \mid \mu^{**}, \boldsymbol{\beta}^{**}, \boldsymbol{\gamma}^{**}). \tag{18}$$

The equivalence noted in (17) and (18) is convenient for the development in this paper, but it is emphasized that Conditions (3) and (5) are rarely used with Procedure 2. Rather, the sums of squares are generated by (14) and (15) applied to the reduced models. Condition 1 is most commonly used with Procedure 2, but Conditions 2 and 4 are found in some references. In the following, only Procedure 2 will be used and $R(\boldsymbol{\beta}_1 \mid \boldsymbol{\beta}_2)$ will be used to denote (13).

The analysis of the two-fold nested model (9) is also complicated by unequal frequencies, but in this case the zero cell problem is not critical. The common nonestimable conditions for this model are listed in Table 2. It should be noted that Conditions 1 and 2 must be accompanied by one additional, but arbitrary, linear restriction on the α_i. Condition 3 is not used with the $R(\ )$ notation but yields a model which is equivalent to (6).

TABLE 2

Nonestimable Conditions for the Two-Fold Nested Model[a]

1.	$\gamma_{i\cdot} = 0$
2.	$\sum_{j=1}^{b} n_{ij}\gamma_{ij} = 0$
3.	$\mu = \alpha_i = 0$

[a] All conditions hold for $i = 1, \ldots, a$.

The choice of nonestimable conditions does not affect the hypotheses associated with the two-way model without interaction as long as one condition is imposed on the α_i and one on the β_j.

3. TWO-WAY CLASSIFICATION WITH INTERACTION

3.1 Main Effect Hypotheses

In this section, three theorems which determine the main effect hypotheses tested by application of Procedure 2 for general sets of nonestimable conditions are stated and proved. The theorems are for general cell frequencies, including the case $n_{ij} = 0$. The number and location of the zero cell frequencies may lead to designs which are not connected. This possibility is included in the theorems although the discussion of connectedness is deferred to Section 3.2.

For the first theorem, assume that the nonestimable conditions on the γ_{ij} are

$$\sum_{i=1}^{a} c_{ij}\gamma_{ij} = \sum_{j=1}^{b} c_{ij}\gamma_{ij} = 0 \tag{19}$$

for $i = 1, \ldots, a, j = 1, \ldots, b$. The requirement on the c_{ij} is that (19) represents $a + b - 1$ linearly independent conditions.

THEOREM 1 In the two-way model, (1), suppose the design is connected and that conditions (19) are imposed. In addition, if $n_{ij} = 0$, set $\gamma_{ij} = 0$ for a total of $a + b - 1 + m$ linearly independent conditions, where m is the number of empty cells. If one condition on the α_i and one on the β_j are added to obtain a full-rank model, then the hypotheses tested by the sums of squares generated by the $R(\ )$ procedure are as follows:

(1) $R(\boldsymbol{\alpha} \mid \mu, \boldsymbol{\beta}, \boldsymbol{\gamma})$

$$H_\alpha\colon\ \sum_{j=1}^{b} d_{ij}\mu_{ij} = \sum_{j=1}^{b} \sum_{i'=1}^{a} d_{ij}d_{i'j}\mu_{i'j}/d_{.j} \qquad i = 1, \ldots, a-1. \tag{20}$$

(2) $R(\beta \mid \mu, \alpha, \gamma)$

$$H_\beta\colon\ \sum_{i=1}^{a} d_{ij}\mu_{ij} = \sum_{i=1}^{a} \sum_{j'=1}^{b} d_{ij}d_{ij'}\mu_{ij'}/d_{i.} \qquad j = 1, \ldots, b-1. \tag{21}$$

Here

$$d_{ij} = \begin{cases} c_{ij} & \text{if} \quad n_{ij} \neq 0 \\ 0 & \text{if} \quad n_{ij} = 0. \end{cases}$$

Proof (The proof is given for H_α only, as the proof of H_β is identical.) In view of the definition (13), it is sufficient to show that under the stated conditions, the hypothesis (20) reduces to $\alpha_i = 0$. To see this, substitute

$$\mu_{ij} = \mu + \alpha_i + \beta_j + \gamma_{ij} \tag{22}$$

into (20) and observe that the terms in μ, β_j, and γ_{ij} vanish leaving the expression

$$d_{i.}\alpha_i - \sum_{j=1}^{b} \sum_{i'=1}^{a} d_{ij}d_{i'j}\alpha_{i'}/d_{.j} = 0 \qquad i = 1, \ldots, a-1. \tag{23}$$

Letting $\mathbf{G}$ denote the appropriate matrix of coefficients of dimension $(a-1) \times a$, (23) is written as

$$\mathbf{G}\boldsymbol{\alpha} = \mathbf{0}. \tag{24}$$

A convenient expression for $\mathbf{G}$ is obtained by letting $i = 1, \ldots, a$ in (23), yielding the $a \times a$ matrix,

$$\mathbf{G}^* = \mathrm{Diag}(d_{i\cdot}) - \mathbf{D}\ \mathrm{Diag}(1/d_{\cdot j})\mathbf{D}', \tag{25}$$

where $\mathbf{D} = (d_{ij})$ is the incidence matrix of the d_{ij}. Note that $\mathbf{J}'\mathbf{G}^* = \mathbf{0}$, where $\mathbf{J}$ is the vector of 1's, hence the last row of $\mathbf{G}^*$ is redundant.

From Theorem 13.2 in Graybill [5], and the assumption that the design is connected, the rank of $\mathbf{G}$ is $a - 1$. Also, from (25) note that $\mathbf{GJ} = \mathbf{0}$, and hence the system of equations (24) is equivalent to

$$(\mathbf{J} \mid -\mathbf{I})\boldsymbol{\alpha} = \mathbf{0}. \tag{26}$$

Thus, (26) and hence (23) is equivalent to the set of conditions $\alpha_i = \alpha_{i'}$ for i, $i' = 1, \ldots, a$. It then follows that any nonestimable condition on the α_i reduces this to the condition $\alpha_i = 0$, $i = 1, \ldots, a$, and the desired result is established.

To extend to the case of nonconnected designs, note that if $\mathbf{G}$ has rank $a - 1$, then it follows from (26) that the design is connected. For nonconnected designs, it is possible to relabel rows and columns such that $\mathbf{D}$, and hence $\mathbf{G}^*$, are block diagonal with s blocks of size a_i and rank $a_i - 1$, where s is the number of connected subsets of the original design. These blocks have the same properties as $\mathbf{G}^*$ and hence the hypothesis (20) is equivalent to $R(\boldsymbol{\alpha} \mid \mu, \boldsymbol{\beta}, \boldsymbol{\gamma})$ if s nonestimable conditions are imposed on the α_i, one for each subset.

Referring to Table 1, note that Theorem 1 includes Conditions 1–3 as special cases. For emphasis, we spell out these hypotheses in detail. With Condition 1, if all $n_{ij} > 0$, then (20) reduces to

$$H_{\alpha 1}\colon\ \mu_{i\cdot} = \mu_{i'\cdot} \qquad \text{all} \quad i, i'. \tag{27}$$

For Condition 2, assuming $n_{i1} > 0$ and $n_{1j} > 0$ for $i = 1, \ldots, a$, $j = 1, \ldots, b$, we obtain

$$H_{\alpha 2}\colon\ \mu_{i1} = \mu_{i'1} \qquad \text{all} \quad i, i'. \tag{28}$$

With Condition 3, the hypothesis is given directly by (20) with d_{ij} replaced by n_{ij} and no restriction on the n_{ij}.

Since (28) compares only the cell means in the first column, it is of little general interest as a main effect hypothesis. The case, $c_{ij} = 1$ which leads to (27) or $c_{ij} = n_{ij}$ are of more general interest, but (27) seems to have the advantage of ease of interpretation.

As noted earlier, Condition 3 is rarely used in practice, but with $c_{ij} = n_{ij}$ yields results which are equivalent to Procedure 1 as indicated by (17). Searle [9] provides a statement of the hypothesis associated with $R_1(\boldsymbol{\alpha} \mid \mu, \boldsymbol{\beta})$

and Hackney [6] develops a general procedure for determining hypotheses tested under $R(\ )$ procedures.

To illustrate the hypotheses which are tested with some $n_{ij} = 0$, we give two examples.

Example 1 With $a = b = 3$, consider the design with cell frequencies shown in Table 3. The hypotheses corresponding to Conditions 1 and 3 in Table 1 are, respectively,

$$\begin{aligned} H_{\alpha 1}:\quad & \mu_{12} - \mu_{32} = \mu_{23} - \mu_{13}, \\ & \mu_{21} - \mu_{31} = \mu_{13} - \mu_{23}, \end{aligned} \tag{29}$$

$$\begin{aligned} H_{\alpha 3}:\quad & (\mu_{12} - \mu_{32})/2 = 2(\mu_{23} - \mu_{13})/3, \\ & \mu_{21} - \mu_{31} = \mu_{13} - \mu_{23}. \end{aligned} \tag{30}$$

The hypothesis (29) provides a rather curious measure of the row effect. The hypothesis (30) is similar, with weights depending on the nonzero cell frequencies.

TABLE 3

Cell Frequencies for Example 1

	j		
i	1	2	3
1	0	1	2
2	1	0	1
3	2	1	0

Example 2 Consider the cell frequencies shown in Table 4. The hypotheses corresponding to Conditions 1 and 3 in Table 1 are, after some simplification,

$$\begin{aligned} H_{\alpha 1}:\quad \mu_{1\cdot} &= \mu_{21} + \mu_{23} + (\mu_{12} + \mu_{32})/2 \\ &= \mu_{31} + \mu_{32} + (\mu_{13} + \mu_{23})/2, \end{aligned} \tag{31}$$

$$\begin{aligned} H_{\alpha 3}:\quad \mu_{1\cdot} &= \mu_{21} + \mu_{23} + (\mu_{12} + \mu_{32})/2 + (\mu_{23} - \mu_{13})/2 \\ &= \mu_{31} + \mu_{32} + (\mu_{13} + \mu_{23})/2 \\ &\quad + (\mu_{12} - \mu_{32})/4 + (\mu_{23} - \mu_{13})/6. \end{aligned} \tag{32}$$

Here $H_{\alpha 3}$ has been written in a form allowing comparison with $H_{\alpha 1}$. The question of whether $H_{\alpha 1}$, much less $H_{\alpha 3}$, is the appropriate measure of the row effect will not be considered in this paper.

TABLE 4

Cell Frequencies for Example 2

	j		
i	1	2	3
1	1	1	2
2	1	0	1
3	2	1	0

The fourth condition in Table 1 is not covered by Theorem 1; hence we consider a second set of nonestimable conditions

$$\sum_{i=1}^{a} v_i \gamma_{ij} = \sum_{j=1}^{b} w_j \gamma_{ij} = 0 \tag{33}$$

for $i = 1, \ldots, a$, $j = 1, \ldots, b$. The requirement on the v_i and w_j is that (33) represents $a + b - 1$ linearly independent conditions. These conditions are discussed by Elston and Bush [2] and Gosslee and Lucas [4]. In the case of zero cell frequencies, Elston and Bush suggested some main effect hypotheses which they felt were reasonable. We shall relate them to the results of the following theorem which establishes the hypotheses tested by the $R(\;)$ procedure.

THEOREM 2 In the two-way model, (1), for connected designs, consider the set of $a + b - 1 + m$ linearly independent conditions consisting of (33) and $\gamma_{ij} = 0$ if $n_{ij} = 0$ plus one condition each on the α_i and β_j. Then the hypotheses tested by the $R(\;)$ procedure are

(1) $R(\boldsymbol{\alpha} \mid \mu, \boldsymbol{\beta}, \boldsymbol{\gamma})$

$$H_\alpha\colon \quad \sum_{j=1}^{b} w_j \delta_{ij} \mu_{ij} = \sum_{j=1}^{b} \sum_{i'=1}^{a} v_{i'} w_j \delta_{i'j} \delta_{ij} \mu_{i'j} \Big/ \left(\sum_{i=1}^{a} v_i \delta_{ij} \right) \qquad i = 1, \ldots, a-1. \tag{34}$$

(2) $R(\boldsymbol{\beta} \mid \mu, \boldsymbol{\alpha}, \boldsymbol{\gamma})$

$$H_\beta\colon \quad \sum_{i=1}^{a} v_i \delta_{ij} \mu_{ij} = \sum_{i=1}^{a} \sum_{j'=1}^{b} v_i w_{j'} \delta_{ij'} \delta_{ij} \mu_{ij'} \Big/ \left(\sum_{j=1}^{b} w_j \delta_{ij} \right) \qquad j = 1, \ldots, b-1. \tag{35}$$

Here

$$\delta_{ij} = \begin{cases} 1 & \text{if} \quad n_{ij} \neq 0 \\ 0 & \text{if} \quad n_{ij} = 0. \end{cases}$$

The proof of Theorem 2 is essentially the same as that for Theorem 1. The essential relation is established by setting $d_{ij} = v_i w_j \delta_{ij}$ and using the same argument as in Theorem 1. The extension to nonconnected designs is also the same.

Some special cases of Theorem 2 are of interest. If all $n_{ij} > 0$, then (34) reduces to

$$\sum_{j=1}^{b} w_j \mu_{ij} = \sum_{j=1}^{b} w_j \mu_{i'j} \qquad \text{all} \quad i, i'. \tag{36}$$

If $w_j = 1$, $j = 1, \ldots, b$, the (36) reduces to (27). The analogous results hold for (35).

To illustrate the hypotheses for missing cell situations, consider the following two examples.

Example 3 Using the cell frequencies from Table 3, the hypotheses (34), for unspecified weights, v_i and w_i, are

$$\begin{aligned} H_\alpha:\quad & v_3 w_2(\mu_{12} - \mu_{32})/(v_1 + v_3) = v_2 w_3(\mu_{23} - \mu_{13})/(v_1 + v_2), \\ & v_3 w_1(\mu_{21} - \mu_{31})/(v_2 + v_3) = v_1 w_3(\mu_{13} - \mu_{23})/(v_1 + v_2). \end{aligned} \tag{37}$$

With $v_i = n_{i.}$ and $w_j = n_{.j}$ as in Condition 4 of Table 1, (34) yields

$$\begin{aligned} H_{\alpha 4}:\quad & \mu_{12} - \mu_{32} = 6(\mu_{23} - \mu_{13})/5, \\ & \mu_{21} - \mu_{31} = \mu_{13} - \mu_{23}. \end{aligned} \tag{38}$$

The hypothesis (38) should be contrasted with (29) and (30).

Elston and Bush [4] suggested hypotheses that might be appropriate for various missing cell combinations. Their hypotheses for one missing cell or two missing cells in the same row agree with the general results of Theorem 2 for H_α but not, in the latter case, for H_β. For more complex situations, they make some recommendations which do not cover Example 3 because they require at least one complete row. The following example illustrates their recommendations and contrasts them with Theorem 2.

Example 4 Using the cell frequencies from Table 4, (34) yields

$$\begin{aligned} H_\alpha:\quad \sum_{j=1}^{3} w_j \mu_{1j} &= w_1 \mu_{21} + w_3 \mu_{23} + w_2(v_1 \mu_{12} + v_3 \mu_{32})/(v_1 + v_3) \\ &= w_1 \mu_{31} + w_2 \mu_{32} + w_3(v_1 \mu_{13} + v_2 \mu_{23})/(v_1 + v_2). \end{aligned} \tag{39}$$

Note that (39) is similar to (31) apart from the weights.

Elston and Bush [4] suggested two alternatives for this situation. Their equation (20) leads to

$$\begin{aligned} w_1 \mu_{11} + w_2 \mu_{12} &= w_1 \mu_{31} + w_2 \mu_{32}, \\ w_1 \mu_{11} + w_3 \mu_{13} &= w_1 \mu_{21} + w_3 \mu_{23}. \end{aligned} \tag{40}$$

This hypothesis seems to be quite reasonable and might well be more appropriate than (39). Our only point here is that (39) arises by application of the $R(\ )$ procedure with conditions (33), and the corresponding sum of squares is easily generated. The second alternative suggested by Elston and Bush [4] consists of a weighted average of the two statements in (40) plus the additional requirement $\mu_{31} = \mu_{21}$. This hypothesis is distinctly different from either (39) or (40) but is a valid hypothesis and should be judged on its own merit.

Finally, consider a third set of nonestimable conditions leading to yet another set of potential main effect hypotheses. The motivation for this set of conditions is to allow us to generate sums of squares corresponding to (15). Specifically, we are interested in the generalization of Condition 5 in Table 1, written as,

$$\sum_{i=1}^{a} c_{ij}(\alpha_i + \gamma_{ij}) = \sum_{j=1}^{b} c_{ij}(\beta_j + \gamma_{ij}) = 0, \tag{41}$$

$$\sum_{i=1}^{a} c_{i\cdot}\alpha_i = \sum_{j=1}^{b} c_{\cdot j}\beta_j = 0, \tag{42}$$

for $i = 1, \ldots, a$, $j = 1, \ldots, b$. The requirements on the c_{ij} are as in Theorem 1, but here, because of the inclusion of α_i and β_j, (41) represents $a + b$ linearly independent conditions. Only one of the conditions in (42) is essential, the other is then implied by (41).

THEOREM 3 In the two-way model, (1), for connected designs, impose the conditions (41) and (42), and if $n_{ij} = 0$, set $\alpha_i + \gamma_{ij} = \beta_j + \gamma_{ij} = 0$ for a total of $a + b + 1 + m$ linearly independent conditions. Then the hypotheses tested by the $R(\ )$ procedure are

(1) $R(\boldsymbol{\alpha} \mid \mu, \boldsymbol{\beta}, \boldsymbol{\gamma})$

$$H_\alpha: \quad \sum_{j=1}^{b} d_{ij}\mu_{ij}/d_{i\cdot} = \sum_{j=1}^{b} d_{i'j}\mu_{i'j}/d_{i'\cdot} \qquad \text{all} \quad i, i'. \tag{43}$$

(2) $R(\beta \mid \mu, \alpha, \gamma)$

$$H_\beta: \quad \sum_{i=1}^{a} d_{ij}\mu_{ij}/d_{\cdot j} = \sum_{i=1}^{a} d_{ij'}\mu_{ij'}/d_{\cdot j'} \qquad \text{all} \quad j, j'. \tag{44}$$

Here

$$d_{ij} = \begin{cases} c_{ij} & \text{if} \quad n_{ij} \neq 0 \\ 0 & \text{if} \quad n_{ij} = 0. \end{cases}$$

The proof of the theorem is straightforward upon substituting $\mu_{ij} = \mu + \alpha_i + \beta_j + \gamma_{ij}$. Note that in this case, (43) reduces directly to $\alpha_i = 0$ quite apart from any consideration of connectedness.

Referring to Table 1, note that the hypothesis being tested under Condition 5 is just (43) with $d_{ij} = n_{ij}$. This is just the hypothesis tested under $R(\boldsymbol{\alpha} \mid \mu)$ using $R(\ )$, Procedure 1. As a second illustration, if $c_{ij} = 1$ and $n_{ij} > 0$ for all i, j, then the hypothesis (43) is identical to (27). With $c_{ij} = 1$, and some $n_{ij} = 0$, (43) reduces to a comparison of the average of the cell means in each row for all nonempty cells. The following example illustrates this point.

Example 5 For the cell frequencies shown in Table 3, the hypothesis (43) with $c_{ij} = 1$ yields

$$\mu_{12} + \mu_{13} = \mu_{21} + \mu_{23} = \mu_{31} + \mu_{32}. \tag{45}$$

For the cell frequencies in Table 4 we obtain,

$$(\mu_{11} + \mu_{12} + \mu_{13})/3 = (\mu_{21} + \mu_{23})/2 = (\mu_{31} + \mu_{32})/2. \tag{46}$$

Note that (45) and (46) are quite reasonable representatives of the row effect and more appealing to the intuition than $H_{\alpha 1}$ for these two examples as illustrated by (29) and (31).

To summarize, Theorems 1–3 provide a means of generating sums of squares for a wide variety of main effect hypotheses. The choice of the weights c_{ij}, v_i, and w_j should be justified in terms of the desired hypothesis rather than a computational procedure. The classical hypothesis $\mu_{i\cdot} = \mu_{i'\cdot}$, for all $n_{ij} > 0$, arises from any of the theorems with equal weights. Hypotheses for the case $n_{ij} = 0$ may be derived from any of the theorems, although the most commonly used procedures are either Theorem 1 with $c_{ij} = 1$ or Theorems 1 and 3 with $c_{ij} = n_{ij}$. Curiously, the most natural missing cell hypotheses arise from Theorem 3 with $c_{ij} = 1$, an option which is rarely, if ever, used.

3.2 Interaction Hypotheses

In this section, the interaction hypotheses associated with the $R(\ )$ notation are developed. An important concept with zero cell frequencies is that of connectedness. John [8] defined a connected, two-factor design as one in which every pair of rows (and columns) can be connected by a row–column–row–column . . . chain. That is, it is possible to move from any row i to row i' along a series of cells with nonzero frequencies by alternately changing the row and column index but not both. A numerical test for connectedness is developed and related to the concept of "estimating" missing data in the model (8).

The following theorem develops the hypothesis tested by $R(\gamma \mid \mu, \alpha, \beta)$.

THEOREM 4 Given any set of $a + b - 1$ linearly independent, nonestimable conditions on the γ_{ij}, say, of the form of (19) or (33), along with $\gamma_{ij} = 0$ if $n_{ij} = 0$, the hypothesis tested by $R(\boldsymbol{\gamma} \mid \mu, \boldsymbol{\alpha}, \boldsymbol{\beta})$ may be obtained as follows:

(1) Let $\mathbf{H}\boldsymbol{\mu} = \mathbf{0}$ denote any set of $(a - 1)(b - 1)$ linearly independent interaction constraints obtained by eliminating redundancies from (7). For example,

$$\mu_{11} - \mu_{i1} - \mu_{1j} + \mu_{ij} = 0 \qquad i = 2, \ldots, a, \quad j = 2, \ldots, b. \tag{47}$$

(2) Assume that the components of μ are ordered so that the m empty cells occur first and apply Gauss reduction to the corresponding columns of $\mathbf{H}$ to obtain the equivalent set of constraints,

$$\begin{bmatrix} \mathbf{H}_{11} & \mathbf{H}_{12} \\ 0 & \mathbf{H}_{22} \end{bmatrix} \boldsymbol{\mu} = \mathbf{0} \tag{48}$$

where $\mathbf{H}_{11}$ has m columns. Then the hypothesis tested by $R(\boldsymbol{\gamma} \mid \mu, \boldsymbol{\alpha}, \boldsymbol{\beta})$ is

$$H_\gamma: \quad [\mathbf{0} \quad \mathbf{H}_{22}]\boldsymbol{\mu} = [\mathbf{0} \quad \mathbf{H}_{22}]\boldsymbol{\gamma} = \mathbf{0}. \tag{49}$$

To prove this theorem, it is necessary to establish some properties of $\mathbf{H}_{11}$ and of the matrix describing the conditions on $\boldsymbol{\gamma}$. The following lemmas contain the necessary results. The proofs are a direct consequence of the definition of connectedness.

LEMMA 1 (1) If the design is connected, $\mathbf{H}_{11}$ is an identity of size m.

(2) If the design is not connected, $\mathbf{H}_{11} = (\mathbf{I} \; \mathbf{P})$ where the dimension of $\mathbf{I}$ is $m - p$ with p being the minimum number of cells to be filled to connect the design.

LEMMA 2 Writing the conditions on $\boldsymbol{\gamma}$ in matrix form as

$$\mathbf{C}\boldsymbol{\gamma} = \mathbf{0}, \tag{50}$$

where $\mathbf{C}$ is $(a + b - 1) \times ab$ of rank $a + b - 1$ and partitioning $\mathbf{C} = (\mathbf{C}_1 \; \mathbf{C}_2)$ according to the partitioning in (48), it follows that:

(1) For connected designs, it is possible to perform row operations on $\mathbf{C}_2$ to generate an identity of size $a + b - 1$.

(2) Otherwise, row operations will reduce $\mathbf{C}_2$ to a matrix of rank $(a + b - 1) - p$.

Proof of Theorem

(1) *Connected designs* Since $\mathbf{H}_{11} = \mathbf{I}_m$, the hypothesis (49) has rank $(a - 1)(b - 1) - m$. Recalling the $R(\;)$ procedure for missing cells, (50) is

augmented by $\gamma_{ij} = 0$ if $n_{ij} = 0$. Thus the nonestimable conditions may be written as

$$\begin{bmatrix} \mathbf{C}_1 & \mathbf{C}_2 \\ \mathbf{I}_m & \mathbf{0} \end{bmatrix} \boldsymbol{\gamma} = \mathbf{0}. \tag{51}$$

By Lemma 2, the matrix in (51) has full row rank of $a + b - 1 + m$. Further, the matrix

$$\mathbf{Q} = \begin{bmatrix} \mathbf{0} & \mathbf{H}_{22} \\ \mathbf{C}_1 & \mathbf{C}_2 \\ \mathbf{I}_m & \mathbf{0} \end{bmatrix} \tag{52}$$

has rank $((a - 1)(b - 1) - m) + (a + b - 1 + m) = ab$ since, by the definition of estimable functions, the rows of (49) are independent of those in (51).

Thus, the hypothesis (49) in conjunction with the nonestimable conditions is $\mathbf{Q}\boldsymbol{\gamma} = \mathbf{0}$, but $\mathbf{Q}$ is nonsingular, and hence this is equivalent to $\gamma_{ij} = 0$ as prescribed by $R(\boldsymbol{\gamma} \mid \mu, \boldsymbol{\alpha}, \boldsymbol{\beta})$.

(2) *Unconnected designs.* In this case, by Lemma 1, $\mathbf{H}_{11} = [\mathbf{I} \quad \mathbf{P}]$ and hence the hypothesis in (49) has rank $(a - 1)(b - 1) - (m - p)$. Further, the matrix in (51) has rank $a + b - 1 + m - p$ in accordance with Lemma 2. Thus the matrix $\mathbf{Q}$ in (52) has rank $((a - 1)(b - 1) - (m - p)) + (a + b - 1 + m - p) = ab$ and hence $\mathbf{Q}\boldsymbol{\gamma} = \mathbf{0}$ is equivalent to $\gamma_{ij} = 0$.

Comments on Theorem 4 (1) If all $n_{ij} \neq 0$, the theorem states that the hypothesis is (7) or, equivalently, (47), independent of the conditions on $\boldsymbol{\gamma}$, subject to the requirement that the rank of $\mathbf{C}$ is $a + b - 1$.

(2) The decomposition of $\mathbf{H}$ as in (48) provides a test for connectedness indicated by $\mathbf{H}_{11} = \mathbf{I}_m$.

(3) The equations

$$[\mathbf{H}_{11} \quad \mathbf{H}_{12}]\boldsymbol{\mu} = \mathbf{0} \tag{53}$$

provide a set of "defining" relations for the cell means of the missing cells in terms of the cells for which $n_{ij} \neq 0$. In the case of connected designs, it is possible to solve uniquely for the cell means corresponding to H_{11} in terms of those corresponding to H_{12}. This is just the basis for "estimation" of missing data in the model (8). That is, if it is assumed, a priori, that the cell means satisfy (7), then the estimate of the missing value is just the estimate of the cell mean as obtained from (53) in terms of the estimates of the non-empty cell means. If the design is not connected, this is not possible and hence the missing data procedures will fail. Note that if (7) is not assumed, then there is no information on which to base an estimate of the cell means for empty cells.

Example 5 Using the cell frequencies from Table 3, the matrix $\mathbf{H}$ of interaction constraints as defined by (47) is

$$\mathbf{H} = \left[\begin{array}{ccc|cccccc} 1 & 1 & 0 & -1 & 0 & -1 & 0 & 0 & 0 \\ 1 & 0 & 0 & 0 & -1 & -1 & 1 & 0 & 0 \\ 1 & 0 & 0 & -1 & 0 & 0 & 0 & -1 & 1 \\ 1 & 0 & 1 & 0 & -1 & 0 & 0 & -1 & 0 \end{array}\right] \tag{54}$$

where the column indices are ordered as (11, 22, 33, 12, 13, 21, 23, 31, 32). After reduction, (48) yields

$$\left[\begin{array}{ccc|cccccc} 1 & 0 & 0 & 0 & -1 & -1 & 1 & 0 & 0 \\ 0 & 1 & 0 & -1 & 1 & 0 & -1 & 0 & 0 \\ 0 & 0 & 1 & 0 & 0 & 1 & -1 & -1 & 0 \\ 0 & 0 & 0 & 1 & -1 & -1 & 1 & 1 & -1 \end{array}\right] \tag{55}$$

In this case, $\mathbf{H}_{11} = \mathbf{I}_3$ and the design is connected. The last row of (55) defines the hypothesis tested by $R(\boldsymbol{\gamma} \mid \mu, \boldsymbol{\alpha}, \boldsymbol{\beta})$, namely

$$\mu_{12} - \mu_{13} - \mu_{21} + \mu_{23} + \mu_{31} - \mu_{32} = 0. \tag{56}$$

The first three rows of (55) define the cell means for the empty cells in terms of the remaining cell means. In particular,

$$\begin{aligned} \mu_{11} &= \mu_{13} + \mu_{21} - \mu_{23}, \\ \mu_{22} &= \mu_{12} - \mu_{13} + \mu_{23}, \\ \mu_{33} &= -\mu_{21} + \mu_{23} + \mu_{31}. \end{aligned} \tag{57}$$

Note that Theorem 4 does not include nonestimable conditions of the type (41) used in Theorem 3. These conditions, although valid, do not lead to a test of the interaction hypothesis via $R(\boldsymbol{\gamma} \mid \mu, \boldsymbol{\alpha}, \boldsymbol{\beta})$, except in the special case $c_{ij} = 1$, for all i, j. In this case the results are the same as in Theorem 4.

4. TWO-WAY CLASSIFICATION WITHOUT INTERACTION

The analysis of the two-way classification model, without interaction, is much simpler. The hypotheses associated with $R(\ )$ notation under either Procedure 1 or Procedure 2 are easily obtained. The model is given by (6) in association with the constraints (7). In the case of missing cells, the appropriate constraints, in terms of only the observed cells, are given by (49). Imposing the constraints on the cell means allows us to write the model in the form (8), and the following theorem results.

THEOREM 5 In the two-way model, (8) for connected designs, the hypotheses tested by the $R(\ )$ procedure are

1. $R(\boldsymbol{\alpha} \mid \mu, \boldsymbol{\beta})$

$$H: \quad \mu_{ij} = \mu_{i'j}$$

for all i, i', j combinations with $n_{ij} \neq 0$. (58)

2. $R(\boldsymbol{\beta} \mid \mu, \boldsymbol{\alpha})$

$$H_\beta: \quad \mu_{ij} = \mu_{ij'}$$

for all i, j, j' combinations with $n_{ij} \neq 0$. (59)

The proof of Theorem 5 follows immediately by noting that, with $\mu_{ij} = \mu + \alpha_i + \beta_j$, eq. (58) reduces to $\alpha_i = \alpha_{i'}$ for all i, i', and hence any set of nonestimable conditions on the α_i reduces to $\alpha_i = 0$. Thus the hypothesis associated with $R(\boldsymbol{\alpha} \mid \mu, \boldsymbol{\beta})$ is independent of the choice of nonestimable conditions. It is of interest to note that this result can also be established by taking the main effect hypotheses from either Theorem 1 or Theorem 2 and imposing the constraints (7). (See Hackney [6].) This follows since the interaction constraints, in conjunction with either of the sets of nonestimable conditions (19) or (33), are equivalent to $\gamma_{ij} = 0$ in (1). This is not true for Theorem 3. For designs which are not connected, the same hypotheses hold. The only distinction is in the proof where it is neccessary to impose a set of nonestimable conditions for each connected design.

The concept of estimating missing data, so as to balance the design, will not be considered here. The reader is referred to the recent papers by Rubin [13] and Cook and Weisberg [1]. The relation of the latter paper to the defining relations (53) should be noted.

Example 6 Using the cell frequencies in Figure 1, Theorem 5 states that the row hypothesis is

$$H: \quad \mu_{21} = \mu_{31}, \qquad \mu_{12} = \mu_{32}, \qquad \mu_{13} = \mu_{23}.$$

As indicated, this hypothesis also arises by imposing (56) on any of (29), (30), or (37).

5. TWO-FOLD NESTED MODEL

In Section 1, it was noted that the model (6) may be used to describe the two-fold nested model (9). The analysis of this model, for unbalanced data, also warrants attention. Letting $j = 1, \ldots, b_i$ in (9) removes the problem of missing data, but the hypotheses tested by $R(\ )$ procedures with unequal n_{ij} are of interest. Typically, the sums of squares used for analysis are (1)

$R(\boldsymbol{\gamma} \mid \mu, \boldsymbol{\alpha})$ and (2) $R(\boldsymbol{\alpha} \mid \mu)$, where the latter represents an application of $R(\ )$, Procedure 1. It may be seen, using the results of Hackney [6], that $R(\boldsymbol{\alpha} \mid \mu)$ is given by $R(\boldsymbol{\alpha} \mid \mu, \boldsymbol{\gamma})$ using Procedure 2, if Condition 2 in Table 2 is imposed to achieve full rank. Generalizing on this, consider the set of nonestimable conditions

$$\sum_{j=1}^{b_i} c_{ij} \gamma_{ij} = 0 \qquad i = 1, \ldots, a, \tag{60}$$

where the c_{ij} are such that the set of equations (60) has rank a.

THEOREM 6 The hypotheses tested by the $R(\ )$ procedure applied to the two-fold nested model (9) are

1. $R(\boldsymbol{\gamma} \mid \mu, \boldsymbol{\alpha})$

$$H_\gamma: \ \mu_{ij} = \mu_{ij'} \qquad \text{for all} \quad i, i', j, j'. \tag{61}$$

2. $R(\alpha \mid \mu, \gamma)$

$$H_\alpha: \ \sum_{j=1}^{b_i} c_{ij} \mu_{ij} / c_{i\cdot} = \sum_{j=1}^{b_i} c_{i'j} \mu_{i'j} / c_{i'} \qquad \text{for all} \quad i, i'. \tag{62}$$

The proof follows by substituting $\mu_{ij} = \mu + \alpha_i + \gamma_{ij}$ and observing that (61) in conjunction with (60) is equivalent to $\gamma_{ij} = 0$ and that (62) is similarly equivalent to $\alpha_i = \alpha_{i'}$. The imposition of any nonestimable condition on the α_i, necessary to achieve full rank in (9), then shows that (62) is equivalent to $\alpha_i = 0$. Reference is made to Hocking and Speed [7] for a discussion of the special case $c_{ij} = n_{ij}$.

6. SUMMARY

The intent of this paper has been to demonstrate the hypotheses under test when standard computing routines are applied to some of the common models. The analyst is urged to study these hypotheses to see if they are reasonable before submitting his data for evaluation. Ideally, the computer program being used should be sufficiently flexible to allow the specification of any linear hypothesis rather than restricting the user to one or more of those cited here. Clearly, this places a burden on the preparation of programs but if the alternative is testing inappropriate hypotheses, it seems that the additional effort is justified.

REFERENCES

[1] COOK, R. D. and WEISBERG, S. (1975). Missing values in unreplicated orthogonal designs. Technical Report 253r, Dept. Appl. Statist., Univ. of Minnesota Press, Minneapolis.

[2] ELSTON, R. C. and BUSH, N. (1964). The hypotheses that can be tested when there are interactions in an analysis of variance model. *Biometrics* 681–699.
[3] FINNEY, D. J. (1948). Main effects and interactions. *J. Amer. Statist. Assoc.* **43** 566–571.
[4] GOSSLEE, D. G. and LUCAS, H. L. (1965). Analysis of variance of disproportionate data when interactions are present. *Biometrics* **21** 115–133.
[5] GRAYBILL, F. A. (1961). *An Introduction to Linear Statistical Models* **I** McGraw-Hill, New York.
[6] HACKNEY, O. P. (1976). Hypothesis testing in general linear model. Unpublished Ph.D. dissertation, Emory Univ.
[7] HOCKING, R. R. and SPEED, F. M. (1975). A full rank analysis of some linear model problems. *J. Amer. Statist. Assoc.* **70** 706–712.
[8] JOHN, P. W. M. (1971). *Statistical Design and Analysis of Experiments*. Macmillan, New York.
[9] SEARLE, S. R. (1967). *Linear Models*. Wiley, New York.
[10] SEARLE, S. R. (1972). Using the $R(\ )$-notation for reductions in sums of squares when fitting linear models. Presented at the Spring Regional Meetings of ENAR, Ames, Iowa.
[11] SPEED, F. M. and HOCKING, R. R. (1976). The use of the $R(\ )$-notation with unbalanced data. *Amer. Statist.* **30** 30–33.
[12] SPEED, F. M., HOCKING, R. R., and HACKNEY, O. P. (1976). Methods of analysis of unbalanced data. Submitted for publication.
[13] RUBIN, D. B. (1972). A non-iterative algorithm for least squares estimation of missing values in any analysis of variance design. *Appl. Statist.* **21** 136–141.

Nonhomogeneous Variances in the Mixed AOV Model; Maximum Likelihood Estimation

William J. Hemmerle *Brian W. Downs*

UNIVERSITY OF RHODE ISLAND

This paper considers maximum likelihood estimation for the general mixed analysis of variance model when the variances associated with the random errors are unequal between groups of observations. A general computational algorithm is given to obtain estimators of the fixed effects, variance components, and error variances in which nonnegative constraints are imposed upon all of the variances involved. Special algorithms are then developed for models used by Grubbs in determining the imprecision of measuring instruments extended to include unbalanced data. The efficiency of the algorithms is demonstrated through examples which are given.

1. INTRODUCTION

This paper considers the problem of obtaining maximum likelihood estimates for the general mixed analysis of variance model when the variances associated with the random errors are unequal between groups of observations. Hartley and Jayattilake [7] considered a similar situation for the fixed effects linear regression or analysis of variance model, obtaining maximum likelihood estimates of the model parameters including the unequal error

Research supported in part by National Science Foundation Grant No. DCR74-15331.

Key words and phrases maximum likelihood estimation, mixed AOV model, nonhomogeneous error variances, Grubbs's measurement models variances constrained to being nonnegative. Hartley and Rao [6] laid the foundation for obtaining maximum likelihood estimates for the general mixed analysis of variance model with equal error variances. The W transformation suggested by Hemmerle and Hartley [8] made it practical computationally to use iterative optimization methods to obtain these estimates. An algorithm was given in [8] to demonstrate feasibility and use of the W transformation; other algorithms employing this transformation to obtain maximum likelihood estimates for the mixed model have subsequently appeared (Jennrich and Sampson [11], Hemmerle and Lorens [9]). It has also been applied by Corbeil and Searle [1] to obtain restricted maximum likelihood (REML) estimates.

The early parts of this paper treat the general mixed model with nonhomogeneous group variances. Theory and methodology are established which permit use of the previously cited algorithms, with little or no modification, to obtain maximum likelihood estimates when the error variances are unequal. An important consideration in this development is the ability to constrain the estimators of the variance components and error variances to being nonnegative. The latter parts of the paper consider the models studied by Grubbs [3,4] and others to determine the imprecision of measuring instruments. One recent paper by Jaech [10] develops approximate likelihood ratios for these models, using estimated variances obtained by least squares (which may be negative), to avoid the presumed difficulty of application of maximum likelihood. We will show that, in fact, maximum likelihood may be applied effectively to these models. Two algorithms are developed, one for balanced data and one for missing data, which exploit the model's inherent structure to yield efficient computation of the estimators and the likelihood function.

2. THE MIXED AOV MODEL WITH UNEQUAL ERROR VARIANCES

Following Hartley and Rao [6] and Hemmerle and Hartley [8] we write the general mixed AOV model as

$$\mathbf{y} = \mathbf{X}\boldsymbol{\alpha} + \mathbf{U}_1\mathbf{b}_1 + \mathbf{U}_2\mathbf{b}_2 + \cdots + \mathbf{U}_c\mathbf{b}_c + \mathbf{e}, \tag{2.1}$$

where $\mathbf{y}$ is an n vector of observations; $\mathbf{X}$ is an $n \times p$ matrix of known fixed numbers, $p \leq n$; $\mathbf{U}_i$ is an $n \times m_i$ matrix of known fixed numbers, $m_i \leq n$; $\boldsymbol{\alpha}$ is a p vector of unknown constants; and $\mathbf{b}_i$ is an m_i vector of independent variables from $N(0, \sigma_i^2)$. In the equal variance case, we assume that $\mathbf{e}$ is an

$n \times 1$ vector from $N(0, \sigma^2)$. We now depart from this assumption with respect to the elements of the vector $\mathbf{e}$ and partition the model (2.1) such that

$$\mathbf{y} = \begin{bmatrix} \mathbf{y}_1 \\ \hline \mathbf{y}_2 \\ \hline \vdots \\ \hline \mathbf{y}_k \end{bmatrix}, \quad \mathbf{X} = \begin{bmatrix} \mathbf{X}_1 \\ \hline \mathbf{X}_2 \\ \hline \vdots \\ \hline \mathbf{X}_k \end{bmatrix}, \quad \mathbf{U}_i = \begin{bmatrix} \mathbf{U}_{i1} \\ \hline \mathbf{U}_{i2} \\ \hline \vdots \\ \hline \mathbf{U}_{ik} \end{bmatrix}, \quad \mathbf{e} = \begin{bmatrix} \mathbf{e}_1 \\ \hline \mathbf{e}_2 \\ \hline \vdots \\ \hline \mathbf{e}_k \end{bmatrix}. \tag{2.2}$$

where $\mathbf{y}_i$ is an n_i vector with $\sum_{i=1}^{k} n_i = n$. We then assume that $\mathbf{e}_i$ has the multivariate normal distribution

$$\mathbf{e}_i \sim \text{MVN}(0, s_i^2\mathbf{I}) \qquad i = 1, \ldots, k \tag{2.3}$$

and that the random vectors $\mathbf{b}_1, \mathbf{b}_2, \ldots, \mathbf{b}_c, \mathbf{e}_1, \mathbf{e}_2, \ldots, \mathbf{e}_k$ are mutually independent. We let

$$\sum_{i=1}^{c} m_i = m \tag{2.4}$$

and also make certain rank assumptions with respect to $\mathbf{X}$ and the $\mathbf{U}_i$ which will be discussed later in this section.

From the model (2.1) and the assumptions related to it, the n vector $\mathbf{y}$ must have the multivariate normal distribution or likelihood function

$$L = \frac{1}{(2\pi)^{n/2}\,|\boldsymbol{\Sigma}|^{1/2}} \exp\{-\tfrac{1}{2}(\mathbf{y} - \mathbf{X}\boldsymbol{\alpha})'\boldsymbol{\Sigma}^{-1}(\mathbf{y} - \mathbf{X}\boldsymbol{\alpha})\} \tag{2.5}$$

with variance–covariance matrix $\boldsymbol{\Sigma}$ given by

$$\boldsymbol{\Sigma} = [\mathbf{D}^* + \sigma_1^2\mathbf{U}_1\mathbf{U}_1' + \sigma_2^2\mathbf{U}_2\mathbf{U}_2' + \cdots + \sigma_c^2\mathbf{U}_c\mathbf{U}_c'], \tag{2.6}$$

where $\mathbf{D}^*$ is the diagonal matrix whose ith diagonal block, D_{ii}^*, is given by

$$\mathbf{D}_{ii}^* = s_i^2\mathbf{I}_{n_i} \qquad i = 1, 2, \ldots, k. \tag{2.7}$$

We now note that the matrix $\boldsymbol{\Sigma}$ given by (2.6) may be represented as

$$\boldsymbol{\Sigma} = [s_1^2\mathbf{E}_1\mathbf{E}_1' + \cdots + s_k^2\mathbf{E}_k\mathbf{E}_k' + \sigma_1^2\mathbf{U}_1\mathbf{U}_1' + \cdots + \sigma_c^2\mathbf{U}_c\mathbf{U}_c'], \tag{2.8}$$

where $\mathbf{E}_i$ is the $n \times n_i$ matrix

$$\mathbf{E}_i = (0\,|\,\cdots\,|0\,|\,\mathbf{I}_{n_i}|\,0\,|\,\cdots\,|0)' \tag{2.9}$$

such that $\mathbf{I}_{n_i}$ is the ith submatrix and

$$(\mathbf{E}_1\,|\,\mathbf{E}_2\,|\,\cdots\,|\,\mathbf{E}_k) = \mathbf{I}_n. \tag{2.10}$$

Next, we write the expression for $\boldsymbol{\Sigma}$ given by (2.8) in the form

$$\begin{aligned}\boldsymbol{\Sigma} = [&s_1^2\mathbf{I} + (s_1^2 - s_1^2)\mathbf{E}_1\mathbf{E}_1' + \cdots \\ &+ (s_k^2 - s_1^2)\mathbf{E}_k\mathbf{E}_k' + \sigma_1^2\mathbf{U}_1\mathbf{U}_1' + \cdots + \sigma_c^2\mathbf{U}_c\mathbf{U}_c']\end{aligned} \tag{2.11}$$

or

$$\Sigma = [s_1^2\mathbf{I} + (s_2^2 - s_1^2)\mathbf{E}_2\mathbf{E}_2' + \cdots + (s_k^2 - s_1^2)\mathbf{E}_k\mathbf{E}_k' + \sigma_1^2\mathbf{U}_1\mathbf{U}_1' + \cdots + \sigma_c^2\mathbf{U}_c\mathbf{U}_c']. \tag{2.12}$$

If we now let

$$\delta_i = (s_i^2 - s_1^2)/s_1^2 \qquad i = 2, \ldots, k, \tag{2.13}$$

$$\gamma_i = \sigma_i^2/s_1^2 \qquad i = 1, 2, \ldots, c, \tag{2.14}$$

and factor out s_1^2 in (2.12) we obtain

$$\Sigma = s_1^2\mathbf{H}, \tag{2.15}$$

where

$$\mathbf{H} = [\mathbf{I} + \delta_2\mathbf{E}_2\mathbf{E}_2' + \cdots + \delta_k\mathbf{E}_k\mathbf{E}_k' + \gamma_1\mathbf{U}_1\mathbf{U}_1' + \cdots + \gamma_c\mathbf{U}_c\mathbf{U}_c']. \tag{2.16}$$

The structure of (2.15) and (2.16) in terms of the new variables $s_1, \delta_2, \ldots, \delta_k$, $\gamma_1, \ldots, \gamma_c$ now conforms exactly with the structure of the variance–covariance matrix for the general mixed model with *equal* variances considered in [6] and [8] with s_1^2 playing the role of σ^2.

For the general mixed AOV model with equal variances, Hartley and Rao [6] initially made the assumption that the design matrices $\mathbf{X}$ and $\mathbf{U}_i$ are all of full rank and that, as a condition for estimability, a basis for the matrix

$$(\mathbf{X}\,|\,\mathbf{U}_1\,|\,\mathbf{U}_2\,|\,\cdots\,|\,\mathbf{U}_c) \tag{2.17}$$

may be formed from the columns of $\mathbf{X}$ and at least one column from each $\mathbf{U}_i$. However, Miller [12] gives a counterexample to show that these conditions are not sufficient for estimability. He replaces them with the following assumptions:

The matrix $\mathbf{X}$ has full rank p; (2.18)

$n \geq p + c + 1$ (i.e., we have at least as many observations as parameters); (2.19)

The matrix $(\mathbf{X}\,|\,\mathbf{U}_i)$ has rank greater than p for $i = 1, 2, \ldots, c$; (2.20)

The matrices $\mathbf{I}_n, \mathbf{U}_1\mathbf{U}_1', \ldots, \mathbf{U}_c\mathbf{U}_c'$ are linearly independent. (2.21)

We will apply these same assumptions to the nonhomogeneous variance model, with variance–covariance matrix in the form given by (2.15) and (2.16), by considering the matrices $\mathbf{E}_2, \ldots, \mathbf{E}_k$ as $\mathbf{U}_{c+1}, \ldots, \mathbf{U}_{c+k-1}$ in (2.18–2.21).

3. CONSTRAINING THE ESTIMATORS

When we consider constraining the maximum likelihood estimators $\hat{s}_i^2$ and $\hat{\sigma}_i^2$ of s_i^2 and σ_i^2, respectively, such that

$$\hat{s}_i^2 \geq 0 \qquad i = 1, 2, \ldots, k, \tag{3.1}$$

$$\hat{\sigma}_i^2 \geq 0 \qquad i = 1, 2, \ldots, c, \tag{3.2}$$

we observe that the transformation (2.13) is unsatisfactory. This is due to the fact that the constraints (3.1) imply that

$$\hat{\delta}_i \geq -1 \qquad i = 2, 3, \ldots, k, \tag{3.3}$$

where $\hat{\delta}_i$ is the maximum likelihood estimator of δ_i. The constraints (3.3) are difficult to handle in the context of the equal variance mixed AOV model. The computational techniques discussed in [8] and in subsequent papers [9,11] may not be applied directly to compute nonnegative maximum likelihood estimators.

We have found that a simple and effective technique for imposing the constraints (3.1) is to introduce a dummy parameter σ^2 into the computations. Proceeding as in Section 2 we let

$$\begin{aligned}\boldsymbol{\Sigma} = [\sigma^2\mathbf{I} + (s_1^2 - \sigma^2)\mathbf{E}_1\mathbf{E}_1' + \cdots \\ + (s_k^2 - \sigma^2)\mathbf{E}_k\mathbf{E}_k' + \sigma_1^2\mathbf{U}_1\mathbf{U}_1' + \cdots + \sigma_c^2\mathbf{U}_c\mathbf{U}_c']\end{aligned} \tag{3.4}$$

similar to (2.11). The expressions corresponding to (2.13–2.16) then become:

$$\delta_i = (s_i^2 - \sigma^2)/\sigma^2 \qquad i = 1, 2, \ldots, k; \tag{3.5}$$

$$\gamma_i = \sigma_i^2/\sigma^2 \qquad i = 1, 2, \ldots, c; \tag{3.6}$$

$$\boldsymbol{\Sigma} = \sigma^2\mathbf{H}; \tag{3.7}$$

$$\mathbf{H} = [\mathbf{I} + \delta_1\mathbf{E}_1\mathbf{E}_1' + \cdots + \delta_k\mathbf{E}_k\mathbf{E}_k' + \cdots + \gamma_1\mathbf{U}_1\mathbf{U}_1' + \cdots + \gamma_c\mathbf{U}_c\mathbf{U}_c']. \tag{3.8}$$

The maximum likelihood estimators of (3.5) and (3.6) are then constrained such that

$$\hat{\delta}_i \geq 0 \qquad i = 1, 2, \ldots, k, \tag{3.9}$$

$$\hat{\gamma}_i \geq 0 \qquad i = 1, 2, \ldots, c, \tag{3.10}$$

and we maximize the log-likelihood λ given by

$$2\lambda = -n \ln 2\pi - n \ln \sigma^2 - \ln|\mathbf{H}| - (\mathbf{y} - \mathbf{X}\boldsymbol{\alpha})'\mathbf{H}^{-1}(\mathbf{y} - \mathbf{X}\boldsymbol{\alpha})/\sigma^2 \tag{3.11}$$

subject to these constraints. The square root transformation suggested in [6] and applied in [8] may be used to assure that the constraints (3.9) and (3.10) are satisfied. We let

$$\omega_i^2 = \delta_i \qquad i = 1, 2, \ldots, k, \tag{3.12}$$

$$\tau_i^2 = \gamma_i \qquad i = 1, 2, \ldots, c, \tag{3.13}$$

and work with the transformed variables in the maximization process. The quantity $\hat{\sigma}^2$ may be evaluated as a quadratic form which is always nonnegative.

Clearly, we have overparametrized in introducing the dummy parameter σ^2 and there is an infinite number of solutions that yield the same maximum value for λ given by (3.11); however, we need only find one of these solutions in order to find the corresponding unique solution for the $\hat{s}_i^2$ and $\hat{\sigma}_i^2$ via the transformations (3.5) and (3.6).

In the cases we have examined, we have found that different starting values for the maximization algorithm produce different final values for the $\hat{\delta}_i$, $\hat{\gamma}_i$, and $\hat{\sigma}^2$. The latter, however, yielded the same value for the log-likelihood (3.11) and transformed into the same values for the $\hat{s}_i^2$ and $\hat{\sigma}_i^2$. Although we have had success with this procedure, we do not claim that it is foolproof. If one of the $\hat{s}_i^2$ wants to converge to zero, then $\hat{\sigma}^2$ should also approach zero since the constraint (3.9) implies

$$\hat{s}_i^2 \geq \hat{\sigma}^2 \qquad i = 1, 2, \ldots, k. \tag{3.14}$$

When this occurs we could apply the deletion or boundary techniques discussed in [8]; however, we have not experimented along these lines. In our selection of examples, we have perhaps been fortunate in not encountering this potentially troublesome situation. The estimates of the variance components, the $\hat{\sigma}_i^2$, may of course converge to zero without $\hat{\sigma}^2$ approaching zero as the example in the following section demonstrates.

4. THE GENERAL ALGORITHM—AN EXAMPLE

We present here an extremely simple example of a mixed AOV model with unequal variances which we first solve analytically and then numerically in order to illustrate the preceding material and the general algorithm. The model for the example is

$$y_{ij} = \mu + b_j + e_{ij} \qquad i = 1, 2 \quad j = 1, 2, \tag{4.1}$$

where μ is fixed, b_j is random with variance component σ_1^2, and

$$\mathrm{var}(e_{1j}) = s_1^2, \qquad \mathrm{var}(e_{2j}) = s_2^2. \tag{4.2}$$

The vector of 4 observations for the model (4.1) is taken to be

$$\mathbf{y} = (y_{11}, y_{12}, y_{21}, y_{22})' = (0, 2, 2, 0)'. \tag{4.3}$$

As an algebraic simplification we let

$$l_i = 1 + \delta_i \qquad i = 1, 2. \tag{4.4}$$

The matrix $\mathbf{H}$ given by (3.8) may then be written as

$$\mathbf{H} = \begin{bmatrix} l_1 + \gamma_1 & 0 & \gamma_1 & 0 \\ 0 & l_1 + \gamma_1 & 0 & \gamma_1 \\ \gamma_1 & 0 & l_2 + \gamma_1 & 0 \\ 0 & \gamma_1 & 0 & l_2 + \gamma_1 \end{bmatrix}, \tag{4.5}$$

with inverse and determinant given by

$$\mathbf{H}^{-1} = 1/q \begin{bmatrix} l_2 + \gamma_1 & 0 & -\gamma_1 & 0 \\ 0 & l_2 + \gamma_1 & 0 & -\gamma_1 \\ -\gamma_1 & 0 & l_1 + \gamma_1 & 0 \\ 0 & -\gamma_1 & 0 & l_1 + \gamma_1 \end{bmatrix}, \tag{4.6}$$

$$|\mathbf{H}| = q^2, \tag{4.7}$$

where

$$q = \gamma_1 l_1 + \gamma_1 l_2 + l_1 l_2. \tag{4.8}$$

Now the first order partials of the log-likelihood (3.11) are:

$$\partial\lambda/\partial\boldsymbol{\alpha} = (1/\sigma^2)\{\mathbf{X}'\mathbf{H}^{-1}\mathbf{y} - (\mathbf{X}'\mathbf{H}^{-1}\mathbf{X})\boldsymbol{\alpha}\}, \tag{4.9}$$

$$\partial\lambda/\partial\sigma^2 = (-n/2\sigma^2) + (1/2\sigma^4)(\mathbf{y} - \mathbf{X}\boldsymbol{\alpha})'\mathbf{H}^{-1}(\mathbf{y} - \mathbf{X}\boldsymbol{\alpha}), \tag{4.10}$$

$$\partial\lambda/\partial\delta_i = -\tfrac{1}{2}\,\mathrm{tr}\{\mathbf{E}_i'\mathbf{H}^{-1}\mathbf{E}_i\} + (1/2\sigma^2)(\mathbf{y} - \mathbf{X}\boldsymbol{\alpha})'\mathbf{H}^{-1}\mathbf{E}_i\mathbf{E}_i'\mathbf{H}^{-1}(\mathbf{y} - \mathbf{X}\boldsymbol{\alpha}), \tag{4.11}$$

$$\partial\lambda/\partial\gamma_i = -\tfrac{1}{2}\,\mathrm{tr}\{\mathbf{U}_i'\mathbf{H}^{-1}\mathbf{U}_i\} + (1/2\sigma^2)(\mathbf{y} - \mathbf{X}\boldsymbol{\alpha})'\mathbf{H}^{-1}\mathbf{U}_i\mathbf{U}_i'\mathbf{H}^{-1}(\mathbf{y} - \mathbf{X}\boldsymbol{\alpha}). \tag{4.12}$$

If we equate (4.9) to zero for our example, we obtain $\hat{\mu} = 1$ and

$$(\mathbf{y} - \mathbf{X}\hat{\boldsymbol{\mu}}) = (-1, 1, 1, -1)' \tag{4.13}$$

for all values of the other variables. Thus, equating (4.10) to zero yields

$$\hat{\sigma}^2 = (l_1 + l_2 + 4\gamma_1)/2q. \tag{4.14}$$

Substituting these values (4.13) and (4.14) into the equations (4.11) and (4.12) yields

$$\partial\lambda/\partial\delta_1 = -(l_2 + \gamma_1)/q + 2(l_2 + 2\gamma_1)^2/q(l_1 + l_2 + 4\gamma_1), \tag{4.15}$$

$$\partial\lambda/\partial\delta_2 = -(l_1 + \gamma_1)/q + 2(l_1 + 2\gamma_1)^2/q(l_1 + l_2 + 4\gamma_1), \tag{4.16}$$

$$\partial\lambda/\partial\gamma_1 = -(l_1 + l_2)/q + 2(l_1 - l_2)^2/q(l_1 + l_2 + 4\gamma_1). \tag{4.17}$$

If we now constrain our estimates such that $\hat{\delta}_1 \geq 0$, $\hat{\delta}_2 \geq 0$, and $\hat{\gamma}_1 \geq 0$, the δ_i partials, (4.15) and (4.16), vanish when $\delta_1 = \delta_2$. Now when $\delta_1 = \delta_2$, the log-likelihood (3.11) reduces to

$$\lambda = \text{constant} + \ln\{(1 + \delta_1)/(1 + \delta_1 + 2\gamma_1)\} \tag{4.18}$$

which attains its constrained maximum for $\hat{\gamma}_1 = 0$. Applying the transformations (3.12) and (3.13) and solving

$$\partial\lambda/\partial\omega_1 = 0, \qquad \partial\lambda/\partial\omega_2 = 0, \qquad \partial\lambda/\partial\tau_1 = 0 \tag{4.19}$$

yields the equivalent solution $\omega_1 = \omega_2$ and $\tau_1 = 0$. Thus the constrained solution for our example is $\hat{\delta}_1 = \hat{\delta}_2$, $\hat{\gamma}_1 = 0$, and

$$\hat{\sigma}^2 = 1/(1 + \hat{\delta}_1) \tag{4.20}$$

as a result of (4.14). Consequently, $\hat{\sigma}^2 + \hat{\delta}_1\hat{\sigma}^2 = 1$ so that from (3.5) we obtain

$$\hat{s}_1^2 = \hat{s}_2^2 = 1, \qquad \hat{\sigma}_1^2 = 0 \tag{4.21}$$

as the unique maximum likelihood estimators for the parameters of the model (4.1).

Two different iterative solutions for the example are presented in Table 1. The general equal variance computer algorithm used for this example (as well as subsequent examples) is essentially the two-stage algorithm described in [8] with the modifications discussed in [9]. It uses the square root transformation to constrain the δ_i and γ_i; it defaults to a Fisher scoring step in the event that a Newton–Raphson step decreases the likelihood or yields a matrix of second order partials that is not negative definite. The first solution in Table 1 was obtained using the starting values 2, 3, and 4 for $\hat{\delta}_1$, $\hat{\delta}_2$, and $\hat{\gamma}_1$, respectively. The second was obtained from the starting values 5, 10, and 5. All results have been rounded. We have perturbed the solution given as further assurance that it is (at least) a local maximum of the likelihood function. We note in passing that all of the rank conditions (2.18–2.21) are satisfied for the example since

$$\operatorname{rank}(\mathbf{X} \mid \mathbf{E}_2) = 3, \qquad \operatorname{rank}(\mathbf{X} \mid \mathbf{U}_1) = 2, \tag{4.22}$$

and the matrices $\mathbf{I}_4$, $\mathbf{E}_2$, and $\mathbf{U}_1$ are linearly independent.

TABLE 1

Iterative Solution of Example

Step	λ	$\hat{\delta}_1$	$\hat{\delta}_2$	$\hat{\gamma}_1$	$\hat{\sigma}^2$	$\hat{s}_1$	$\hat{s}_2$
0	−6.87157	2.00000	3.00000	4.00000	.287500	0.86250	1.15000
1[a]	−5.68531	7.93133	6.68494	0.01701	.121156	1.07989	0.93107
2	−5.67576	7.16521	7.20129	0.00002	.122201	0.99780	1.00221
3	−5.67575	7.18314	7.18314	0.00000	.122202	1.00000	1.00000
0	−6.49401	5.0000	10.0000	5.00000	.122517	0.73510	1.34768
1[a]	−5.73981	20.6539	16.0330	0.49996	.052408	1.13484	0.89267
2	−5.67940	16.5585	17.7361	0.02355	.055162	0.96857	1.03353
3	−5.67575	17.0887	17.0740	0.00000	.055306	1.00041	0.99959
4	−5.67575	17.0813	17.0813	0.00000	.055306	1.00000	1.00000

[a] Fisher scoring was applied for this step.

5. THE CASE OF PROPORTIONAL VARIANCES

When the variances are proportional such that

$$s_i^2 = l_i\sigma^2 \qquad i = 1, 2, \ldots, k \tag{5.1}$$

and the l_i are known, the variance–covariance matrix for the model (2.1) becomes

$$\sigma^2\mathbf{H} = \sigma^2(\mathbf{L} + \mathbf{VDV}'), \tag{5.2}$$

where

$$\mathbf{V} = (\mathbf{U}_1 | \mathbf{U}_2 | \cdots | \mathbf{U}_c), \tag{5.3}$$

and **D** and **L** are diagonal matrices whose ith diagonal blocks, $\mathbf{D}_{ii}$ and $\mathbf{L}_{ii}$, respectively, are given by

$$\mathbf{D}_{ii} = \gamma_i \mathbf{I}_{m_i} \qquad i = 1, 2, \ldots, c, \tag{5.4}$$

$$\mathbf{L}_{ii} = l_i \mathbf{I}_{n_i} \qquad i = 1, 2, \ldots, k, \tag{5.5}$$

with γ_i defined as in (3.6). Then, appealing to the development of the W transformation in [8] we have that

$$(\mathbf{L} + \mathbf{VDV}')^{-1} = \mathbf{L}^{-1} - \mathbf{L}^{-1}\mathbf{V}(\mathbf{D}^{-1} + \mathbf{V}'\mathbf{L}^{-1}\mathbf{V})^{-1}\mathbf{V}'\mathbf{L}^{-1} \tag{5.6}$$

and

$$\mathbf{V}'(\mathbf{L} + \mathbf{VDV}')^{-1}\mathbf{V} = \mathbf{V}'\mathbf{L}^{-1}\mathbf{V} - \mathbf{V}'\mathbf{L}^{-1}\mathbf{V}(\mathbf{D}^{-1} + \mathbf{V}'\mathbf{L}^{-1}\mathbf{V})^{-1}\mathbf{V}'\mathbf{L}^{-1}\mathbf{V}, \tag{5.7}$$

$$\mathbf{V}'(\mathbf{L} + \mathbf{VDV}')^{-1}\mathbf{X} = \mathbf{V}'\mathbf{L}^{-1}\mathbf{X} - \mathbf{V}'\mathbf{L}^{-1}\mathbf{V}(\mathbf{D}^{-1} + \mathbf{V}'\mathbf{L}^{-1}\mathbf{V})^{-1}\mathbf{V}'\mathbf{L}^{-1}\mathbf{X}, \tag{5.8}$$

and so forth.

In the equal variance case we form the matrix

$$\mathbf{W}_0 = \left[\begin{array}{c|c|c} \mathbf{V}'\mathbf{V} & \mathbf{V}'\mathbf{X} & \mathbf{V}'\mathbf{y} \\ \hline \mathbf{X}'\mathbf{V} & \mathbf{X}'\mathbf{X} & \mathbf{X}'\mathbf{y} \\ \hline \mathbf{y}'\mathbf{V} & \mathbf{y}'\mathbf{X} & \mathbf{y}'\mathbf{y} \end{array}\right] \tag{5.9}$$

initially and then apply the W transformation at each iteration to simply compute the forms

$$\mathbf{U}_i'\mathbf{H}^{-1}\mathbf{U}_j, \quad \mathbf{U}_i'\mathbf{H}^{-1}\mathbf{X}, \quad \mathbf{U}_i'\mathbf{H}^{-1}\mathbf{y}, \quad \mathbf{X}'\mathbf{H}^{-1}\mathbf{X}, \quad \mathbf{X}'\mathbf{H}^{-1}\mathbf{y}, \quad \mathbf{y}'\mathbf{H}^{-1}\mathbf{y} \tag{5.10}$$

needed in the calculation of first order partials, second order partials, or expected values of second order partials for the optimization step.

Now the partial derivatives of the log-likelihood have the same mathematical form in the case when $\mathbf{H} = (\mathbf{L} + \mathbf{VDV}')$, for $\mathbf{L}$ known, as when $\mathbf{H} = (\mathbf{I} + \mathbf{VDV}')$. Consequently, for the proportional variance case we first form the matrix

$$\mathbf{W}_0^* = \left[\begin{array}{c|c|c} \mathbf{V}'\mathbf{L}^{-1}\mathbf{V} & \mathbf{V}'\mathbf{L}^{-1}\mathbf{X} & \mathbf{V}'\mathbf{L}^{-1}\mathbf{y} \\ \hline \mathbf{X}'\mathbf{L}^{-1}\mathbf{V} & \mathbf{X}'\mathbf{L}^{-1}\mathbf{X} & \mathbf{X}'\mathbf{L}^{-1}\mathbf{y} \\ \hline \mathbf{y}'\mathbf{L}^{-1}\mathbf{V} & \mathbf{y}'\mathbf{L}^{-1}\mathbf{X} & \mathbf{y}'\mathbf{L}^{-1}\mathbf{y} \end{array}\right] \tag{5.11}$$

and then proceed exactly as in the equal variance case in applying the W transformation to obtain the forms (5.10) required for the partials. In order to evaluate the log-likelihood in the equal variance case, $|\mathbf{H}|$ is evaluated as

$$|\mathbf{H}| = |\mathbf{D}| \cdot |\mathbf{D}^{-1} + \mathbf{V}'\mathbf{V}|, \tag{5.12}$$

where $|\mathbf{D}^{-1} + \mathbf{V}'\mathbf{V}|$ is a by-product of the W transformation. In the proportional variance case we must compute $|\mathbf{H}|$ as

$$|\mathbf{H}| = |\mathbf{L}| \cdot |\mathbf{D}| \cdot |\mathbf{D}^{-1} + \mathbf{V}'\mathbf{L}^{-1}\mathbf{V}|, \tag{5.13}$$

where, in this case, $|\mathbf{D}^{-1} + \mathbf{V}'\mathbf{L}^{-1}\mathbf{V}|$ would be a by-product of the W transformation.

In essence what this says is that if we weight the observations, the rows of $\mathbf{X}$, and the rows of indicator variables in $\mathbf{V}$ by the reciprocal of the square root of the corresponding l_i before we process the data, we will obtain the proper maximum likelihood estimators for σ^2 and the γ_i as well as for $\boldsymbol{\alpha}$. Later in the sequel, we will use the relationships given in this section in developing algorithms for cases when the l_i are unknown.

6. MEASURING INSTRUMENT MODELS

In the remaining sections of this paper we will concentrate on special cases of the model (2.1) with nonhomogeneous variances that have received attention in the literature. In particular, we consider those models studied by Grubbs [3,4] and others for assessing the imprecision of measuring instruments. Algorithms for both balanced and unbalanced data will be developed which capitalize upon the structure of these models to provide more efficient computation than the general algorithm discussed in the previous sections.

The model considered by Grubbs [4] in using 3 instruments to make simultaneous measurements on each of a series of N items is the mixed model

$$y_{ij} = \alpha_i + \beta_j + e_{ij}, \tag{6.1}$$

where: $i = 1, 2, 3$ and $j = 1, 2, \ldots, N$; α_i is fixed; β_j is random with variance component σ_1^2; and $\mathrm{var}(e_{ij}) = s_i^2$. We will consider the same model for the more general case when

$$i = 1, 2, \ldots, M. \tag{6.2}$$

It can be verified that the rank conditions (2.18–2.21) are satisfied for this model provided that we have $N \geq 3$, $M \geq 2$.

Application of the general nonhomogeneous variances algorithm described in Sections 2, 3, and 4 consists essentially of augmenting the design matrix for the variance components with a partitioned identity matrix. As a consequence, the $n \times n$ inverse matrix $\mathbf{H}^{-1}$ is in fact computed explicitly even though the W transformation is applied. Since n is the number of observations [$n = M \cdot N$ for the model (6.1)], the computational requirements can become excessive. We intend to exploit the balanced data structure of the model (6.1), with the generalization specified by (6.2), to avoid explicit calculation of $\mathbf{H}^{-1}$ for both balanced and unbalanced data. Towards this end, we present without proof in the remainder of this section some relationships for the model (6.1) which are easily confirmed.

Using the same notation as (5.1) and (5.2), the variance–covariance matrix $\mathbf{H}$ may be written as

$$\mathbf{H} = \begin{bmatrix} (\gamma_1 + l_1)\mathbf{I}_N & \gamma_1\mathbf{I}_N & \cdots & \gamma_1\mathbf{I}_N \\ \gamma_1\mathbf{I}_N & (\gamma_1 + l_2)\mathbf{I}_N & \cdots & \gamma_1\mathbf{I}_N \\ \vdots & \vdots & \ddots & \vdots \\ \gamma_1\mathbf{I}_N & \gamma_1\mathbf{I}_N & \cdots & (\gamma_1 + l_M)\mathbf{I}_N \end{bmatrix} \tag{6.3}$$

The inverse of (6.3), $\mathbf{H}^{-1}$, is similarly composed of blocks of $N \times N$ diagonal matrices. Let us denote the ijth block of $\mathbf{H}^{-1}$ corresponding to the ijth block of $\mathbf{H}$ by

$$(ij\text{th block of } \mathbf{H}^{-1}) = a_{ij}\mathbf{I}_N. \tag{6.4}$$

Then if we let

$$u = \prod_{i=1}^{M} l_i, \qquad v = \sum_{i=1}^{M} (1/l_i), \tag{6.5}$$

the a_{ij} in (6.4) are given by

$$a_{ij} = [\delta_{ij}/l_i] - [\gamma_1/(1 + v\gamma_1)l_i l_j], \tag{6.6}$$

where δ_{ij} is the Kronecker delta, and the determinant of (6.3) may be obtained as

$$|\mathbf{H}| = q^N, \tag{6.7}$$

where

$$q = u(1 + v\gamma_1). \tag{6.8}$$

For example, when $M = 3$ the variance–covariance matrix (6.3) has as its inverse

$$\mathbf{H}^{-1} = 1/q \begin{bmatrix} (\gamma_1 l_2 + \gamma_1 l_3 + l_2 l_3)\mathbf{I}_N & -\gamma_1 l_3 \mathbf{I}_N & -\gamma_1 l_2 \mathbf{I}_N \\ -\gamma_1 l_3 \mathbf{I}_N & (\gamma_1 l_1 + \gamma_1 l_3 + l_1 l_3)\mathbf{I}_N & -\gamma_1 l_1 \mathbf{I}_N \\ -\gamma_1 l_2 \mathbf{I}_N & -\gamma_1 l_1 \mathbf{I}_N & (\gamma_1 l_1 + \gamma_1 l_2 + l_1 l_2)\mathbf{I}_N \end{bmatrix}, \tag{6.9}$$

where

$$q = \gamma_1 l_1 l_2 + \gamma_1 l_1 l_3 + \gamma_1 l_2 l_3 + l_1 l_2 l_3. \tag{6.10}$$

Using the solution given by (6.4–6.6) for $\mathbf{H}^{-1}$, one can show that the maximum likelihood estimators for the α_i in (6.1) are the analysis of variance estimators

$$\hat{\alpha}_i = (1/N) \sum_{j=1}^{N} y_{ij} \qquad i = 1, 2, \ldots, M. \tag{6.11}$$

An additional relationship which we will use in what follows is the explicit determination of $\mathbf{W}_0^*$ given by (5.11) for the model (6.1). This matrix is given by

$$\mathbf{W}_0^* = \left[\begin{array}{cccc|cccc|c}
\sum 1/l_i & 0 & \cdots & 0 & 1/l_1 & 1/l_2 & \cdots & 1/l_M & \sum y_{i1}/l_i \\
 & \sum 1/l_i & \cdots & 0 & 1/l_1 & 1/l_2 & \cdots & 1/l_M & \sum y_{i2}/l_i \\
 & & \ddots & \vdots & \vdots & \vdots & & \vdots & \vdots \\
 & & & \sum 1/l_i & 1/l_1 & 1/l_2 & \cdots & 1/l_M & \sum y_{iN}/l_i \\
\hline
 & & & & N/l_1 & 0 & \cdots & 0 & (1/l_1)\sum y_{1j} \\
 & & & & & N/l_2 & \cdots & 0 & (1/l_2)\sum y_{2j} \\
 & \text{(SYMMETRIC)} & & & & & \ddots & \vdots & \vdots \\
 & & & & & & & N/l_M & (1/l_M)\sum y_{Mj} \\
\hline
 & & & & & & & & \sum\sum y_{ij}^2/l_i
\end{array}\right] \tag{6.12}$$

7. THE l_i ALGORITHM FOR BALANCED DATA

The concept of this algorithm is very simple. We appeal to the case of proportional variances treated in Section 5 and consider the l_i as fixed in obtaining new approximations for $\hat{\sigma}^2$ and $\hat{\gamma}_1$. (In the unbalanced case to follow we must also obtain new approximations for $\hat{\alpha}$.) The matrix $\mathbf{W}_0^*$ given by (6.12) is easily formed for a given set of l_i and the W transformation may then be utilized by using $\mathbf{W}_0^*$ in obtaining the new values for $\hat{\sigma}^2$ and $\hat{\gamma}_1$. In order to obtain new approximations for the $\hat{l}_i$, for use in the next iteration, we initially compute the a_{ij} of $\mathbf{H}^{-1}$ given by (6.6) and store these $\frac{1}{2}M(M-1)$ coefficients. These coefficients are then utilized in computing the first order l_i partials as well as the second order l_i partials (or expected values of these second order partials) for a Newton–Raphson step (or Fisher scoring step) to improve the l_i approximations. Both $\hat{\sigma}^2$ and $\hat{\gamma}_1$ are included in the l_i partials.

Due to the structure of $\mathbf{H}^{-1}$ given by (6.4), these l_i partials are easily computed. The form of the first order l_i partial is the same as (4.11) since $\delta_i = l_i - 1$. We have then that

$$-\frac{1}{2}\operatorname{tr}\{\mathbf{E}_i'\mathbf{H}^{-1}\mathbf{E}_i\} = -\frac{N}{2}a_{ii} \tag{7.1}$$

with a_{ii} given by (6.6). If we let

$$\mathbf{d} = \mathbf{y} - \mathbf{X}\boldsymbol{\alpha}, \tag{7.2}$$

we obtain

$$\mathbf{d}'\mathbf{H}^{-1}\mathbf{E}_i = \left(\sum_{k=1}^{M} d_{k1} a_{ki}, \quad \sum_{k=1}^{M} d_{k2} a_{ki}, \quad \ldots, \quad \sum_{k=1}^{M} d_{kN} a_{ki} \right) \tag{7.3}$$

with the a_{ki} terms given by (6.6). Consequently, the first order l_i partials are given by

$$\frac{\partial \lambda}{\partial l_i} = -\frac{N}{2} a_{ii} + \frac{1}{2\sigma^2} \left\{ \sum_{j=1}^{N} \left(\sum_{k=1}^{M} d_{kj} a_{ki} \right)^2 \right\}. \tag{7.4}$$

We have found that (7.4), when equated to zero, can be manipulated algebraically to obtain a solution for l_i in terms of the remaining l_j, $j \neq i$; however, using these resulting equations to update the l_i produces only first order convergence. For a Fisher scoring step for the l_i we must compute

$$E\left(\frac{\partial^2 \lambda}{\partial l_i \, \partial l_j} \right) = -\frac{1}{2} \operatorname{tr}\{\mathbf{E}_i'\mathbf{H}^{-1}\mathbf{E}_j \mathbf{E}_j'\mathbf{H}^{-1}\mathbf{E}_i\} \tag{7.5}$$

which simplifies in this case to

$$E\left(\frac{\partial^2 \lambda}{\partial l_i \, \partial l_j} \right) = -\frac{N}{2} a_{ij}^2, \tag{7.6}$$

where a_{ij} is again given by (6.6). For the Fisher scoring step we would then solve the equations

$$\frac{N}{2} \begin{bmatrix} a_{11}^2 & a_{12}^2 & \cdots & a_{1M}^2 & \Delta l_1 \\ a_{12}^2 & a_{22}^2 & \cdots & a_{2M}^2 & \Delta l_2 \\ \vdots & \vdots & & \vdots & \vdots \\ a_{1M}^2 & a_{2M}^2 & \cdots & a_{MM}^2 & \Delta l_M \end{bmatrix} = \begin{bmatrix} \frac{\partial \lambda}{\partial l_1} \\ \frac{\partial \lambda}{\partial l_2} \\ \vdots \\ \frac{\partial \lambda}{\partial l_M} \end{bmatrix} \tag{7.7}$$

to obtain the increments Δl_i to add to the current l_i values. In actuality, we apply the square root transformation to constrain the $\hat{l}_i$ to nonnegative values and form the previous equations (7.7) in terms of the transformed variables.

The full second order l_i partials for a Newton–Raphson step are given by

$$\frac{\partial^2 \lambda}{\partial l_i \, \partial l_j} = \frac{1}{2} \operatorname{tr}\{\mathbf{E}_i'\mathbf{H}^{-1}\mathbf{E}_j\mathbf{E}_j'\mathbf{E}_j'\mathbf{H}^{-1}\mathbf{E}_j\} - \frac{1}{\sigma^2} \mathbf{d}'\mathbf{H}^{-1}\mathbf{E}_i \mathbf{E}_i'\mathbf{H}^{-1}\mathbf{E}_j \mathbf{E}_j'\mathbf{H}^{-1}\mathbf{d}. \tag{7.8}$$

These reduce to

$$\frac{\partial^2 \lambda}{\partial l_i \, \partial l_j} = \frac{N}{2} a_{ij}^2 - \frac{a_{ij}}{\sigma^2}\left(\sum_{k=1}^{N} \sum_{l=1}^{M} d_{lk} a_{li} \sum_{m=1}^{M} d_{mk} a_{mj}\right) \tag{7.9}$$

for the instrumentation model (6.1). The matrix of these partials would replace the coefficient matrix of (7.7) for a Newton–Raphson step to obtain new $\hat{l}_i$ values.

As an example of the balanced l_i algorithm, we use the data given in Grubbs [4], reproduced in Table 2, in which he records the measurements taken by three velocity chronographs on each of twelve successive rounds fired from a 155 mm gun. The model is (6.1) with $M = 3$ and $N = 12$. Using his estimation procedure, Grubbs obtains the values 0.0065, 0.0525, 0.2186, and 2.0164 as estimates of s_1^2, s_2^2, s_3^2, and σ_1^2, respectively. A summary of our computations to obtain the maximum likelihood estimators for these quantities is given in Table 3 for the set of starting values $\hat{l}_1 = 1$, $\hat{l}_2 = 1$, $\hat{l}_3 = 1$, and $\hat{\gamma}_1 = 100$.

TABLE 2

Velocity Data

j	y_{1j}	y_{2j}	y_{3j}	j	y_{1j}	y_{2j}	y_{3j}
1	793.8	794.6	793.2	7	791.7	792.4	791.6
2	793.1	793.9	793.3	8	792.3	792.8	792.4
3	792.4	793.2	792.6	9	789.6	790.2	788.5
4	794.0	794.0	793.8	10	794.4	795.0	794.7
5	791.4	792.2	791.6	11	790.9	791.6	791.3
6	792.4	793.1	791.6	12	793.5	793.8	793.5

Notice that we work with the l_i and σ^2 in doing the calculations and have the same overparametrization here as in the general algorithm. Nevertheless, we converged to the same values for $\hat{s}_1^2$, $\hat{s}_2^2$, $\hat{s}_3^2$, and $\hat{\sigma}_1^2$ for a variety of different starting values. We also used the general computer algorithm discussed earlier to confirm the final values given in Table 3 to 12 significant digits of agreement in the log-likelihood. To improve the precision of our computations we subtracted 792.2, the approximate mean, from each of the y_{ij} values prior to processing the data. The computations were then performed on an IBM 370/155 using double precision arithmetic (as were all of the computations for this paper).

For this example, the balanced l_i algorithm requires less than 1/10th of the amount of array storage used by the general algorithm. Furthermore, it also

TABLE 3

Summary of Calculations for the Balanced Velocity Chronograph Data

Fixed effects		
$\hat{\alpha}_1$	$\hat{\alpha}_2$	$\hat{\alpha}_3$
792.4583333	793.0666666	792.3416666

Step	$\hat{\sigma}^2$	$\hat{\gamma}_1$	$\hat{l}_1$	$\hat{l}_2$	$\hat{l}_3$
0	.0628182294	100.00000000	1.0000000000	1.0000000000	1.000000000
1[a]	.0465849312	42.02936857	.2895803462	.6992443331	4.621304986
2[b]	.0466757995	37.30136948	.3228890941	.8360312287	4.279427970
3	.0467126746	37.88541045	.2879708705	.8592276851	4.262282065
6	.0467163814	37.96310735	.2843958388	.8612274862	4.260314670
8	.0467163814	37.96311035	.2843958036	.8612274782	4.260314499
10	.0467163814	37.96311035	.2843958036	.8612274782	4.260314499

Step	$\hat{\sigma}_1^2$	$\hat{s}_1^2$	$\hat{s}_2^2$	$\hat{s}_3^2$	λ
0	6.281822936	.0628182294	.0628182294	.0628182294	−35.509269302009
1[a]	1.957133243	.0134845548	.0325609062	.2151949916	−27.712724236744
2[b]	1.741071242	.0150711066	.0390224258	.1997457218	−27.595715027062
3	1.769728850	.0134518896	.0401368232	.1991025951	−27.593245080587
6	1.773499000	.0132859445	.0402334317	.1990264848	−27.593217095291
8	1.773499142	.0132859428	.0402334313	.1990264770	−27.593217095291
10	1.773499142	.0132859428	.0402334313	.1990264770	−27.593217095291

[a] Fisher scoring for γ_1 and the l_i.
[b] Fisher scoring for the l_i only.

requires less than 1/10th of the amount of computer time used by the general algorithm to achieve the same degree of precision in the results (execution time for the l_i algorithm to complete 10 iterations was approximately 6 seconds).

8. THE MISSING DATA ALGORITHM

Although the examples of the model (6.1) appearing in the literature deal primarily with balanced data, some provision should realistically be made for missing data values. Suppose that, for notational simplicity, we partition the $\mathbf{y}$ vector for the model (6.1) such that

$$\mathbf{y} = \begin{bmatrix} \mathbf{y}_1 \\ \hline \mathbf{y}_2 \end{bmatrix} \tag{8.1}$$

where $\mathbf{y}_1$ is a q vector corresponding to q missing observations and $\mathbf{y}_2$ is an $(n-q)$ vector of observed data. In a similar manner we partition

$$\mathbf{X} = \left[\frac{\mathbf{X}_1}{\mathbf{X}_2}\right], \qquad \mathbf{E}_i = \left[\frac{\mathbf{E}_{i1}}{\mathbf{E}_{i2}}\right], \qquad \mathbf{U}_1 = \left[\frac{\mathbf{U}_{11}}{\mathbf{U}_{12}}\right], \tag{8.2}$$

$$\mathbf{H} = \left[\begin{array}{c|c} \mathbf{H}_{11} & \mathbf{H}_{12} \\ \hline \mathbf{H}'_{12} & \mathbf{H}_{22} \end{array}\right], \qquad \mathbf{H}^{-1} = \left[\begin{array}{c|c} \mathbf{B}_{11} & \mathbf{B}_{12} \\ \hline \mathbf{B}'_{12} & \mathbf{B}_{22} \end{array}\right]. \tag{8.3}$$

The l_i partials, in the form (4.11), for the observed data are then given by

$$\begin{aligned} \partial\lambda/\partial l_i = & -\tfrac{1}{2}\operatorname{tr}\{\mathbf{E}'_{i2}\mathbf{H}_{22}^{-1}\mathbf{E}_{i2}\} \\ & + (1/2\sigma^2)(\mathbf{y}_2 - \mathbf{X}_2\boldsymbol{\alpha})'\mathbf{H}_{22}^{-1}\mathbf{E}_{i2}\mathbf{E}'_{i2}\mathbf{H}_{22}^{-1}(\mathbf{y}_2 - \mathbf{X}_2\boldsymbol{\alpha}), \end{aligned} \tag{8.4}$$

and it is well known that we may express $\mathbf{H}_{22}^{-1}$ in (8.4) in terms of the elements of $\mathbf{H}^{-1}$ as

$$\mathbf{H}_{22}^{-1} = \mathbf{B}_{22} - \mathbf{B}'_{12}\mathbf{B}_{11}^{-1}\mathbf{B}_{12}. \tag{8.5}$$

We first attempted to employ the procedure suggested initially by Hartley [5] of replacing missing quantities in the data with manufactured values; these values are manufactured at each iterative step in such a manner that the first order partial derivatives for the balanced data (with the dummy values included) equal those for the given, unbalanced data. This procedure has recently been cast in a more general framework by Dempster *et al.* [2] and referred to as the *EM* algorithm (*E* for expectations, since the substituted quantities are conditional expectations, and *M* for maximize). Unfortunately, we soon discovered that application of this procedure to the model (6.1) required manufacturing values for E_{i1}, $i = 1, 2, \ldots, M$ and U_{11} in (8.2) as well as for the missing observations y_1 in (8.1). In so doing, we destroy the structure of $\mathbf{H}$ and $\mathbf{H}^{-1}$ in (8.3) that permits the balanced problem to be solved easily.

We finally took a more direct approach to the problem which proved to be reasonably effective. Essentially, we use the relationship (8.5) to obtain $\mathbf{H}_{22}^{-1}$ from the balanced $\mathbf{H}^{-1}$ which we already know; but we do this in a way that ties in nicely with the balanced l_i algorithm and, for a small amount of missing data, requires approximately $(1/N)$th of the storage required by the general algorithm in containing $\mathbf{H}_{22}^{-1}$. Furthermore, the amount of computation is reduced substantially.

Initially, we store the N diagonal elements of the ijth diagonal blocks of $\mathbf{H}^{-1}$ in row major ordering for $i \le j$, in a linear array as

$$\underbrace{\overbrace{a_{11}, a_{11}, \ldots, a_{11}}^{N}}_{11\text{ block}},\ \underbrace{\overbrace{a_{12}, a_{12}, \ldots, a_{12}}^{N}}_{12\text{ block}},\ \ldots,\ \underbrace{\overbrace{a_{MM}, a_{MM}, \ldots, a_{MM}}^{N}}_{MM\text{ block}}$$

where the a_{ij} are given by (6.6). We then modify these values in accordance with the missing data in this manner.

Consider first one missing value, say y_{ij}. In order to obtain $\mathbf{H}_{22}^{-1}$ we need only perform an in-place Gauss–Jordan reduction or sweep of $\mathbf{H}^{-1}$ given by (6.4), pivoting on the jth diagonal element in the ith diagonal block of $\mathbf{H}^{-1}$ [iith block of (8.6)]. Only those elements in the jth position in each diagonal block are affected (changed) by this reduction. If we then zero out the jth diagonal element in the ikth and kith blocks of (8.6) for $k = 1, \ldots, i$ and $k = i + 1, \ldots, M$, respectively, what remains is $\mathbf{H}_{22}^{-1}$ arrayed in storage corresponding to the observed data; that is, the row and column corresponding to the missing observation have been replaced by zeros. This process may be repeated for each additional missing value which results in the formation of $\mathbf{H}_{22}^{-1}$ computed in the positions corresponding to the observed data and zeros elsewhere.

We also store zeros in $\mathbf{y}$ corresponding to the missing data and modify the appropriate elements of $\mathbf{W}_0^*$ to reflect the different frequencies. That is, if y_{ij} is the only missing value, $\mathbf{W}_0^*$ would be modified so that the value of the jth diagonal element of $\mathbf{U}_1'\mathbf{L}^{-1}\mathbf{U}_1$ is $\sum_{k \neq i} 1/l_k$, the jith element of $\mathbf{U}_1'\mathbf{L}^{-1}\mathbf{X}$ is 0, and the ith diagonal element of $\mathbf{X}'\mathbf{L}^{-1}\mathbf{X}$ is $(N - 1)/l_i$. Using the previous configurations of $\mathbf{H}_{22}^{-1}$ and the $\mathbf{y}$ vector, the l_i partials or their expected values can then be computed easily in an orderly manner. For example, the quantity $\operatorname{tr}\{\mathbf{E}_{i2}'\mathbf{H}_{22}^{-1}\mathbf{E}_{i2}\}$ in (8.4) is obtained by summing the N elements which have overwritten the iith block of (8.6); the quantity $\operatorname{tr}\{\mathbf{E}_{i2}'\mathbf{H}_{22}^{-1}\mathbf{E}_{j2}\mathbf{E}_{j2}'\mathbf{H}_{22}^{-1}\mathbf{E}_{i2}\}$, applicable to (7.5) and (7.8) is obtained by summing the squares of the N elements which have overwritten the ijth block of (8.6).

In order to exhibit the unbalanced l_i algorithm we consider again the example given in the previous section, but delete the six observations $y_{1,2}$, $y_{2,4}$, $y_{2,6}$, $y_{3,3}$, $y_{3,7}$, and $y_{3,10}$, and treat these values as missing data. A summary of our computations appears in Table 4. These final values were also confirmed by way of the general algorithm. For this example, the unbalanced l_i algorithm requires less than 1/6th of the array storage used by the general algorithm and less than 1/10th of the amount of computer time (execution time for the l_i algorithm was approximately 9 seconds).

Table 4 (and Table 3 in the previous section) presents many more significant digits for the estimators, as well as more steps, than one is likely to require. This information is provided to give the reader a better appreciation of the numerical process. In both examples, the last step given represents the point at which no further increases in the log-likelihood could be obtained using double precision arithmetic. All first order partials with respect to γ_1 and the l_i were then less than 10^{-11}. We also experimented with using second order crossed partials (or their expected values) for all of the

TABLE 4

Summary of Calculations for the Unbalanced Velocity Chronograph Data

Step	$\hat{\alpha}_1$	$\hat{\alpha}_2$	$\hat{\alpha}_3$
0	792.4836615	793.0844856	792.2657578
1[a]	792.4730846	793.1073418	792.2598219
2[b]	792.4726491	793.1129143	792.2580932
3	792.4724770	793.1140898	792.2571620
5	792.4727034	793.1148954	792.2548044
8	792.4727051	793.1148993	792.2547932
11	792.4727051	793.1148993	792.2547932

Step	$\hat{\sigma}^2$	$\hat{\gamma}_1$	$\hat{l}_1$	$\hat{l}_2$	$\hat{l}_3$
0	.0558909111	100.00000000	1.0000000000	1.0000000000	1.000000000
1[a]	.0258141251	44.59714570	.9721551766	1.6454117993	9.496625137
2[b]	.0257212813	59.54429813	.3039525835	.9044378324	9.075683245
3	.0268541880	68.59720869	.1682388337	.6961923631	8.800563109
5	.0265734777	68.92396767	.0652626508	.8540719476	8.541872204
8	.0265737410	68.91649182	.0647522348	.8550312088	8.538965799
11	.0265737410	68.91649287	.0647522096	.8550312240	8.538965726

Step	$\hat{\sigma}_1^2$	$\hat{s}_1^2$	$\hat{s}_2^2$	$\hat{s}_3^2$	λ
0	5.589091113	.0558909111	.0558909111	.0558909111	−32.334098961167
1[a]	1.151236299	.0250953354	.0424748661	.2451470697	−25.419290200588
2[b]	1.531555639	.0078180498	.0232632999	.2334382013	−23.142307210287
3	1.842122344	.0045179173	.0186956807	.2363319770	−22.927474069614
5	1.831549518	.0017342555	.0226956619	.2269872505	−22.909450473464
8	1.831369004	.0017207091	.0227213779	.2269122655	−22.909449403642
11	1.831369032	.0017207084	.0227213783	.2269122636	−22.909449403643

[a] Fisher scoring for γ_1 and the l_i.
[b] Fisher scoring for the l_i only.

variables included in the general case as suggested by Jennrich and Sampson [11]. For the l_i algorithms we used these second order quantities for the α_i, σ^2, and γ_1. In addition, we applied their constraining techniques in order to obtain nonnegative estimators. These procedures did yield further verification of the results for our examples; however, we did not make a thorough attempt to assess potential efficiency differences.

REFERENCES

[1] Corbeil, R. R. and Searle, S. R. (1976). Restricted maximum likelihood (REML) estimation of variance components in the mixed model. *Technometrics* **18** 31–38.

[2] Dempster, A. P., Laird, N. M., and Rubin, D. M. (1976). Maximum likelihood from incomplete data via the EM algorithm. Research Reports S-38 NS-320, Dept. of Statistics, Harvard Univ.

[3] GRUBBS, F. E. (1948). On estimating precision of measuring instruments and product variability. *J. Amer. Statist. Assoc.* **43** 243–264.

[4] GRUBBS, F. E. (1973). Errors of measurement, precision, accuracy and the statistical comparison of measuring instruments. *Technometrics* **15** 53–66.

[5] HARTLEY, H. O. (1958). Maximum likelihood estimation from incomplete data. *Biometrics* **14** 174–194.

[6] HARTLEY, H. O. and RAO, J. N. K. (1967). Maximum likelihood estimation for the mixed analysis of variance model. *Biometrika*, **54** 93–108.

[7] HARTLEY, H. O. and JAYATTILAKE, K. S. E. (1973). Estimation for linear models with unequal variances. *J. Amer. Statist. Assoc.* **68** 185–189.

[8] HEMMERLE, W. J. and HARTLEY, H. O. (1973). Computing maximum likelihood estimates for the mixed A.O.V. model using the W transformation. *Technometrics* **15** 819–831.

[9] HEMMERLE, W. J. and LORENS, J. A. (1976). Improved algorithm for the W transformation in variance component estimation. *Technometrics* **18** 207–212.

[10] JAECH, J. L. (1976). Large sample tests for Grubbs estimators of instrument precision with more than two instruments. *Technometrics* **18** 127–133.

[11] JENNRICH, R. I. and SAMPSON, P. F. (1976). Newton–Raphson and related algorithms for maximum likelihood variance component estimation. *Technometrics* **18** 11–18.

[12] MILLER, J. J. (1973). Asymptotic properties and computation of maximum likelihood estimates in the mixed model of the analysis of variance. Tech. Rep. 12, Dept. of Statistics, Stanford Univ.

Concurrency of Regression Equations with k Regressors

A. M. Kshirsagar *Violeta Sonvico*†

TEXAS A & M UNIVERSITY

Some exact tests of significance for testing the goodness of fit of a proposed vector of coordinates for the concurrency of m regression equations with k independent variables are derived in this paper. The tests are based on the Bartlett decomposition of a $(k + 1) \times (k + 1)$ noncentral Wishart matrix and provide a generalization of Williams's tests. A method for estimating the point of concurrence is also provided.

1. INTRODUCTION

Statistical tests about the point of concurrence of several regression equations have been considered so far by Tocher (1952), Williams (1953), and Saw (1966). Both Williams and Saw have based their tests on the roots of a 2×2 noncentral Wishart matrix. Saw avoids the exact distribution of the roots by using an upper bound on their percentage points, while Williams was able to get some exact tests by using a hypothetical point of concurrence and testing its goodness of fit. The present paper is a generalization of Williams's work where the regression of y on more than one regressor is considered.

This generalization cannot be achieved by extending Williams's method, as this would involve the distribution of the roots of a $(k + 1) \times (k + 1)$ noncentral Wishart matrix which is unmanageable. However, we consider

† Present address: Dto. Estadistica, INTA—Castelar—Pcia Bs. As. Argentina.

Key words and phrases Regression, concurrency, noncentral Wishart matrix, Bartlett decomposition.

its Bartlett decomposition to obtain the tests. Incidentally when $k = 1$, this provides an easier and more satisfactory alternative derivation of Williams's tests. There are many obscure and unexplained steps in Williams's paper; this method throws more light on these.

2. GOODNESS OF FIT OF A HYPOTHETICAL POINT OF CONCURRENCE

Consider m regression equations of a dependent variable y on k independent variables $x_1, \ldots, x_k$. If there are n_i observations in the ith group ($i = 1, \ldots, m$) and if the regression equations are concurrent with $(\eta, \xi_1, \ldots, \xi_k)$ as the point of concurrence, the setup will be

$$Y_{ij} = \eta + \beta_{1i}(x_{1ij} - \xi_1) + \cdots + \beta_{ki}(x_{kij} - \xi_k) + \varepsilon_{ij}$$
$$i = 1, \ldots, m; \quad j = 1, \ldots, n_i. \tag{2.1}$$

We assume $m > k + 1$. We shall use the following vectors and matrices:

$$\mathbf{Y}_i' = [y_{i1}, \ldots, y_{in_i}], \qquad \boldsymbol{\beta}_i' = [\beta_{1i}, \ldots, \beta_{ki}], \qquad \boldsymbol{\xi}' = [\xi_1, \ldots, \xi_k],$$

$\mathbf{X}_i$ is the $k \times n_i$ matrix $[X_{lij}]$ $\quad l = 1, \ldots, k; \quad j = 1, \ldots, n_i,$

$$N = \sum_1^m n_i,$$

$$\mathbf{q}_i = (1/n_i)\mathbf{X}_i'\mathbf{Y}_i, \qquad \mathbf{q} = \sum_1^m n_i \mathbf{q}_i/N, \qquad \bar{\boldsymbol{\beta}} = \sum n_i \boldsymbol{\beta}_i/N,$$

$$\bar{y}_i = \sum_{j=1}^{n_i} y_{ij}/n_i, \qquad \bar{y} = \sum n_i \bar{y}_i/N.$$

$\mathbf{E}_{ab}$ denotes the $a \times b$ matrix of all unit elements. Further we assume that the observations x_{lij} are so chosen that

$$\mathbf{X}_i \mathbf{E}_{n_i 1} = \mathbf{0} \qquad i = 1, \ldots m \tag{2.2}$$

and

$$(1/n_i)\mathbf{X}_i \mathbf{X}_i' = \mathbf{G}, \tag{2.3}$$

the same matrix for all the groups.

The ε_{ij} are assumed to be independent normal variables with zero means and a common variance σ^2. This will be denoted by

$$\varepsilon_{ij} \sim \mathrm{NI}(0, \sigma^2). \tag{2.4}$$

The model (2.1) can now be written in matrix notation as

$$E(\mathbf{Y}_i) = \eta \mathbf{E}_{n_i 1} + \mathbf{X}_i' \boldsymbol{\beta}_i - (\boldsymbol{\xi}' \boldsymbol{\beta}_i)\mathbf{E}_{n_i 1}, \tag{2.5}$$

$$V(\mathbf{Y}_i) = \sigma^2 \mathbf{I}_{n_i}, \tag{2.6}$$

where E stands for expectations, V for the variance–covariance matrix, and $\mathbf{I}_a$ denotes the identity matrix of order a. Let $\mathbf{G}^{-1/2}$ denote a $n_i \times n_i$ nonsingular matrix, such that $\mathbf{G}^{-1/2}\mathbf{G}^{-1/2} = \mathbf{G}^{-1}$.

From the standard least squares theory about the estimates of $\boldsymbol{\beta}_i$, it follows easily that

$$\mathbf{Z}_i = \begin{bmatrix} \sqrt{n_i}\,\mathbf{G}^{-1/2}\mathbf{q}_i \\ \sqrt{n_i}(\bar{y}_i - \eta) \end{bmatrix} \qquad i = 1, \ldots, m \tag{2.7}$$

has a $(k + 1)$-variate normal distribution with means

$$E(\mathbf{Z}_i) = \begin{bmatrix} \sqrt{n_i}\,\mathbf{G}^{1/2}\boldsymbol{\beta}_i \\ \sqrt{n_i}\,\boldsymbol{\xi}'\boldsymbol{\beta}_i \end{bmatrix} \tag{2.8}$$

and variance–covariance matrix $\sigma^2\mathbf{I}$. The $\mathbf{Z}_1, \ldots, \mathbf{Z}_m$ are independent, and further they are also independent of the residual sum of squares

$$\phi = \sum_i \sum_j (y_{ij} - \bar{y}_i)^2 - \sum_{i=1}^{m} n_i \mathbf{q}_i' \mathbf{G}^{-1} \mathbf{q}_i \tag{2.9}$$

which has a $\chi^2\sigma^2$ distribution with $f = N - m(k + 1)$ degrees of freedom (d.f.). Let

$$Z = [\mathbf{Z}_1, \ldots, \mathbf{Z}_m] = [Z_{ir}] \qquad i = 1, \ldots m, \quad r = 1, \ldots k + 1 \tag{2.10}$$

and let $\bar{Z}_r = \sum_{i=1}^{m} n_i Z_{ir}/N$. Then one can easily show that the matrix

$$\mathbf{W} = \left[\sum_{i=1}^{m} (Z_{ir} - \bar{Z}_r)(Z_{is} - \bar{Z}_s) \right] \qquad r, s = 1, \ldots, k + 1 \tag{2.11}$$

is a noncentral Wishart matrix. Using (2.7), we see that

$$\begin{aligned} \mathbf{W} &= \left[\begin{array}{c|c} \mathbf{G}^{-1/2} \sum_i n_i(\mathbf{q}_i - \mathbf{q})(\mathbf{q}_i - \mathbf{q})'\mathbf{G}^{-1/2} & -\mathbf{G}^{-1/2} \sum n_i \bar{y}_i(\mathbf{q}_i - \mathbf{q}) \\ \hline -\sum n_i \bar{y}_i (\mathbf{q}_i - \mathbf{q})'\mathbf{G}^{-1/2} & \sum n_i(\bar{y}_i - \bar{y})^2 \end{array} \right] \\ &= \left[\begin{array}{c|c} \mathbf{A} & -\mathbf{B} \\ \hline -\mathbf{B}' & \mathbf{C} \end{array} \right]\begin{matrix} k \\ 1 \end{matrix} . \\ &\qquad\;\; k \qquad\; 1 \end{aligned} \tag{2.12}$$

From (2.8), one can see that the noncentrality matrix is $(1/\sigma^2)\boldsymbol{\Lambda}$, where

$$\boldsymbol{\Lambda} = \begin{bmatrix} \mathbf{G}^{1/2} \\ \boldsymbol{\xi}' \end{bmatrix} \sum_{i=1}^{m} n_i(\boldsymbol{\beta}_i - \bar{\boldsymbol{\beta}})(\boldsymbol{\beta}_i - \bar{\boldsymbol{\beta}})' \begin{bmatrix} \mathbf{G}^{1/2} \\ \boldsymbol{\xi}' \end{bmatrix}' . \tag{2.13}$$

The parameter matrix in the noncentral Wishart distribution is $\sigma^2\mathbf{I}$. This shows that

$$\mathbf{\Lambda l} = \mathbf{0}, \tag{2.14}$$

where the vector $\mathbf{l}$ is defined by

$$\mathbf{l}' = (1 + \boldsymbol{\xi}'\mathbf{G}^{-1}\boldsymbol{\xi})^{-1/2}[\boldsymbol{\xi}'\mathbf{G}^{-1/2} \mid -1]. \tag{2.15}$$

Thus the vector $\boldsymbol{\xi}$ of the coordinates of the point of concurrence defines a vector $\mathbf{l}$ which is orthogonal to $\mathbf{\Lambda}$. We employ this to obtain a test of goodness of fit of a proposed point of concurrence. First we construct a $(k + 1) \times (k + 1)$ orthogonal matrix $\mathbf{L}$ whose last row is $\mathbf{l}'$, i.e.,

$$\mathbf{L} = \left[\frac{\mathbf{L}_1}{\mathbf{l}'}\right] \tag{2.16}$$

where $\mathbf{L}_1\mathbf{l} = \mathbf{0}$, $\mathbf{L}_1\mathbf{L}_1' = \mathbf{I}$. Then make the transformation

$$\mathbf{U} = \mathbf{LWL}'. \tag{2.17}$$

It follows that $(1/\sigma^2)\mathbf{U}$ has a noncentral Wishart distribution with noncentrality matrix

$$\frac{1}{\sigma^2}\mathbf{L\Lambda L}' = \frac{1}{\sigma^2}\begin{bmatrix} \mathbf{L}_1\mathbf{\Lambda L}_1' & \mathbf{0} \\ \mathbf{0} & \mathbf{0} \end{bmatrix}\begin{matrix} k \\ 1 \end{matrix} \tag{2.18}$$

with column orders k, 1,

d.f. $m - 1$ and parameter matrix $\mathbf{I}$. Now consider the Bartlett decomposition (see Kshirsagar, 1963) of the matrix $\mathbf{U}$; i.e., let $\mathbf{T}$ be a lower triangular matrix of order $k + 1$, with nonzero elements t_{ij} $(i > j)$ such that

$$\mathbf{LWL}' = \mathbf{U} = \mathbf{TT}'. \tag{2.19}$$

Then on account of the null elements in the last row of (2.18) and from standard results about the decomposition matrix $\mathbf{T}$ it follows that $t_{k+1,1}$, $t_{k+1,2}, \ldots, t_{k+1,k}$ are $\mathrm{NI}(0, \sigma^2)$ and $t_{k+1,k+1}^2$ is a $\chi^2\sigma^2$ with $m - 1 - k$ d.f. All these are independent and also independent of ϕ of (2.9). Hence if we set

$$T_0^2 = \sum_{r=1}^{k+1} t_{k+1,r}^2 = T_1^2 + T_2^2, \tag{2.20}$$

$$T_1^2 = t_{k+1,k+1}^2, \tag{2.21}$$

and

$$T_2^2 = T_0^2 - T_1^2, \tag{2.22}$$

it follows that

$$\frac{f}{m-1-k} \cdot \frac{T_1^2}{\phi} \sim F_{m-1-k,\, f} \tag{2.23}$$

$$\frac{f}{k} \cdot \frac{T_0^2}{\phi} \sim F_{k,\, f} \tag{2.24}$$

and

$$\frac{f}{m-1} \cdot \frac{T_0^2}{\phi} \sim F_{m-1,\, f}, \tag{2.25}$$

where $F_{a,\, b}$ denotes an F distribution with a and b d.f. These tests are useful in testing the following hypothesis H.

H: The m regression equations of y on $x_1, \ldots, x_k$ as defined earlier are concurrent and the $x_1, x_2, \ldots, x_k$ coordinates of the point of concurrence are $\xi_1, \ldots, \xi_k$ (specified).

The overall test criterion for this hypothesis H is provided by T_0^2 and the test by (2.25). However, this hypothesis H comprises two parts: (a) the regression equations are concurrent and (b) the point of concurrence has specified coordinates $\xi_1, \ldots, \xi_k$. A significant T_0^2 could be due to (a) or (b) or both being not true. If one is interested in testing (a) or (b) separately one could use T_1^2 and T_2^2 and the tests (2.23) and (2.24).

Strictly speaking, a test for concurrency alone should not depend on the assigned coordinates $\xi_1, \ldots, \xi_k$, which is why Williams describes it as a "fluid" test. This entire argument is very similar to one employed by Williams (1952a) and Bartlett (1951) for testing the goodness of fit of a hypothetical discriminant function, where the hypothesis consists of two parts: (a) a single function is adequate and (b) that single function is the proposed one. Bartlett and Williams factorized Wilks' Λ criterion corresponding to these parts. In fact, Williams (1952b,c, 1955) has used this type of logic and factorization of an overall criterion in several other problems successfully; $t_{k+1,1,\ldots}, t_{k+1,k}$ are like orthogonalized regression coefficients in the multiple regression of the variable which corresponds to the last row of $\mathbf{U}$, on the preceding k variables and $t_{k+1,k+1}^2$ is the residual sum of squares in this regression. From the way $\mathbf{U}$ is constructed one can see readily that this last variable and hence $t_{k+1,1,\ldots}, t_{k+1,k}$ depend more heavily on the vector $\boldsymbol{\xi}$ than does T_1^2, which is why Williams calls T_2^2 the test statistic for testing the concordance of the data with the proposed $\boldsymbol{\xi}$, and T_1^2 the test statistic for testing departure from concurrency. (Of course, in Williams's paper $k = 1$, T_2^2 is t_{21}^2, and T_1^2 is t_{22}^2, but the logic is the same.) Further he obtains them through the eigenvalues of his 2×2 matrix $\mathbf{W}$. It only remains now to express these statistics in terms of the original quantities. This is accomplished in the next section. For more details see Sonvico (1976).

3. TEST STATISTICS T_0^2, T_1^2, T_2^2

From (2.19) and (2.15) we have

$$\begin{aligned} T_0^2 &= t_{k+1,1}^2 + \cdots + t_{k+1,k+1}^2 \\ &= \text{last element of } \mathbf{TT}' = \text{last element of } \mathbf{U} = \mathbf{LWL}' = \mathbf{l}'\mathbf{W}\mathbf{l} \\ &= \frac{\begin{aligned}[\boldsymbol{\xi}'\mathbf{G}^{-1} \textstyle\sum_1^m n_i(\mathbf{q}_i - \mathbf{q})(\mathbf{q}_i - \mathbf{q})'\mathbf{G}^{-1}\boldsymbol{\xi} \\ + 2\boldsymbol{\xi}'\mathbf{G}^{-1} \textstyle\sum_i n_i \bar{y}_i(\mathbf{q}_i - \mathbf{q}) + \sum n_i(\bar{y}_i - \bar{y})^2]\end{aligned}}{1 + \boldsymbol{\xi}'\mathbf{G}^{-1}\boldsymbol{\xi}}. \end{aligned} \tag{3.1}$$

Again since $\mathbf{L}$ is orthogonal, we obtain from (2.19)

$$(\mathbf{TT}')^{-1} = \mathbf{LW}^{-1}\mathbf{L}'.$$

Now equate the last elements on both sides, remembering that $\mathbf{T}$ is lower triangular. We obtain

$$\frac{1}{t_{k+1,k+1}^2} = \mathbf{l}'\mathbf{W}^{-1}\mathbf{l}. \tag{3.2}$$

From (2.12), we have

$$\mathbf{W}^{-1} = \left[\begin{array}{c|c} \mathbf{A}^{-1} + \dfrac{\mathbf{A}^{-1}\mathbf{BB}'\mathbf{A}^{-1}}{\mathbf{C} - \mathbf{B}'\mathbf{A}^{-1}\mathbf{B}} & \dfrac{\mathbf{A}^{-1}\mathbf{B}}{(\mathbf{C} - \mathbf{B}'\mathbf{A}^{-1}\mathbf{B})} \\ \hline \dfrac{\mathbf{B}'\mathbf{A}^{-1}}{(\mathbf{C} - \mathbf{B}'\mathbf{A}^{-1}\mathbf{B})} & (\mathbf{C} - \mathbf{B}'\mathbf{A}^{-1}\mathbf{B})^{-1} \end{array}\right] \tag{3.3}$$

From (3.2), (3.3), and (2.16) one can now easily express $T_1^2 = t_{k+1,k+1}^2$ in terms of $\boldsymbol{\xi}$, $\mathbf{A}$, $\mathbf{B}$, and $\mathbf{C}$ of (2.12).

4. ESTIMATION OF ξ AND η

From (2.15) we find that the vector

$$\left[\begin{array}{c} \mathbf{G}^{-1/2}\boldsymbol{\xi} \\ \hline -1 \end{array}\right] \tag{4.1}$$

is an eigenvector of $\boldsymbol{\Lambda}$, the corresponding eigenvalue being zero. One can verify that

$$E(\mathbf{W}) = \boldsymbol{\Lambda} + (m-1)\sigma^2\mathbf{I} \tag{4.2}$$

and also note that $E(\phi/f) = \sigma^2$. Therefore

$$E\left(\mathbf{W} - \frac{(m-1)\phi}{f}\mathbf{I}\right) = \mathbf{\Lambda} \tag{4.3}$$

Hence we naturally expect that

$$\begin{bmatrix} \mathbf{G}^{-1/2}\hat{\xi} \\ \hline -1 \end{bmatrix}$$

(where $\hat{\xi}$ is an estimate of ξ) will be the eigenvector of $\hat{\mathbf{\Lambda}}$, i.e.,

$$\mathbf{W} - \frac{(m-1)\phi}{f}\mathbf{I},$$

corresponding to its smallest root. So if θ_m is the smallest root of $\mathbf{W}$, we obtain

$$\left(\mathbf{W} - \frac{(m-1)\phi}{f}\mathbf{I} - \theta_m\mathbf{I}\right)\begin{bmatrix} \mathbf{G}^{-1/2}\hat{\xi} \\ \hline -1 \end{bmatrix} = \mathbf{0}. \tag{4.4}$$

Using (2.13), this yields

$$\left\{\mathbf{A} - \left[\frac{(m-1)}{f}\phi + \theta_m\right]\mathbf{I}\right\}\mathbf{G}^{-1/2}\hat{\xi} + \mathbf{B} = \mathbf{0}$$

and therefore

$$\hat{\xi} = -\mathbf{G}^{1/2}\left\{\mathbf{A} - \left[\frac{m-1}{f}\phi + \theta_m\right]\mathbf{I}\right\}^{-1}\mathbf{B}. \tag{4.5}$$

This estimate will differ from that of Williams's in the particular case $k = 1$ because he does not use the adjustment $(m-1)\sigma^2\mathbf{I}$ in (4.2) to get an unbiased estimate of $\mathbf{\Lambda}$. The Y coordinate of the point of concurrence is η and it can be estimated by standard least squares theory (when ξ is known) by

$$\xi'\mathbf{G}^{-1}\mathbf{q}. \tag{4.6}$$

We now substitute $\hat{\xi}$ for ξ as given by (4.5) in (4.6) to obtain $\hat{\eta}$, an estimate of η.

5. TEST OF GOODNESS OF FIT OF A PROPOSED ξ WHEN η IS KNOWN

When η is known, one need not consider

$$\sum_{i=1}^{m}(Z_{ir} - \bar{Z}_r)(Z_{is} - \bar{Z}_s)$$

in forming **W** of (2.11). Instead one can define

$$\mathbf{W} = \left[\sum_{i=1}^{m} Z_{ir} Z_{is}\right] \tag{5.1}$$

and then **W** will have m d.f. instead of $m - 1$. If these changes are made, the rest of the derivation will be the same as in Sections 2–4.

REFERENCES

BARTLETT, M. S. (1951). The goodness of fit of a single hypothetical discriminant function in the case of several groups. *Ann. Eugen.* **16** 199–214.

KSHIRSAGAR, A. M. (1963). Effect of non-centrality on the Bartlett decomposition of a Wishart matrix. *Ann. Inst. Statist. Math.* **14** 208–217.

SAW, J. G. (1966). A conservative test for the concurrence of several regression lines and related problems. *Biometrika* **53** 272–275.

SONVICO, Violeta (1976). Some problems of statistical inference in regression and discriminant analysis. Ph.D. Dissertation, Texas A & M University, College Station.

TOCHER, K. D. (1952). On the concurrence of a set of regression lines. *Biometrika* **39** 109–112.

WILLIAMS, E. J. (1952a). Some exact tests in multivariate analysis. *Biometrika* **39** 17–31.

WILLIAMS, E. J. (1952b). The interpretation of interaction in factorial experiments. *Biometrika* **39** 65–81.

WILLIAMS, E. J. (1952c). Use of scores for the analysis of association in contingency tables. *Biometrika* **39** 274–289.

WILLIAMS, E. J. (1953). Tests of significance for concurrent regression lines. *Biometrika* **40** 297–305.

WILLIAMS, E. J. (1955). Significance tests for discriminant functions and linear functional relationships. *Biometrika* **42** 360–381.

A Univariate Formulation of the Multivariate Linear Model

S. R. Searle

CORNELL UNIVERSITY

The vec operator of matrix algebra is exploited to extend the univariate formulation of the multivariate linear model that is in the literature to deal not only with estimation but also with hypothesis testing. Hotelling's generalized T_0^2 statistic can then be derived from the numerator sum of squares of the univariate F statistic. Application of the vec operator to Jacobian results is also explored.

1. THE VEC OPERATOR AND SOME ASSOCIATED RESULTS

Let $\mathbf{A}$, $\mathbf{B}$, and $\mathbf{C}$ be matrices such that the products and inverses used in the sequel exist. Then, following Neudecker (1969), the vec operator is such that vec $\mathbf{A}$ is the vector formed by writing the columns of $\mathbf{A}$ one under the other in sequence. Thus if $\mathbf{a}_j$ is the jth column of $\mathbf{A}_{r\times c}$, of order $r \times c$

$$\mathbf{A}_{r\times c} = [\mathbf{a}_1 \quad \mathbf{a}_2 \quad \cdots \quad \mathbf{a}_c] \text{ and vec } \mathbf{A} = \begin{bmatrix} \mathbf{a}_1 \\ \mathbf{a}_2 \\ \vdots \\ \mathbf{a}_c \end{bmatrix} \text{ of order } rc \times 1.$$

Paper No. BU-318 in the Biometrics Unit, Cornell University. Thanks go to several colleagues for helpful discussions, and to Florida State University for partially supporting the work during the course of a sabbatical leave.

Key words and phrases Vec and vech operators, Hotelling's generalized T_0^2, Jacobian, linear model.

For example, with

$$\mathbf{A} = \begin{bmatrix} 1 & 7 & 5 \\ 2 & 6 & 8 \end{bmatrix} \quad \text{then} \quad \text{vec}\,\mathbf{A} = \begin{bmatrix} 1 \\ 2 \\ 7 \\ 6 \\ 5 \\ 8 \end{bmatrix}$$

In passing, note that $(\text{vec}\,\mathbf{A}')'$ is the row vector consisting of the rows of $\mathbf{A}$ set alongside one another and is to be distinguished from $(\text{vec}\,\mathbf{A})'$, the transpose of vec $\mathbf{A}$.

We use results given by Neudecker (1969) concerning the vec of a product matrix and relating the vec operator to the trace operator. The first is

$$\text{vec}(\mathbf{ABC}) = (\mathbf{C}' \otimes \mathbf{A})\,\text{vec}\,\mathbf{B}, \tag{1}$$

where, in general, $\mathbf{A} \otimes \mathbf{B} = \{a_{ij}\mathbf{B}\}$ for $i = 1, \ldots, r$ and $j = 1, \ldots, c$ is the Kronecker product of $\mathbf{A}$ and $\mathbf{B}$. A useful special case of (1) is

$$\text{vec}(\mathbf{AB}) = (\mathbf{I} \otimes \mathbf{A})\,\text{vec}\,\mathbf{B}. \tag{2}$$

In introducing the Kronecker product we recall some of its useful properties:

$$(\mathbf{A} \otimes \mathbf{B})' = \mathbf{A}' \otimes \mathbf{B}', \qquad (\mathbf{A} \otimes \mathbf{B})(\mathbf{C} \otimes \mathbf{D}) = \mathbf{AC} \otimes \mathbf{BD}$$

and (3)

$$(\mathbf{A} \otimes \mathbf{B})^{-1} = \mathbf{A}^{-1} \otimes \mathbf{B}^{-1};$$

also, that for $\mathbf{A}$ and $\mathbf{B}$ square of order a and b, respectively,

$$|\mathbf{A} \otimes \mathbf{B}| = |\mathbf{A}|^b |\mathbf{B}|^a. \tag{4}$$

Trace operations using the vec operator are based on Neudecker's (1969)

$$\text{tr}(\mathbf{AB}) = (\text{vec}\,\mathbf{A}')'\,\text{vec}\,\mathbf{B} \tag{5}$$

and an extension of this

$$\text{tr}(\mathbf{AD}'\mathbf{BDC}) = (\text{vec}\,\mathbf{D})'(\mathbf{A}'\mathbf{C}' \otimes \mathbf{B})\,\text{vec}\,\mathbf{D} \tag{6}$$

that is easily established using (1).

Variance–covariance properties of a matrix of random variables in multivariate analysis can be succinctly stated using the vec operator. We establish the following lemma.

LEMMA Suppose $\mathbf{x}_j$ and $\mathbf{x}_k$ are columns of a matrix $\mathbf{X}$ of random variables such that the covariance matrix of $\mathbf{x}_j$ and $\mathbf{x}_k'$ defined as

$$\text{cov}(\mathbf{x}_j, \mathbf{x}_k') = \mathscr{E}[\mathbf{x}_j - \mathscr{E}(\mathbf{x}_j)][\mathbf{x}_k - \mathscr{E}(\mathbf{x}_k)]'$$

for the expectation operator $\mathscr{E}$ is

$$\operatorname{cov}(\mathbf{x}_j, \mathbf{x}_k') = m_{jk}\mathbf{\Sigma}. \tag{7}$$

Then for $\mathbf{M} = \{m_{jk}\}$ it follows that

$$\operatorname{var}(\operatorname{vec} \mathbf{X}) = \mathbf{M} \otimes \mathbf{\Sigma}, \tag{8}$$

$$\operatorname{var}(\operatorname{vec} \mathbf{TX}) = \mathbf{M} \otimes \mathbf{T\Sigma T}', \tag{9}$$

$$\operatorname{var}(\operatorname{vec} \mathbf{X}') = \mathbf{\Sigma} \otimes \mathbf{M}, \tag{10}$$

where $\operatorname{var}(\mathbf{y})$ is the variance–covariance matrix of the vector $\mathbf{y}$, and in (9) $\mathbf{TX}$ represents any linear transformation of the variables in $\mathbf{X}$.

Proof Result (8) is just a composite restatement of (7); utilizing $\operatorname{cov}[\mathbf{Tx}_j, (\mathbf{Tx}_k)'] = m_{jk}\mathbf{T\Sigma T}'$ from (7) gives (9) analogous to (8); and (10) is true because if x_{ij} is the ith element of $\mathbf{x}_j$ then (7) means that $\operatorname{cov}(x_{ij}, x_{rk}) = m_{jk}\sigma_{ir}$ for $\mathbf{\Sigma} = \{\sigma_{ir}\}$. Then if $(\mathbf{x}')_i$ and $(\mathbf{x}')_r$ are the ith and rth rows of $\mathbf{X}$ their covariance matrix is $\operatorname{cov}[(\mathbf{x}')_i', (\mathbf{x}')_r] = \sigma_{ir}\mathbf{M}$, so leading to (10).

2. THE MODEL

The usual multivariate linear model is

$$\mathscr{E}(\mathbf{Y}_{N\times p}) = \mathbf{X}_{N\times q}\mathbf{B}_{q\times p}, \tag{11}$$

where $\mathscr{E}$ is expectation and $\mathbf{Y}_{N\times p}$ is a matrix of N observations on each of p random variables. Then on using (2)

$$\mathscr{E}(\operatorname{vec} \mathbf{Y}) = \operatorname{vec}(\mathbf{XB}) = (\mathbf{I} \otimes \mathbf{X}) \operatorname{vec} \mathbf{B}. \tag{12}$$

The model customarily includes the property that the covariance structure of the $\mathbf{y}_j$'s is $\operatorname{cov}(\mathbf{y}_j, \mathbf{y}_k') = \sigma_{jk}\mathbf{I}_N$ for $j, k = 1, 2, \ldots, p$. Hence from (8)

$$\operatorname{var}(\operatorname{vec} \mathbf{Y}) = \mathbf{\Sigma} \otimes \mathbf{I}_N, \tag{13}$$

where $\mathbf{\Sigma} = \{\sigma_{ij}\}$ for $i, j = 1, 2, \ldots, p$. The multivariate linear model is now represented in (12) and (13) in a univariate manner, similar, for example, to Zellner (1962), Goldberger (1964), and Eaton (1970).

3. ESTIMATION

Generalized least squares estimation of vec $\mathbf{B}$ based on (12) and (13) gives

$$(\mathbf{I}_p \otimes \mathbf{X})'(\mathbf{\Sigma} \otimes \mathbf{I}_N)^{-1}(\mathbf{I}_p \otimes \mathbf{X}) \operatorname{vec} \hat{\mathbf{B}} = (\mathbf{I} \otimes \mathbf{X})'(\mathbf{\Sigma} \otimes \mathbf{I}_N)^{-1} \operatorname{vec} \mathbf{Y}. \tag{14}$$

From properties of Kronecker products given in (3) this reduces to

$$\operatorname{vec} \hat{\mathbf{B}} = [\mathbf{I} \otimes (\mathbf{X}'\mathbf{X})^{-1}\mathbf{X}'] \operatorname{vec} \mathbf{Y} \tag{15}$$

and using (2) again gives

$$\hat{\mathbf{B}} = (\mathbf{X}'\mathbf{X})^{-1}\mathbf{X}'\mathbf{Y}, \tag{16}$$

the familiar result. The sampling variance of vec $\hat{\mathbf{B}}$ from (13) and (15) is

$$\text{var}(\text{vec } \hat{\mathbf{B}}) = [\mathbf{I} \otimes (\mathbf{X}'\mathbf{X})^{-1}\mathbf{X}'](\mathbf{\Sigma} \otimes \mathbf{I}_N)[\mathbf{I} \otimes (\mathbf{X}'\mathbf{X})^{-1}\mathbf{X}']' = \mathbf{\Sigma} \otimes (\mathbf{X}'\mathbf{X})^{-1} \tag{17}$$

so that $\text{cov}(\hat{\mathbf{b}}_i, \hat{\mathbf{b}}_j') = \sigma_{ij}(\mathbf{X}'\mathbf{X})^{-1}$. This familiar result can also be derived in the form (17) by direct application of (9)–(13) and (15).

To obtain an estimate of $\mathbf{\Sigma}$ we use $\mathbf{\Delta}$ as the analogue of $\mathbf{y} - \hat{\mathbf{y}}$ in the univariate case defined by

$$\text{vec } \mathbf{\Delta} \equiv \text{vec } \mathbf{Y} - \text{vec } \hat{\mathbf{Y}} = \text{vec } \mathbf{Y} - (\mathbf{I} \otimes \mathbf{X}) \text{ vec } \hat{\mathbf{B}} = (\mathbf{I} \otimes \mathbf{P}) \text{ vec } \mathbf{Y} \tag{18}$$

on using (15) and on defining

$$\mathbf{P} = [\mathbf{I} - \mathbf{X}(\mathbf{X}'\mathbf{X})^{-1}\mathbf{X}'] = \mathbf{P}' = \mathbf{P}^2, \quad \text{with} \quad \mathbf{PX} = \mathbf{0}. \tag{19}$$

Hence from applying (2) to (18)

$$\mathbf{\Delta} = \mathbf{PY}. \tag{20}$$

Now consider

$$\mathscr{E}(\mathbf{\Delta}'\mathbf{\Delta}) = \mathscr{E}(\mathbf{Y}'\mathbf{PY}) = \{\mathscr{E}(\mathbf{y}_j'\mathbf{P}\mathbf{y}_k)\} = \left\{\mathscr{E}\tfrac{1}{2}[\mathbf{y}_j' \quad \mathbf{y}_k'] \begin{bmatrix} \mathbf{0} & \mathbf{P} \\ \mathbf{P} & \mathbf{0} \end{bmatrix} \begin{bmatrix} \mathbf{y}_j \\ \mathbf{y}_k \end{bmatrix}\right\} \tag{21}$$

for $j, k = 1, 2, \ldots, p$, $\mathbf{y}_j$ and $\mathbf{y}_k$ being columns of $\mathbf{Y}$. Then (11) and (13) give

$$\mathscr{E}\begin{bmatrix} \mathbf{y}_j \\ \mathbf{y}_k \end{bmatrix} = \begin{bmatrix} \mathbf{X}\mathbf{b}_j \\ \mathbf{X}\mathbf{b}_k \end{bmatrix} \quad \text{and} \quad \text{var}\begin{bmatrix} \mathbf{y}_j \\ \mathbf{y}_k \end{bmatrix} = \begin{bmatrix} \sigma_{jj}\mathbf{I} & \sigma_{jk}\mathbf{I} \\ \sigma_{jk}\mathbf{I} & \sigma_{kk}\mathbf{I} \end{bmatrix} \tag{22}$$

and on using the well-known univariate theorem [e.g., Searle (1971, Section 2.5*a*)] that when, for some general vector of variables $\mathbf{z}$ say, $\mathscr{E}(\mathbf{z}) = \boldsymbol{\mu}$ and $\text{var}(\mathbf{z}) = \mathbf{V}$ then $\mathscr{E}(\mathbf{z}'\mathbf{A}\mathbf{z}) = \text{tr}(\mathbf{AV}) + \boldsymbol{\mu}'\mathbf{A}\boldsymbol{\mu}$, we get from (21) and (22), using (19)

$$\mathscr{E}(\mathbf{Y}'\mathbf{PY}) = \{\sigma_{jk} \text{ tr}(\mathbf{P})\} = (N - r)\mathbf{\Sigma}$$

for $r = r(\mathbf{X})$, the rank of $\mathbf{X}$, in this case $r = q$. Hence an unbiased estimator of $\mathbf{\Sigma}$ is the familiar

$$\hat{\mathbf{\Sigma}} = \mathbf{Y}'\mathbf{PY}/(N - q) = \mathbf{Y}'[\mathbf{I} - \mathbf{X}(\mathbf{X}'\mathbf{X})^{-1}\mathbf{X}']\mathbf{Y}/(N - q). \tag{23}$$

4. INDEPENDENCE UNDER NORMALITY

To show that $\hat{\mathbf{B}}$ of (16) and $\hat{\mathbf{\Sigma}}$ of (23) are independent under normality assumptions we use the theorem (*loc. cit.*) that when $\mathbf{z}$ is normally distributed with mean $\boldsymbol{\mu}$ and variance-covariance matrix $\mathbf{V}$, i.e., $\mathbf{z} \sim N(\boldsymbol{\mu}, \mathbf{V})$, then the linear and quadratic forms $\mathbf{Hz}$ and $\mathbf{z'Az}$ respectively are independent if $\mathbf{HVA} = \mathbf{0}$. Assuming normality for the columns of $\mathbf{Y}$, (12) and (13) yield

$$\text{vec } \mathbf{Y} \sim N[(\mathbf{I}_p \otimes \mathbf{X}) \text{ vec } \mathbf{B}, \quad \mathbf{\Sigma} \otimes \mathbf{I}_N]. \tag{24}$$

Now if $\mathbf{e}'_i$ is the ith row of $\mathbf{I}_N$, similarly write

$$\mathbf{E}_i = [0 \quad \cdots \quad 0 \quad \mathbf{I}_N \quad 0 \quad \cdots \quad 0]_{N \times Np},$$

a partitioned matrix which is null except for the ith $N \times N$ submatrix being $\mathbf{I}_N$. Then (23) implies $(N - q)\hat{\sigma}_{ij} = (\text{vec } \mathbf{Y})'\mathbf{E}_i\mathbf{PE}'_j \text{ vec } \mathbf{Y}$, and from (15) we establish independence of vec $\hat{\mathbf{B}}$ and $\hat{\sigma}_{ij}$ by considering

$$\begin{aligned} &[\mathbf{I} \otimes (\mathbf{X'X})^{-1}\mathbf{X'}](\mathbf{\Sigma} \otimes \mathbf{I})\mathbf{E}_i\mathbf{PE}'_j \\ &= [\mathbf{\Sigma} \otimes (\mathbf{X'X})^{-1}\mathbf{X'}]\begin{bmatrix}\text{A matrix that is null except} \\ \text{for } \mathbf{P} \text{ as its } ij\text{th submatrix}\end{bmatrix} \\ &= \mathbf{0} \quad \text{because} \quad \mathbf{X'P} = \mathbf{0}. \end{aligned}$$

Hence $\hat{\mathbf{B}}$ and $\hat{\sigma}_{ij}$ are independent; and so therefore are $\hat{\mathbf{B}}$ and $\hat{\mathbf{\Sigma}}$.

5. HYPOTHESIS TESTING

When, in univariate analysis $\mathbf{y} \sim N(\mathbf{Xb}, \sigma^2\mathbf{I})$, the F statistic for testing the hypothesis H: $\mathbf{K'b} = \mathbf{m}$ for $\mathbf{K}'$ being of full row rank q is $F = Q/q\hat{\sigma}^2$ with $Q = (\mathbf{K'\hat{b}} - \mathbf{m})'[\mathbf{K'(X'X)^{-1}K}]^{-1}(\mathbf{K'\hat{b}} - \mathbf{m})$ and

$$\hat{\sigma}^2 = \mathbf{y}'[\mathbf{I} - \mathbf{X(X'X)^{-1}X'}]\mathbf{y}/[N - r(\mathbf{X})],$$

where $\mathbf{X}$ has full column rank. Adapted to our situation here Q is

$$Q = (\mathbf{K}'\text{vec } \hat{\mathbf{B}} - \mathbf{m})'\{\mathbf{K}'[\mathbf{\Sigma} \otimes (\mathbf{X'X})^{-1}]\mathbf{K}\}^{-1}(\mathbf{K}'\text{vec } \hat{\mathbf{B}} - \mathbf{m}). \tag{25}$$

Further development of Q depends on the form of $\mathbf{K}'$. Consider the hypothesis discussed in Anderson (1956) where, on partitioning $\mathbf{B}$ as

$$\mathbf{B} = \begin{bmatrix}\mathbf{B}_1 \\ \mathbf{B}_2\end{bmatrix} \begin{matrix}\text{with } \mathbf{B}_1 \text{ of order } q_1 \times p \\ \text{and } \mathbf{B}_2 \text{ of order } q_2 \times p\end{matrix} \tag{26}$$

where $q_1 + q_2 = q$, we test H: $\mathbf{B}_1 = \mathbf{B}_{10}$, equivalent to H: vec $\mathbf{B}_1 =$ vec $\mathbf{B}_{10}$. Then H can also be expressed as

$$H\colon \quad \mathbf{K}'\text{vec } \mathbf{B} = \text{vec } \mathbf{B}_{10} \tag{27}$$

for appropriately chosen $\mathbf{K}'$, which turns out to be

$$(\mathbf{K}')_{q_1p \times qp} = \mathbf{I} \otimes \mathbf{L} \quad \text{for} \quad \mathbf{L} = [\mathbf{I}_{q_1} \quad \mathbf{0}_{q_1 \times q_2}]. \tag{28}$$

Then in (25)

$$\mathbf{K}'[\mathbf{\Sigma} \otimes (\mathbf{X}'\mathbf{X})^{-1}]\mathbf{K} = \mathbf{\Sigma} \otimes \mathbf{L}(\mathbf{X}'\mathbf{X})^{-1}\mathbf{L}'. \tag{29}$$

Partitioning $\mathbf{X} = [\mathbf{X}_1 \quad \mathbf{X}_2]$ conformably with (26) and substituting (28) into (29) gives

$$\mathbf{K}'[\mathbf{\Sigma} \otimes (\mathbf{X}'\mathbf{X})^{-1}]\mathbf{K} = \mathbf{\Sigma} \otimes \left\{[\mathbf{I} \quad \mathbf{0}]\begin{bmatrix} \mathbf{X}_1'\mathbf{X}_1 & \mathbf{X}_1'\mathbf{X}_2 \\ \mathbf{X}_2'\mathbf{X}_1 & \mathbf{X}_2'\mathbf{X}_2 \end{bmatrix}^{-1}\begin{bmatrix} \mathbf{I} \\ \mathbf{0} \end{bmatrix}\right\} = \mathbf{\Sigma} \otimes \mathbf{P}_{11\cdot 2}^{-1} \tag{30}$$

where $\mathbf{P}_{11\cdot 2}$ is, from the inverse of a partitioned matrix,

$$\mathbf{P}_{11\cdot 2} = \mathbf{X}_1'\mathbf{X}_1 - \mathbf{X}_1'\mathbf{X}_2(\mathbf{X}_2'\mathbf{X}_2)^{-1}\mathbf{X}_2'\mathbf{X}_1 \tag{31}$$

of order $q_1 \times q_1$. Hence (30) and (27) used in (25) give

$$\begin{aligned} Q &= (\mathbf{K}'\text{vec}\,\hat{\mathbf{B}} - \text{vec}\,\mathbf{B}_{10})'(\mathbf{\Sigma} \otimes \mathbf{P}_{11\cdot 2}^{-1})^{-1}(\mathbf{K}'\text{vec}\,\hat{\mathbf{B}} - \text{vec}\,\mathbf{B}_{10}) \\ &= (\text{vec}\,\hat{\mathbf{B}}_1 - \text{vec}\,\mathbf{B}_{10})'(\mathbf{\Sigma}^{-1} \otimes \mathbf{P}_{11\cdot 2})(\text{vec}\,\hat{\mathbf{B}}_1 - \text{vec}\,\mathbf{B}_{10}). \end{aligned} \tag{32}$$

General linear model theory for $\mathbf{\Sigma}$ known tells us that this has a distribution proportional to a χ^2. But it contains $\mathbf{\Sigma}^{-1}$ and so cannot be used when $\mathbf{\Sigma}$ is unknown, as is usually the case. In the univariate case $\mathbf{\Sigma}^{-1} = 1/\sigma^2$ and we estimate σ^2 by $\hat{\sigma}^2$, for which $\hat{\sigma}^2/\sigma^2$ is also proportional to a χ^2 and independent of Q; then the σ^2's cancel in $Q/(q\hat{\sigma}^2/\sigma^2)$ and we are left with $Q/q\hat{\sigma}^2$ as an F statistic. But in (32) no such cancelling of $\mathbf{\Sigma}^{-1}$ can occur. However, suppose in (32) we replace $\mathbf{\Sigma}^{-1}$ by $(\hat{\mathbf{\Sigma}})^{-1}$, which we shall write as $\hat{\mathbf{\Sigma}}^{-1}$, and consider

$$\hat{Q} = (\text{vec}\,\hat{\mathbf{B}}_1 - \text{vec}\,\mathbf{B}_{10})'(\hat{\mathbf{\Sigma}}^{-1} \otimes \mathbf{P}_{11\cdot 2})(\text{vec}\,\hat{\mathbf{B}}_1 - \text{vec}\,\mathbf{B}_{10}). \tag{33}$$

Now define $\mathbf{W}$ by

$$\text{vec}\,\hat{\mathbf{B}}_1 - \text{vec}\,\mathbf{B}_{10} \equiv \text{vec}\,\mathbf{W} \quad \text{for} \quad \mathbf{W} = [\mathbf{w}_1 \quad \cdots \quad \mathbf{w}_p], \tag{34}$$

where $\mathbf{w}_j$ has order $q_1 \times 1$ for $j = 1, 2, \ldots, p$. Then (33) is

$$\hat{Q} = (\text{vec}\,\mathbf{W})'(\hat{\mathbf{\Sigma}}^{-1} \otimes \mathbf{P}_{11\cdot 2})\,\text{vec}\,\mathbf{W}$$

and on using (6)

$$\hat{Q} = \text{tr}(\hat{\mathbf{\Sigma}}^{-1}\mathbf{W}'\mathbf{P}_{11\cdot 2}\mathbf{W}).$$

Because $\mathbf{X}'\mathbf{X}$ has full rank, $\mathbf{P}_{11\cdot 2}$ of (30) is positive-definite and there exists a nonsingular matrix $\mathbf{R}$ such that $\mathbf{P}_{11\cdot 2} = \mathbf{R}'\mathbf{R}$. Let $\mathbf{U} = \mathbf{W}'\mathbf{R}'$. Then

$$\hat{Q} = \text{tr}(\hat{\mathbf{\Sigma}}^{-1}\mathbf{U}\mathbf{U}'). \tag{35}$$

We show that $(N - q)\hat{Q}$ is Hotelling's generalized T_0^2.

The normality property of (24) together with (15) and (17) implies that

$$\text{vec}\,\hat{\mathbf{B}} \sim N[\text{vec}\,\mathbf{B}, \quad \mathbf{\Sigma} \otimes (\mathbf{X}'\mathbf{X})^{-1}]$$

from which it is not difficult to see that under the null hypothesis H: $\mathbf{B}_1 = \mathbf{B}_{10}$ the distribution of vec $\mathbf{W}$ of (34) is

$$\text{vec } \mathbf{W} \sim N[\mathbf{0}, \quad \boldsymbol{\Sigma} \otimes \mathbf{P}_{11\cdot 2}^{-1}].$$

Therefore for $\mathbf{U} = \mathbf{W}'\mathbf{R}'$ result (9) gives

$$\text{vec } \mathbf{U}' = \text{vec}(\mathbf{RW}) \sim N[\mathbf{0}, \quad \boldsymbol{\Sigma} \otimes \mathbf{RP}_{11\cdot 2}^{-1}\mathbf{R}']. \tag{36}$$

But $\mathbf{P}_{11\cdot 2} = \mathbf{R}'\mathbf{R}$ for $\mathbf{R}$ nonsingular, and so on applying (10) to (36)

$$\text{vec } \mathbf{U} \sim N[\mathbf{0}, \mathbf{I} \otimes \boldsymbol{\Sigma}] \sim N \left\{ \begin{bmatrix} \mathbf{0} \\ \mathbf{0} \\ \vdots \\ \mathbf{0} \end{bmatrix}, \begin{bmatrix} \boldsymbol{\Sigma} & \mathbf{0} & \cdots & \mathbf{0} \\ \mathbf{0} & \boldsymbol{\Sigma} & & \mathbf{0} \\ \vdots & & \ddots & \vdots \\ \mathbf{0} & & & \boldsymbol{\Sigma} \end{bmatrix} \right\}. \tag{37}$$

Therefore the columns of $\mathbf{U}$ are independently and identically distributed $N(\mathbf{0}, \boldsymbol{\Sigma})$. Hence $\mathbf{UU}'$ is a Wishart matrix with parameters $(\boldsymbol{\Sigma}, p, q_1)$; and since $\mathbf{U} = \mathbf{W}'\mathbf{R}' = (\hat{\mathbf{B}}_1 - \mathbf{B}_{10})'\mathbf{R}'$ we have $\mathbf{U}$ independent of $\hat{\boldsymbol{\Sigma}}$ because $\hat{\mathbf{B}}_1$ is. Thus on rewriting (23) as $(N - q)\hat{\boldsymbol{\Sigma}} = \mathbf{S}$ where $\mathbf{S} = \mathbf{Y}'\mathbf{PY}$, which we know is also a Wishart matrix, we have (35) as

$$(N - q)\hat{Q} = \text{tr}(\mathbf{S}^{-1}\mathbf{UU}')$$

with $\mathbf{S}$ and $\mathbf{UU}'$ being independent Wishart matrices. Therefore $(N - q)\hat{Q}$ is Hotelling's generalized T_0^2 statistic corresponding to Wilk's Λ statistic, $\Lambda = |\mathbf{S}|/|\mathbf{S} + \mathbf{U}'\mathbf{U}|$ [see, for example, Kshirsagar (1972, p. 331)].

6. JACOBIANS

The 1-to-1 linear transformation $\mathbf{y} = \mathbf{Tx}$ of a vector of variables $\mathbf{x}$ to $\mathbf{y}$ has Jacobian $J_{x,y}$ given by

$$1/J_{x,y} = J_{y,x} = \left\| \frac{\partial \mathbf{y}}{\partial \mathbf{x}} \right\| = \|\mathbf{T}\|, \tag{38}$$

where the notation $\|\mathbf{T}\|$ denotes the absolute value of the determinant of $\mathbf{T}$. The vec operator permits this standard result to be extended quite easily to transformations that involve not just vectors of variables but matrices of them. For if $\mathbf{X}$ represents the initial variables and $\mathbf{Y}$ the transformed variables then, by definition, the Jacobian of the transformation is

$$1/J_{X,Y} = J_{Y,X} = \left\| \frac{\partial \text{ vec } \mathbf{Y}}{\partial \text{ vec } \mathbf{X}} \right\|. \tag{39}$$

Applications of this involve the vec of products and the properties of Kronecker products shown in (1) and (2), and (3) and (4), respectively. Three examples illustrate this:

For the transformation

$$\mathbf{Y}_{N \times p} = \mathbf{A}_{N \times N} \mathbf{X}_{N \times p}$$

(2) yields

$$\text{vec } \mathbf{Y} = (\mathbf{I}_p \otimes \mathbf{A}) \text{ vec } \mathbf{X}$$

so that (38) and (39) give

$$J_{Y,X} = \|\mathbf{I}_p \otimes \mathbf{A}\| = \|\mathbf{A}\|^p$$

from (3). Similarly for

$$\mathbf{Y}_{N \times p} = \mathbf{A}_{N \times N} \mathbf{X}_{N \times p} \mathbf{B}_{p \times p},$$

$$\text{vec } \mathbf{Y} = (\mathbf{B}' \otimes \mathbf{A}) \text{ vec } \mathbf{X} \qquad \text{and} \qquad J_{Y,X} = \|\mathbf{B}' \otimes \mathbf{A}\| = \|\mathbf{A}\|^p \|\mathbf{B}\|^N.$$

And for

$$\mathbf{Y}_{n \times n} = \mathbf{X}_{n \times n}^{-1} \qquad \text{and} \qquad \mathbf{XY} = \mathbf{I},$$

using (1) twice gives

$$(\mathbf{I} \otimes \mathbf{X}) \text{ vec } \mathbf{Y} = (\mathbf{Y}' \otimes \mathbf{I}) \text{ vec } \mathbf{X} = \text{vec } \mathbf{I}.$$

Hence

$$\text{vec } \mathbf{Y} = (\mathbf{I} \otimes \mathbf{X})^{-1} (\mathbf{Y}' \otimes \mathbf{I}) \text{ vec } \mathbf{X} = (\mathbf{Y}' \otimes \mathbf{Y}) \text{ vec } \mathbf{X}$$

giving

$$J_{Y,X} = \|\mathbf{Y}' \otimes \mathbf{Y}\| = \|\mathbf{Y}\|^{2n}.$$

These are the results given variously in Deemer and Olkin (1951), Dwyer (1967), and elsewhere, derived without benefit of the vec operator.

Extension of the vec operator to symmetric matrices in this context is not quite so direct because when $\mathbf{X}$ is symmetric it contains only $\frac{1}{2}n(n+1)$ different variables. To account for this, a new operator is proposed, to be called vech (for "vector, half"), defined solely for symmetric $\mathbf{X}$ so that vech $\mathbf{X}$ is the vector formed by writing the columns of $\mathbf{X}$, starting in each case at the diagonal element, one under the other in sequence. For example, whereas for

$$\mathbf{X} = \begin{bmatrix} a & b & c \\ b & d & e \\ c & e & f \end{bmatrix}$$

$$(\text{vec } \mathbf{X})' = [a \quad b \quad c \quad b \quad d \quad e \quad c \quad e \quad f],$$
$$(\text{vech } \mathbf{X})' = [a \quad b \quad c \quad d \quad e \quad f].$$

Thus vech $\mathbf{X}$ has just $\frac{1}{2}n(n+1)$ different elements in it, and in transformations from $\mathbf{X}$ to $\mathbf{Y}$ for symmetric $\mathbf{Y}$

$$J_{Y,X} = \left\| \frac{\partial \text{ vech } \mathbf{Y}}{\partial \text{ vech } \mathbf{X}} \right\|.$$

Clearly

$$\text{vech } \mathbf{X} = \mathbf{H}_n \text{ vec } \mathbf{X} \qquad \text{and} \qquad \text{vec } \mathbf{X} = \mathbf{G}_n \text{ vech } \mathbf{X}$$

for matrices $\mathbf{H}_n$ and $\mathbf{G}_n$ that can be specified for any order n. Then, for example, in the transformation

$$\mathbf{Y} = \mathbf{AXA}'$$

we have from (1) that

$$\text{vec } \mathbf{Y} = (\mathbf{A} \otimes \mathbf{A}) \text{ vec } \mathbf{X}$$

and so

$$\text{vech } \mathbf{Y} = \mathbf{H}_n(\mathbf{A} \otimes \mathbf{A})\mathbf{G}_n \text{ vech } \mathbf{X}$$

from which

$$J_{Y,X} = \|\mathbf{H}_n(\mathbf{A} \otimes \mathbf{A})\mathbf{G}_n\|.$$

This, we know, from Deemer and Olkin (1951) equals $|\mathbf{A}|^{n+1}$, as can also be shown from properties of $\mathbf{H}_n$ and $\mathbf{G}_n$. Work in progress is concerned with using these properties for extending the available results on Jacobians.

REFERENCES

ANDERSON, T. W. (1956). *An Introduction to Multivariate Statistical Analysis.* Wiley, New York.

DEEMER, W. L. and OLKIN, I. (1951). The Jacobians of certain matrix transformations useful in multivariate analysis. *Biometrika* **38** 345–367.

DWYER, P. S. (1967). Some applications of matrix derivatives in multivariate analysis. *J. Amer. Statist. Assoc.* **62** 607–625.

EATON, M. L. (1970). Gauss–Markov estimation for multivariate linear models: a co-ordinate free approach. *Ann. Math. Statist.* **41** 528–538.

GOLDBERGER, A. S. (1964). *Econometric Theory.* Wiley, New York.

KSHIRSAGAR, A. M. (1972). *Multivariate Analysis.* Marcell Dekker, New York.

NEUDECKER, H. (1969). Some theorems on matrix differentiation with special reference to Kronecker matrix products. *J. Amer. Statist. Assoc.* **64** 953–963.

SEARLE, S. R. (1971). *Linear Models.* Wiley, New York.

ZELLNER, A. (1962). An efficient method of estimating unrelated regressions and tests for aggregation bias. *J. Amer. Statist. Assoc.* **57** 338–342.

Multinomial Selection Index

W. B. Smith *D. M. Scott*†

TEXAS A & M UNIVERSITY

A linear selection index is developed assuming the phenotypic vector is distributed multinomially with both complete and incomplete data records. Utilizing all data, estimates of the phenotypic means and covariances are given, and those individuals with partial records are also indexed. Limited Monte Carlo simulation studies are presented in support of the optimality properties, and numerical comparisons are made between this index and a simpler index developed for multivariate normal phenotypic vectors.

1. INTRODUCTION

The linear selection index developed by H. F. Smith (1936) is a linear combination of the elements of the phenotypic observation vector $\mathbf{X}_j$,

$$\mathbf{I}_j = \mathbf{b}'\mathbf{X}_j, \tag{1}$$

where $\mathbf{I}_j$ denotes the composite index value associated with the jth member of a population and $\mathbf{b}$ is an n vector of unknown coefficients (weights). This index was conceived to aid in discriminating between selection programs among varieties of plants. Assuming $\mathbf{X}_j$ was distributed as a multivariate

† Present address: U.S. Steel, Monroeville, Pennsylvania.

This research was partially supported by the National Aeronautics and Space Administration, Research Grant NGR 44-001-097. The authors thank R. R. Hocking and an anonymous referee for their constructive criticisms of earlier drafts.

Key words and phrases Missing data, multinomial distribution, selection index.

normal with phenotypic covariance matrix **P**, Smith showed that the optimal choice of **b** (i.e., yielding greatest expected genetic advance) is

$$\mathbf{b} = \mathbf{P}^{-1}\mathbf{G}\boldsymbol{\alpha}, \tag{2}$$

where G is the genotypic covariance matrix and $\boldsymbol{\alpha}$ is an n vector of economic weights.

Since Smith's paper, much research has been conducted on the linear index and its nonlinear competitors. Notable contributions are Henderson (1963), Kempthorne and Nordskog (1959), Williams (1962), Hazel (1943), and VanVleck (1970). See Williams (1962) for a thorough review.

Smith and Pfaffenberger (1970) considered index estimation using multivariate normal phenotypic observations, both full and partially complete vectors, assuming **G** and $\boldsymbol{\alpha}$ are known, but **P** unknown. This procedure applied a technique of Hocking and Smith (1968) for estimating the parameters of a multivariate normal distribution in the presence of partial data. All data are used and several alternative methods are presented for indexing those individuals possessing partial records. A contrast between Henderson's techniques and that of Smith and Pfaffenberger (S–P in the sequel) is given.

Indices, whose input variates are distributed discretely, are constructed in many areas. For example, when determining a performance index in poultry, numbers of eggs laid, graded quality, and daily production are used. Economic indicators often utilize discrete data—number unemployed, size of work force, etc. Each index so calculated using standard normal-based procedures violates basic assumptions; thus the need for procedures developed under different assumptions.

The main objection to the use of multivariate normal based selection index techniques rests on the normality assumption, not on continuity, since in most practical situations the distributions of input variates are distinctly asymmetric. In addition, in many applications the input variates follow distributions which are a mixed combination of both continuous and discrete variates.

As an example, Andrus and McGilliard (1975) in selecting dairy cattle for overall economic excellence use a set of input variates that includes both the theoretically continuous measurements of body weight and milk production, and also the number of live freshenings and an indicator of the presence of mastitis.

Nagai, *et al.* (1975) include two variables measuring the mother's nursing ability and litter body weight at the age of 42 days in developing a selection index for choosing breeding stock of research mice. Each nursing mother nursed eight offspring of other mothers, with their nursing ability calculated by the mean 12-day weight of the litter. As the mean is very sensitive to outliers, a possibly better indicator of nursing ability would be a discrete variable calculated on the basis of overall offspring weight ranks. Thus, a performance index so calculated would include both continuous and discrete variates. As a final example, Koch *et al.* (1974a,b) construct an index in selecting beef cattle which includes body weights at various times and a muscling score measurement.

Each of these investigators used normal-based statistical tools in their analysis. The technique described here is based on a multinomial assumption; an extreme case in which input variates are discretely distributed in a specified manner. The effort will be to compare the new procedure with results obtained from using normal-based procedures on the same data.

This paper considers the linear selection index as described by (1) with **b** chosen as in (2) assuming both full and partial records are available and that the phenotypic vectors follow a multinomial distribution. Thus, this index deviates both from the assumption that the phenotypic covariance matrix is known and from the assumption of normality.

An application of the estimation procedure of Hocking and Oxspring (1971) is discussed and certain simulation studies are presented to support claimed optimality properties. In addition, the S–P multivariate normal technique is applied to multinomial data for comparison with the multinomial estimation procedure.

2. ESTIMATION PROCEDURE

Consider a phenotypic observation vector $\mathbf{X}_i' = (X_{i1} \cdots X_{ik})$ which is distributed multinomially with probability parameters $\boldsymbol{\theta}' = (\theta_1 \cdots \theta_k)$. That is,

$$P(\mathbf{X}_i' = \mathbf{x}_k') = M! \prod_{j=1}^{k+1} (\theta_j^{x_{ij}}/x_{ij}!), \tag{3}$$

where $x_{i,k+1} = M - \sum_{j=1}^{k} x_{ij}$ and $\theta_{k+1} = 1 - \sum_{j=1}^{k} \theta_j$. We desire to index each vector from a population distributed as (3); however, some of these vectors have missing elements (recall that any marginal distribution from a multinomial is again multinomial in form). As in Smith and Pfaffenberger (1970) all information, both full and partial vectors, is utilized in estimating $\theta_1, \ldots, \theta_k$ and thus in estimating each individual's index, assuming **G** and $\boldsymbol{\alpha}$ known.

Following the outline of Hocking and Smith (1968) group the data vectors by which elements are missing, estimate within each group the available θ_j, and then optimally combine these estimates. For example, consider a sample of size N, where n_1 individuals have recorded all elements of the phenotypic observation vector while n_2 individuals have only the first (renumbering if necessary) $l < k$ elements recorded, $n_1 + n_2 = N$. Thus, from the full data group each parameter θ_j can be estimated unbiasedly by ${}_1\hat{\theta}_j$, $j = 1, \ldots, k$, whereas from the second group (partial data) only $\theta_1, \ldots, \theta_l$ can be estimated by ${}_2\hat{\theta}_j$, $j = 1, \ldots, l$. In each case, the usual maximum likelihood estimates are used. Combining these estimates as in Hocking and Smith

$$\tilde{\theta}_j = {}_1\hat{\theta}_j + \sum_{r=1}^{l} a_{rj}({}_1\hat{\theta}_r - {}_2\hat{\theta}_r) \qquad j = 1, \ldots, k. \tag{4}$$

Note that $\mathbf{A}_j' = (a_{1j}, \ldots, a_{lj})$ is chosen to minimize the variance of $\hat{\theta}_j$, $j = 1, \ldots, k$. If $\mathbf{A}_j'$ does not depend on the parameters $\boldsymbol{\theta}'$, then $\tilde{\theta}_j$ is minimum variance unbiased. In general, $\tilde{\theta}_j$ is consistent, asymptotically unbiased, and asymptotically efficient when full data estimates of $\boldsymbol{\theta}'$ are used in $\mathbf{A}_j'$.

A general formulation for $\mathbf{A}_j$ can be given. Let $\mathbf{V}$ be the covariance matrix of $(X_1, \ldots, X_k)$. Thus,

$$M^{-1}\mathbf{V} = \mathrm{Diag}(\boldsymbol{\theta}) - \boldsymbol{\theta}\boldsymbol{\theta}'.$$

Then the covariance matrix of ${}_1\hat{\boldsymbol{\theta}}'$ is given by $\mathbf{V}/n_1 M$ and for ${}_2\hat{\boldsymbol{\theta}}$ by $\mathbf{D}_2\mathbf{V}\mathbf{D}_2'/n_2 M$, where $\mathbf{D}_2 = (\mathbf{I}_l \,\vdots\, \mathbf{0})$ and $\mathbf{I}_l$ is an identity matrix of order l. Thus,

$$\mathbf{A} = \begin{bmatrix} \mathbf{A}_1' \\ \cdots \\ \cdots \\ \mathbf{A}_k' \end{bmatrix} = -\frac{n_2}{n_1 + n_2}(\mathbf{D}_2\mathbf{V}\mathbf{D}_2')^{-1}\mathbf{D}_2\mathbf{V}.$$

Note that $\mathbf{D}_2\boldsymbol{\theta} = {}_2\boldsymbol{\theta}$, where ${}_2\boldsymbol{\theta}' = (\theta_1, \ldots, \theta_l)$.

If in addition, there is a third data group of n_3 multinomial vectors with parameters M and $\mathbf{D}_3$, $\mathbf{D}_3$ is a $s \times k$ unitary matrix of 1's and 0's, then new estimates would combine $\tilde{\boldsymbol{\theta}}$ with $\{{}_3\hat{\theta}_j\}$, the estimates from this third group. In such a case, in matrix notation

$$\tilde{\tilde{\boldsymbol{\theta}}} = \tilde{\boldsymbol{\theta}} + \mathbf{B}'(\mathbf{D}_3\tilde{\boldsymbol{\theta}} - {}_3\hat{\boldsymbol{\theta}}).$$

Note that $\mathbf{D}_3$ makes $\tilde{\boldsymbol{\theta}}$ conformable to ${}_3\hat{\boldsymbol{\theta}}$. $\mathbf{B}$ is chosen to minimize the variance of $\tilde{\tilde{\boldsymbol{\theta}}}$ and satisfies

$$[\mathbf{D}_3(\mathbf{W} + \mathbf{V}/n_3 M)\mathbf{D}_3']\mathbf{B} = \mathbf{D}_3\mathbf{W},$$

where $\mathbf{W}$ is the covariance matrix for the combined first two data groups. That is,

$$\begin{aligned} \mathbf{W} &= [(\mathbf{V}/n_1 M)^{-1} + \mathbf{D}_2'(\mathbf{D}_2\mathbf{V}\mathbf{D}_2')/n_2 M)\mathbf{D}_2]^{-1} \\ &= [n_1 + n_2\mathbf{V}\mathbf{D}_2'(\mathbf{D}_2\mathbf{V}\mathbf{D}_2')\mathbf{D}_2]^{-1}\mathbf{V}/M. \end{aligned}$$

To estimate $\boldsymbol{\theta}'$ when more than three groups of data are available, continue in the fashion just outlined, producing at each stage the estimated asymptotic covariance matrix of the combined estimate for use at the next stage.

Now to achieve an estimate based on all data the final estimate of $\boldsymbol{\theta}'$ is used for estimation of the covariance matrix $\mathbf{P}$. That is, in the previous case $\tilde{\tilde{\theta}}_j$ is substituted for θ_j in the formulation of $\mathbf{P}$, yielding a matrix $\tilde{\tilde{\mathbf{P}}}$. Then set

$$\tilde{\tilde{\mathbf{b}}} = \tilde{\tilde{\mathbf{P}}}^{-1}\mathbf{G}.$$

With $\tilde{\tilde{\mathbf{b}}}$ an index for each phenotypic vector can be given; in those cases of missing data the final estimate for the mean of that element is substituted to produce a full vector to index. In general, the procedure yields consistent and asymptotically efficient estimates. Note that estimated phenotypic means, variances, and covariances are available upon termination.

Example 1 Let $k = 3$, $l = 2$, as before. Then

$$\tilde{\boldsymbol{\theta}} = \hat{\boldsymbol{\theta}} + \mathbf{A}(\mathbf{D}_2\hat{\boldsymbol{\theta}} - {}_2\hat{\boldsymbol{\theta}}).$$

In this case

$$\mathbf{D}_2 = \begin{pmatrix} 1 & 0 & 0 \\ 0 & 1 & 0 \end{pmatrix}$$

and

$$\mathbf{A} = \frac{n_2}{n_1 + n_2}\begin{bmatrix} -1 & 0 & \theta_3/(1-\theta_1-\theta_2) \\ 0 & -1 & \theta_3/(1-\theta_1-\theta_2) \end{bmatrix}.$$

Thus, $\tilde{\boldsymbol{\theta}}_1$ and $\tilde{\boldsymbol{\theta}}_2$ are just the weighted sums of ${}_1\hat{\theta}_1$, ${}_2\hat{\theta}_1$ and ${}_1\hat{\theta}_2$, ${}_2\hat{\theta}_2$, respectively, and are minimum variance unbiased. Note, however, that the coefficients for $\tilde{\theta}_3$ depend on $\tilde{\boldsymbol{\theta}}'$; thus, $\tilde{\theta}_3$ will be a consistent, asymptotic efficient estimate when ${}_1\hat{\theta}_j$ is substituted for θ_j, $j = 1, 2, 3$.

Example 2 Consider the same situation as in Example 1, but with a third data group, $s = 1$. Then

$$\tilde{\tilde{\boldsymbol{\theta}}} = \tilde{\boldsymbol{\theta}} + \mathbf{B}(\tilde{\theta}_1 - {}_3\tilde{\theta}_1),$$

where

$$\mathbf{D}_3 = (1 \quad 0 \quad 0)$$

and

$$\mathbf{B} = -\left[\left(\frac{1}{n_1+n_2} + \frac{1}{n_3}\right)\theta_1(1-\theta_1)\right]^{-1}\begin{bmatrix} \theta_1(1-\theta_1)/(n_1+n_2) \\ \theta_1\theta_2/(n_1+n_2) \\ \theta_1\theta_3/(n_1+n_2) \end{bmatrix}$$

$$= \frac{-n_3}{(n_1+n_2+n_3)}\begin{bmatrix} 1 \\ \theta_2/(1-\theta_1) \\ \theta_3/(1-\theta_1) \end{bmatrix}.$$

Again the coefficient depends on the parameters to be estimated, but substitution by the "best" previous estimates (i.e., $\widetilde{\boldsymbol{\theta}}$) yields consistent and asymptotically efficient estimates.

Example 3 Consider the same situation as in Example 1, but with a third data group containing information on X_2 and X_3.

$$\tilde{\tilde{\boldsymbol{\theta}}} = \tilde{\boldsymbol{\theta}} + \mathbf{C}' \begin{bmatrix} \tilde{\theta}_2 - {}_3\hat{\theta}_2 \\ \tilde{\theta}_3 - {}_3\hat{\theta}_3 \end{bmatrix},$$

where

$$\mathbf{D}_3 = \begin{pmatrix} 0 & 1 & 0 \\ 0 & 0 & 1 \end{pmatrix}$$

and

$$\mathbf{C} = -\frac{n_3}{n_1 + n_2 + n_3} \begin{bmatrix} \theta_2(1-\theta_2) & -\theta_2\theta_3 \\ -\theta_2\theta_3 & \theta_3(1-\theta_3) \end{bmatrix}^{-1} \cdot \begin{bmatrix} -\theta_1\theta_2 & \theta_2(1-\theta_2) & -\theta_2\theta_3 \\ -\theta_1\theta_3 & -\theta_2\theta_3 & \theta_3(1-\theta_3) \end{bmatrix}.$$

3. SIMULATION STUDIES

In the following Monte Carlo simulation studies we set $\boldsymbol{\alpha}' = (1\ 1\ 1)$ and

$$\mathbf{G} = \begin{bmatrix} 2 & 0.75 & 2 \\ 0.75 & 3 & 1.5 \\ 2 & 1.5 & 4 \end{bmatrix}$$

with $\boldsymbol{\theta}' = (.15, .25, .40)$ in each case.

Table 1 records a summary of simulation studies (200 runs) of Example 2 where $M_1 = M_2 = M_3 = 20$, $n_1 = 100$, $n_2 = 50$, $n_3 = 25$, and clearly indicates the greater precision achieved by using the partial data vectors. Table 2 summarizes a similar experiment with the same data configuration but in the presence of a much higher percentage (80%) of partial data vectors ($n_1 = 10$, $n_2 = 15$, $n_3 = 25$). Again, reductions in the sample variance of the estimates are noted. In both cases the estimates of $\boldsymbol{\theta}'$ are virtually unbiased but the estimates of $\mathbf{b}' = (b_1, b_2, b_3)$ are biased slightly upward. These examples, of course, are for a situation for which we have "nested" data, and in such situations Hocking and Oxspring (1971) have shown that this technique yields maximum likelihood estimates.

TABLE 1

Simulation of Example 2[a]

Parameter	Mean estimate					
	$\theta_1 = .15$	$\theta_2 = .25$	$\theta_3 = .40$	$b_1 = 2383$	$b_2 = 2169$	$b_3 = 2124$
1st data group	0.1502	0.2507	0.4005	2401.4	2185.0	2141.3
2nd adjoined	0.1496	0.2515	0.4004	2402.9	2184.9	2141.9
3rd adjoined	0.1490	0.2517	0.4006	2404.4	2183.2	2140.2
	Sample variances of estimates					
	$\theta_1(\times 10^{-4})$	$\theta_2(\times 10^{-4})$	$\theta_3(\times 10^{-3})$	b_1	b_2	b_3
1st data group	0.7976	0.9419	1.0223	7151.0	6115.8	6759.9
2nd adjoined	0.4006	0.6386	0.8862	6494.8	5665.9	5725.7
3rd adjoined	0.3705	0.6372	0.8860	6394.8	5585.3	5648.8

[a] $n_1 = 100$, $n_2 = 50$, $n_3 = 25$; 200 repetitions

TABLE 2

Simulation of Example 2[a]

Parameter	Mean estimate					
	$\theta_1 = .15$	$\theta_2 = .25$	$\theta_3 = .40$	$b_1 = 2383$	$b_2 = 2169$	$b_3 = 2124$
1st data group	0.1503	0.2515	0.3994	2455.2	2276.2	2181.5
2nd adjoined	0.1495	0.2515	0.3999	2438.6	2215.2	2171.3
3rd adjoined	0.1493	0.2516	0.4001	2434.4	2213.9	2171.2
	Sample variances of estimates					
	$\theta_1(\times 10^{-4})$	$\theta_2(\times 10^{-4})$	$\theta_3(\times 10^{-3})$	b_1	b_2	b_3
1st data group	0.6766	0.8813	1.939	83,296.2	77,727.3	77,617.3
2nd adjoined	0.2816	0.3781	0.8791	72,107.6	67,039.3	60,450.5
3rd adjoined	0.1325	0.3557	0.8606	70,834.9	65,625.6	59,290.9

[a] $n_1 = 10$, $n_2 = 15$, $n_3 = 25$; 500 repetitions

Tables 3 and 4 summarize simulation conducted on Example 3 but with two different sample sizes. Each of these tables again reflect the consistency of the estimate and the reduction in variances achieved by utilizing the partial data. Again there is some bias noted in the **b** term.

TABLE 3

Simulation of Example 3[a]

Parameter	Mean estimate					
	$\theta_1 = .15$	$\theta_2 = .25$	$\theta_3 = .40$	$b_1 = 2383.3$	$b_2 = 2170$	$b_3 = 2125$
1st data group	0.1486	0.2514	0.3995	2395.1	2171.8	2128.7
2nd adjoined	0.1488	0.2511	0.3995	2391.9	2170.8	2127.8
3rd adjoined	0.1487	0.2518	0.3991	2391.7	2168.9	2127.6
	Sample variances of estimates					
	$\theta_1 (\times 10^{-4})$	$\theta_2 (\times 10^{-4})$	$\theta_3 (\times 10^{-4})$	b_1	b_2	b_3
1st data group	1.063	1.921	1.654	12,832.8	9546.9	10,163.6
2nd adjoined	0.555	0.969	1.449	10,686.5	8569.4	9167.9
3rd adjoined	0.512	0.689	1.042	7407.8	7809.4	7300.4

[a] $n_1 = 50$, $n_2 = 50$, $n_3 = 50$; 100 repetitions

TABLE 4

Simulation of Example 3[a]

	Mean estimate					
	$\theta_1 = .15$	$\theta_2 = .25$	$\theta_3 = .40$	$b_1 = 2383.3$	$b_2 = 2170$	$b_3 = 2125$
1st data group	0.1506	0.2514	0.3992	2454.8	2227.1	2181.9
2nd adjoined	0.1496	0.2512	0.4000	2437.8	2215.1	2171.7
3rd adjoined	0.1495	0.2515	0.3998	2418.7	2195.2	2152.3
	Sample variance of estimates					
	$\theta_1 (\times 10^{-3})$	$\theta_2 (\times 10^{-3})$	$\theta_3 (\times 10^{-3})$	b_1	b_2	b_3
1st data group	0.700	0.882	1.215	82,713.9	77,909.5	77,439.5
2nd adjoined	0.283	0.380	0.883	72,867.4	67,739.1	61,087.3
3rd adjoined	0.243	0.194	0.297	25,302.4	32,804.3	31,159.7

[a] $n_1 = 10$, $n_2 = 15$, $n_3 = 25$; 100 repetitions

TABLE 5

Average Correlation With "True" Ranking (10 Runs each) of Multinomial Data Simulation of Example 2[a]

Estimation index before deletion	0.7448	0.7236
Multinomial estimates	0.7415	0.7175
Smith–Pfaffenberger multivariate normal estimates	0.7356	0.7151
	$\theta' = (.15, .25, .40)$	$\theta' = (.27, .08, .47)$

[a] $n_1 = 100$, $n_2 = 50$, $n_3 = 25$.

Table 5 summarizes an Example 2 experiment using the same parameters as those of Table 1. That is, 175 vectors $\mathbf{X}' = (X_1\ X_2\ X_3)$ were generated to follow a multinomial distribution with parameters $M = 20$, and $\boldsymbol{\theta}' = (\theta_1\ \theta_2\ \theta_3)$. Each of these vectors is indexed by equations (1) and (2) using the population values for $\boldsymbol{\theta}'$. The order resulting is called the "true" order. It is desired to compare this order with the order resulting from estimated indexes in several different cases. First, all 175 full data vectors are used to estimate $\boldsymbol{\theta}'$ and $\mathbf{P}$, thus yielding an estimated index for each. The correlation between the estimated ordering and the "true" ordering is given by the first entry of Table 5. Next, a missing data situation is created by randomly selecting 75 vectors and deleting X_3 from 50 of them, and X_2 and X_3 from 25 of them. Thus, we now have available 100 full vectors and 75 partial vectors of two types. The procedure of Section 2 is applied to estimate $\boldsymbol{\theta}'$ and $\mathbf{P}$ (and thus $\mathbf{b}$). The index order resulting is compared to the "true" ordering yielding the second entry of the table. Finally, the Smith–Pfaffenberger (1970) multivariate normal indexing technique is applied to the partial multinomial data and the final entry is the correlation between the resulting ordering and the true ordering. It should be noted that in the above case the partial data vectors were indexed by means of applying the estimated $\mathbf{b}$ vector to the partial vectors, where the missing element is in turn estimated by its expected value in the multinomial case and by regression estimate in the multivariate normal case. Further explanation of the regression estimate in the multivariate normal case is given in the paper by Smith and Pfaffenberger. It should be noted that the estimate of the population mean is precise as indicated by the simulation of estimates of $\boldsymbol{\theta}_j$.

4. CONCLUSIONS

In this paper we develop a linear selection index using phenotypic observation vectors multinomially distributed and estimate the index value of each vector by estimating in an optimal, sequential fashion the parameters of the parent multinomial distribution. Moreover, the estimation procedure of Section 2 does not require that all data records be full (i.e., have no missing elements). In addition, a vector with missing elements is indexed by multiplying **b** by that vector with its missing elements filled by the combined mean estimate (i.e., M $\tilde{\tilde{\theta}}_j$).

Thus, the index of Section 2 deviates from a "standard" index in that, first, we estimate **b** by estimating **P**, and, second, the parent phenotypic vector distribution is nonnormal. We cite the results tabulated in Section 3 (Tables 1 to 4) as empirical indications of the procedure's properties, viz., consistency and asymptotic efficiency. For further theoretical justifications for a similar technique see Hocking and Oxspring (1971).

Note that in Table 5, the comparison of the ranking resulting from the S–P multivariate normal procedure and the Section 2 procedure indicates that use of an estimated index assuming a multivariate normal configuration does not lead to unwarranted results. Thus, the value of the Section 2 procedure is in the slightly more precise ordering achieved and in providing during the indexing process estimates of both phenotypic means and covariance matrix.

Future problems include combining the multivariate normal and multinomial procedures to yield an index of vectors some of whose elements are continuously distributed, others discrete. In addition, nonlinear competitors for both estimate indices are being considered.

REFERENCES

ANDRUS, D. F. and MCGILLIARD, L. D. (1975). Selection of dairy cattle for overall excellence. *J. Dairy Sci.* **58** 1876–1879.

HAZEL, L. N. (1943). Genetic basis for selection indices. *Genetics* **28** 476–490.

HENDERSON, C. R. (1963). Selection index and expected genetic advance. *NAS-NRC Publication 982* 141–163.

HOCKING, R. R. and OXSPRING, H. H. (1971). Maximum likelihood estimation with incomplete multinomial data. *J. Amer. Statist. Assoc.* **66** 65–70.

HOCKING, R. R. and SMITH, W. B. (1968). Estimation of parameters in the multivariate normal distribution with missing observations. *J. Amer. Statist. Assoc.* **63** 159–173.

KEMPTHORNE, O. and NORDSKOG, A. W. (1959). Restricted selection indices. *Biometrics* **15** 10–19.

KOCH, R. M., GREGORY, K. E., and CUNDIFF, L. V. (1947a). Selection in beef cattle. I. Selection applied and generation interval. *J. Animal Sci.* **39** 449–458.

KOCH, R. M., GREGORY, K. E., and CUNDIFF, L. V. (1974b). Selection in beef cattle. II. Selection response. *J. Animal Sci.* **39** 459–470.

NAGAI, J., HICKMAN, C. G., and BARR, G. R. (1975). Selection index based on the nursing ability of the mother and the mature weight of the offspring in mice. *J. Animal Sci.* **40** 590–597.

SMITH, H. F. (1936). A discriminant function for plant selection. *Ann. Eugen. Long.* **7** 240–250.

SMITH, W. B. and PFAFFENBERGER, R. C. (1970). Selection index estimation from partial multivariate normal data. *Biometrics* **26** 625–639.

VANVLECK, L. D. (1970). Index selection for direct and maternal genetic components of economic traits. *Biometrics* **26** 477–484.

WILLIAMS, J. S. (1962). The evaluation of a selection index. *Biometrics* **18** 375–393.

Part III

TIME SERIES

Applications of Time Series Analysis

G. E. P. Box *G. C. Tiao*

UNIVERSITY OF WISCONSIN
MADISON

This paper illustrates two of the recently developed time series techniques. In the first, a canonical analysis is made on a five-variate time series on U.S. hog supply, hog price, corn supply, corn price, and farm wages. The analysis shows that the five original series can be decomposed into (i) a nonstationary component; (ii) two serially correlated stationary components, and (iii) two white noise components. The white noise components can reflect contemporaneous relationships among the original variables while the nonstationary component characterizes the overall growth pattern of the series. In the second, intervention methods are applied to determine the effects of control measures and changes in measurement techniques on carbon monoxide pollution data in Los Angeles county. In addition, a brief survey of other recent works on time series is also presented.

1. INTRODUCTION

During the past sixteen years, much work has been done in collaboration with Gwilym Jenkins on the modeling of time series for forecasting and process control. It began at Princeton with the problem of finding methods for the adaptive optimization of a chemical reactor. It was a problem whose

This research was partially supported by the Air Force Office of Scientific Research under grant AFOSR 72-2363D and the Army Research Office under grant DAAG29-76-G-0304.

Key words and phrases Multiple time series, transfer function, autoregressive-moving average models, intervention analysis, canonical analysis, nonstationary series.

dynamic and stochastic aspects could not be ignored and it led, by a process of iteration between practice and theory, to some interesting research and a number of publications [1, 2], including a book [3]. These researches have continued both in England and at Wisconsin in the belief that good theory and good practice flourish together as a result of interaction and mutual enrichment. The purpose of this paper is to summarize some of the current research on various aspects of time series analysis and to discuss in some detail two recent studies on (i) a canonical transformation of multiple time series and (ii) intervention techniques.

2. SERIAL AND NONSERIAL MODELS

Suppose $\mathbf{z}_t = (z_{1t}, z_{2t}, \ldots, z_{kt})$ are k *data* streams observable at equally spaced time intervals some of which may be represented by random variables and others not. Then a wide class of *statistical models* may be represented as a transformation of the data to white noise of the form

$$\mathbf{f}(\mathbf{z}_t, \mathbf{z}_{t-1}, \ldots, |\boldsymbol{\theta}) = \mathbf{a}_t, \tag{1}$$

where $\{\mathbf{a}_t\}$ is the vector-valued "white noise" sequence of independently and identically distributed random variables having mean zero and covariance matrix $\boldsymbol{\Sigma}$.

Much statistical analysis may be regarded as an iterative search for such a model. For instance, in standard regression analysis, a linear model of the form

$$z_{1t} - \theta_1 - \theta_2 z_{2t} - \theta_3 z_{3t} - \cdots - \theta_k z_{kt} = a_t \tag{2}$$

is often employed. This can be written as

$$f(\mathbf{z}_t | \boldsymbol{\theta}) = a_t \tag{3}$$

and is seen to be an example of the more general form (1) with the very special property that it *needs to involve only contemporaneous data.* The contemporaneous linear model (2) is used in the analysis of regression problems but (by allowing all or some of the z's to be indicator variables) it is of course also employed in the analysis of variance and covariance of data obtained from standard statistical designs.

Analysis with such a model ordinarily takes the form of trial fitting followed by examination of residuals $\hat{a}_t$. Such examination may point to the existence of bad data values, or (if the overall functional form is to be retained) to the need for further transformation of one or more variables. It may also suggest the presence in $\hat{a}_t$ of components of the other variables $z_{k+1,t}$ etc., which may need to be incorporated in the model. Finally, serial correlation of the residuals $\hat{a}_t$ could indicate that lagged values of the z's might be needed.† For example, a relationship

$$a_t = \phi a_{t-1} + \alpha_t,$$

where $\{\alpha_t\}$ was a white noise sequence, could be written

$$a_t = \frac{\alpha_t}{1 - \phi B},$$

where B was the backshift operator such that $Ba_t = a_{t-1}$. But this implies a model

$$(1 - \phi B) f(\mathbf{z}_t | \boldsymbol{\theta}) = \alpha_t,$$

which is of the *serial* (noncontemporaneous) form

$$f_1(\mathbf{z}_t, \mathbf{z}_{t-1} | \boldsymbol{\theta}) = \alpha_t,$$

a particular case of the general model (1).

A Serial Model of Wide Application

A widely useful form of (1) may be obtained by allowing for possible dynamic (or inertial) relationships between the variables. Dynamic relationships between a k_2-dimensional input sequence $\{\mathbf{x}_t\}$ and a k_1-dimensional output sequence $\{\mathbf{y}_t\}$ is provided by a linear filtering operation parsimoniously represented by a system of difference equations

$$\{\mathbf{I} - \boldsymbol{\delta}_1 B - \boldsymbol{\delta}_2 B^2 - \cdots - \boldsymbol{\delta}_r B^r\}\mathbf{y}_t = (\boldsymbol{\omega}_0 - \boldsymbol{\omega}_1 B - \cdots - \boldsymbol{\omega}_s B^s) B^b \mathbf{x}_t,$$

where $(\mathbf{I}, \boldsymbol{\delta}_1, \ldots, \boldsymbol{\delta}_r)$ and $(\boldsymbol{\omega}_0, \boldsymbol{\omega}_1, \ldots, \boldsymbol{\omega}_s)$ are, respectively, $k_1 \times k_1$ and $k_1 \times k_2$ matrices. This may be written succinctly as

$$\boldsymbol{\delta}(B)\mathbf{y}_t = \boldsymbol{\Omega}(B)\mathbf{x}_t$$

or

$$\mathbf{y}_t = \boldsymbol{\delta}^{-1}(B)\boldsymbol{\Omega}(B)\mathbf{x}_t. \tag{4}$$

When the input is a white noise sequence $\{\mathbf{a}_t\}$ of independent k_1-dimensional random variables distributed as $N(\mathbf{0}, \boldsymbol{\Sigma})$, the same mathematical form provides a model for k_1 dependent time series $\{\mathbf{u}_t\}$

$$\mathbf{u}_t = \boldsymbol{\phi}^{-1}(B)\boldsymbol{\theta}(B)\mathbf{a}_t, \tag{5}$$

† Although it is often done, starting the iterative model building process with a contemporaneous model of the form (2) usually makes little sense for serial data. It is usually more realistic and less a temptation to error to provide for serial dependence from the beginning.

where following tradition $\boldsymbol{\phi}(B)$ may be called the *autoregressive operator* and $\boldsymbol{\theta}(B)$ the *moving average* operator. It may be shown that models of the form of (5) can represent relationships containing *fixed deterministic* functions of time which are combinations of polynomials, exponentials, sines and cosines as well as *evolving* functions of the same form. See, for example [11].

A very general and valuable class of serial models, of the form of (1), may now be obtained by representing the outputs $\mathbf{y}_t$ as partially a response to known inputs $\mathbf{x}_t$ and partially determined by serially correlated noise $\mathbf{u}_t$. Combining (4) and (5) we then obtain

$$\mathbf{y}_t = \boldsymbol{\delta}^{-1}(B)\boldsymbol{\Omega}(B)\mathbf{x}_t + \boldsymbol{\phi}^{-1}(B)\boldsymbol{\theta}(B)\mathbf{a}_t. \tag{6}$$

This is seen to be a special case of (1) where $\mathbf{z}_t$ represents both $\mathbf{y}_t$ and $\mathbf{x}_t$.

We now present two developments of such models which have proved valuable in the analysis of actual data.

3. A CANONICAL ANALYSIS USEFUL FOR DETECTING CONTEMPORANEOUS AND OTHER RELATIONSHIPS

It helps to introduce some specific data at this point. In 1957 Quenouille studies a five-variate time series containing 82 yearly observations for 1867–1948 of hog supply, hog price, corn price, corn supply, and farm wages [4]. He made adjustments where necessary, logarithmically transformed each variable and then linearly coded the logs to produce numbers of comparable magnitude. The series are plotted in Figure 1 and identified in Table 1.

TABLE 1

Variate	Symbol	As logged and linearly coded by Quenouille	Used in our analysis
Hog supply	H_s	$10^3 \log(H_s - 7) = X_{1t}$	$z_{1t} = X_{1t}$
Hog price	H_p	$10^3 \log H_p = X_{2t}$	$z_{2t} = X_{2(t+1)}$
Corn price	R_p	$10^3 \log(R_p - 1) = X_{3t}$	$z_{3t} = X_{3t}$
Corn supply	R_s	$10^3 \log(R_s - 8) = X_{4t}$	$z_{4t} = X_{4t}$
Farm wages	W	$10^3 \log(W - 1) = X_{5t}$	$z_{5t} = X_{5(t+1)}$

(7)

Although Quenouille's motivation was otherwise, regression relationships of the form of (2) have often been sought among time series such as the hog data. Two difficulties are:

(i) while different relationships are obtained depending which variables are classified as "independent" and which as "dependent" variables, for

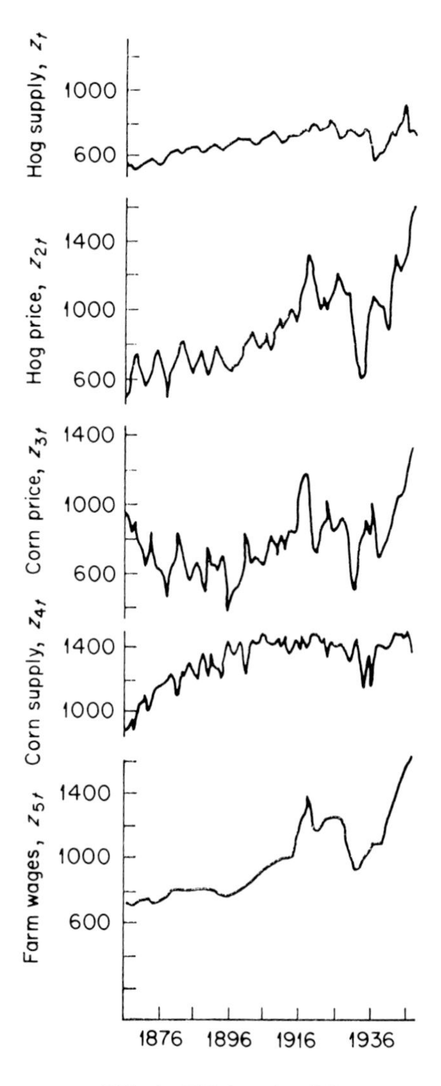

FIG. 1 U.S. hog data [7].

many examples such as this one no *natural* distinction of this kind seems appropriate;

(ii) in any case, regression methods are appropriate only on the assumption that a nonserial relationship of the form of (3) actually exists. When the errors a_t's are serially correlated, results from standard least squares fitting can be extremely misleading. See, for example, [5, 6].

In recent work [7] an attempt is made to overcome both difficulties by approaching the problem differently. The approach is first to build a multivariate stochastic model of the form (5) for the entire system and then to determine objectively whether one or more components appear to exist having the nonserial form (3). Fuller details can be obtained from the above reference. The main ideas are set out in the next section.

A Canonical Transformation of the Series $\{\mathbf{Z}_t\}$

Suppose $\{\mathbf{z}_t\}$ is a $k \times 1$ vector process and let $\mathbf{Z}_t = \mathbf{z}_t - \boldsymbol{\mu}$, where $\boldsymbol{\mu}$ is a convenient origin which is taken to be the mean if the process is stationary.

Write

$$\mathbf{Z}_t = \hat{\mathbf{Z}}_{t-1}(1) + \mathbf{a}_t, \tag{8}$$

where

$$\hat{\mathbf{Z}}_{t-1}(1) = E(\mathbf{Z}_t | \mathbf{Z}_{t-1}, \mathbf{Z}_{t-2}, \ldots) = \sum_{l=1}^{p} \boldsymbol{\pi}_l \mathbf{Z}_{t-l} \tag{9}$$

is the expected value of $\mathbf{Z}_t$, conditional on past history up to time $t-1$, $\{\mathbf{a}_t\}$ is a k-variate sequence of independent random variables distributed as $N(\mathbf{0}, \boldsymbol{\Sigma})$ independent of $\hat{\mathbf{Z}}_{t-1}(1)$.

Then we may write

$$\left(\mathbf{I} - \sum_{l=1}^{p} \boldsymbol{\pi}_l B^l\right) \mathbf{Z}_t = \mathbf{a}_t. \tag{10}$$

Note that with p taken sufficiently large this is an alternative form of (5). In what follows we shall assume that $\mathbf{Z}_t$ is stationary.

Now consider a linear combination

$$\overset{*}{\mathbf{Z}}_t = \mathbf{m}'\mathbf{Z}_t. \tag{11}$$

Then

$$\mathbf{m}'\mathbf{Z}_t = \mathbf{m}'\hat{\mathbf{Z}}_{t-1}(1) + \mathbf{m}'\mathbf{a}_t \tag{12}$$

which may be written

$$\overset{*}{Z}_t = \hat{\overset{*}{Z}}_{t-1}(1) + \overset{*}{a}_t \tag{13}$$

so that

$$\sigma^2(\overset{*}{Z}) = \sigma^2(\hat{\overset{*}{Z}}) + \sigma^2(\overset{*}{a}). \tag{14}$$

Now define a quantity λ measuring the contribution of the past values of $\overset{*}{Z}$ to the present value

$$\lambda = \sigma^2(\hat{\overset{*}{Z}})/\sigma^2(\overset{*}{Z}). \tag{15}$$

If there exists within (10), a contemporaneous linear relationship

$$\theta_1 Z_{1t} + \theta_2 Z_{2t} + \cdots + \theta_k Z_{kt} = \overset{*}{a}_t \tag{16}$$

or

$$\overset{*}{Z}_t = \boldsymbol{\theta}'\mathbf{Z}_t = \overset{*}{a}_t, \tag{17}$$

then for the particular linear combination with $\mathbf{m} = \boldsymbol{\theta}$, $\overset{*}{Z}_t = \overset{*}{a}_t$ and the value of λ is zero.

We can proceed, therefore, by finding that value of $\mathbf{m}$ which minimizes λ. Now, if $\boldsymbol{\Gamma}_j(\mathbf{Z})$ is the jth lag autocovariance matrix of $\mathbf{Z}$, then

$$\lambda = \mathbf{m}'\boldsymbol{\Gamma}_0(\hat{\mathbf{Z}})\mathbf{m}/\mathbf{m}'\boldsymbol{\Gamma}_0(\mathbf{Z})\mathbf{m}. \tag{18}$$

Thus λ is the smallest root of the determinantal equation

$$\det\{\boldsymbol{\Gamma}_0(\hat{\mathbf{Z}}) - \lambda\boldsymbol{\Gamma}_0(\mathbf{Z})\} = 0 \tag{19}$$

and $\mathbf{m}$ is the solution of

$$\{\boldsymbol{\Gamma}_0(\hat{\mathbf{Z}}) - \lambda\boldsymbol{\Gamma}_0(\mathbf{Z})\}\mathbf{m} = \mathbf{0}. \tag{20}$$

It follows that λ is the smallest eigenvalue of the matrix $\boldsymbol{\Gamma}_0^{-1}(\mathbf{Z})\boldsymbol{\Gamma}_0(\hat{\mathbf{Z}})$ and $\mathbf{m}$ is the corresponding eigenvector. Suppose the k individual eigenvalues, ordered with λ_1 the smallest, are $\lambda_1, \lambda_2, \ldots, \lambda_k$ with corresponding eigenvectors $\mathbf{m}_1', \mathbf{m}_2', \ldots, \mathbf{m}_k'$ which form the rows of a matrix $\mathbf{M}$. Then the analysis produces k linear functions

$$\overset{*}{\mathbf{Z}}_t = \mathbf{M}\mathbf{Z}_t \tag{21}$$

and we have a k-variate transformed process

$$\overset{*}{\mathbf{Z}}_t = \hat{\overset{*}{\mathbf{Z}}}_{t-1}(1) + \overset{*}{\mathbf{a}}_t, \tag{22}$$

where $\hat{\overset{*}{\mathbf{Z}}}_{t-1}(1) = \mathbf{M}\hat{\mathbf{Z}}_{t-1}(1)$, and $\overset{*}{\mathbf{a}}_t = \mathbf{M}\mathbf{a}_t$.

Suppose we choose the arbitrary scale factors of the eigenvectors so that $\boldsymbol{\Gamma}_0(\overset{*}{\mathbf{Z}}) = \mathbf{I}$, then it is easily shown that

$$\boldsymbol{\Gamma}_0(\overset{*}{\mathbf{Z}}) = \boldsymbol{\Gamma}_0(\hat{\overset{*}{\mathbf{Z}}}) + \overset{*}{\boldsymbol{\Sigma}} \tag{23}$$

gives

$$\mathbf{I} = \boldsymbol{\Lambda} + (\mathbf{I} - \boldsymbol{\Lambda}), \tag{24}$$

where $\boldsymbol{\Lambda}$ is a diagonal matrix with elements $\lambda_1, \lambda_2, \ldots, \lambda_k$.

Thus, the transformation produces k series $\overset{*}{Z}_{1t}, \overset{*}{Z}_{2t}, \ldots, \overset{*}{Z}_{kt}$ which

(i) are ordered from least predictable to most predictable,

(ii) are contemporaneously independent,

(iii) have predictable components $\{\hat{\overset{*}{Z}}_{1(t-1)}(1), \ldots, \hat{\overset{*}{Z}}_{k(t-1)}(1)\}$ which are contemporaneously independent, and

(iv) have unpredictable components $\{\overset{*}{a}_{1t}, \ldots, \overset{*}{a}_{kt}\}$ which are contemporaneously and serially independent.

Analysis of the Hog Data

This analysis may be applied to any model of the form of (10), or equivalently (5).

As an example, consider the hog data introduced earlier. This may be represented approximately by a five-variate first order† autoregressive model

$$\mathbf{Z}_t = \boldsymbol{\phi}\mathbf{Z}_{t-1} + \mathbf{a}_t. \tag{25}$$

For such a model it is easily shown that $\boldsymbol{\Lambda} = \overset{*}{\boldsymbol{\phi}}\overset{*}{\boldsymbol{\phi}}'$ and the λ's are the eigenvalues of

$$\boldsymbol{\Gamma}_0^{-1}(\mathbf{Z})\boldsymbol{\Gamma}_1'(\mathbf{Z})\boldsymbol{\Gamma}_0^{-1}(\mathbf{Z})\boldsymbol{\Gamma}_1(\mathbf{Z}). \tag{26}$$

Substituting estimates for the covariance matrices, results are obtained as follows.

The estimated eigenvalues and eigenvectors are given in Table 2. The transformed process is $\overset{*}{\mathbf{Z}}_t = \overset{*}{\boldsymbol{\phi}}\overset{*}{\mathbf{Z}}_{t-1} + \overset{*}{\mathbf{a}}_t$ with

$$\hat{\overset{*}{\boldsymbol{\phi}}} = \begin{bmatrix} .1213 & -.0778 & .0465 & -.0110 & .0113 \\ .2215 & .2766 & -.1241 & -.0309 & .0119 \\ -.0321 & .3167 & .6334 & .0444 & -.0404 \\ .0885 & -.0025 & -.0492 & .8235 & .0416 \\ -.0801 & .0378 & .0396 & -.0363 & .9360 \end{bmatrix}. \tag{27}$$

Figure 2 shows the 5 transformed series $\overset{*}{\mathbf{z}}_t = \mathbf{M}\mathbf{z}_t$.

TABLE 2

		$\hat{\mathbf{m}}_j'$					(28)
j	$\hat{\lambda}_j$	H_s	H_p	R_p	R_s	W	
1	0.0232	(1.0000	0.3876	−0.2524	−0.5896	−0.2665) × 0.0284	
2	0.1421	(0.2080	1.0000	−0.8614	−0.3382	−0.3655) × 0.0111	
3	0.5061	(0.8925	−0.6433	−0.8277	−0.4784	1.0000) × 0.0074	
4	0.6901	(−0.9358	−0.2410	−0.4391	−0.5614	1.0000) × 0.0129	
5	0.8868	(0.6687	−0.1206	−0.0134	0.0396	1.0000) × 0.0039	

† Quenouille was doubtful about the fit of such a model. However, we showed in our paper that the fit can be improved by appropriately shifting series 2 and 5 backward by one period as indicated above.

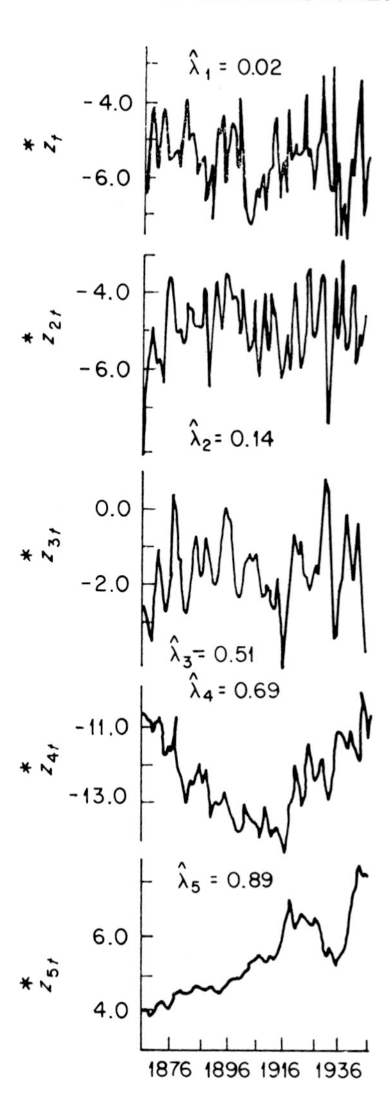

FIG. 2 The transformed series [7].

The estimated (proportional) contributions, $\hat{\overset{*}{\phi}}{}^2_{ji}$ and $1 - \hat{\lambda}_j$, of $\overset{*}{Z}_{1(t-1)}, \ldots, \overset{*}{Z}_{5(t-1)}$ and $\overset{*}{a}_{jt}$ to the variance of $\overset{*}{Z}_{jt}$ are shown in Table 3.

There is evidently very little contribution from past history to either $\overset{*}{Z}_1$ or $\overset{*}{Z}_2$ (omitting the subscript t). Thus, these two series are essentially contemporaneous and contemporaneous relationships between the original variables must lie in the hyperplane

$$\mathscr{L} = \alpha\overset{*}{Z}_1 + \beta\overset{*}{Z}_2 = c_1 Z_1 + c_2 Z_2 + c_3 Z_3 + c_4 Z_4 + c_5 Z_5. \qquad (29)$$

TABLE 3

	$\overset{*}{Z}_{1(t-1)}$	$\overset{*}{Z}_{2(t-1)}$	$\overset{*}{Z}_{3(t-1)}$	$\overset{*}{Z}_{4(t-1)}$	$\overset{*}{Z}_{5(t-1)}$	$\overset{*}{a}_{jt}$
$\overset{*}{Z}_{1t}$	0.015	0.006	0.002	0.000	0.000	0.977
$\overset{*}{Z}_{2t}$	0.049	0.077	0.015	0.001	0.000	0.858
$\overset{*}{Z}_{3t}$	0.001	0.100	0.401	0.002	0.002	0.494
$\overset{*}{Z}_{4t}$	0.008	0.000	0.002	0.678	0.002	0.310
$\overset{*}{Z}_{5t}$	0.006	0.001	0.002	0.001	0.876	0.113

(30)

In considering possibilities it is natural to seek linear combinations that are scientifically meaningful. Now the dollar value of the hogs sold is proportional to $H_p H_s$ and the dollar value of the corn needed to feed them is $R_p R_s$. If then a $\mathscr{Z}$ exists involving these dollar values it will be such that $c_1 = c_2$ and $c_3 = c_4$. By least squares or otherwise it is easy to find the linear combination for which this is nearly true. Specifically, by setting $\alpha = 30.01$ and $\beta = 59.51$ we obtain

$$\mathscr{Z}_1 = Z_1 + Z_2 - 0.78Z_3 - 0.73Z_4 - 0.48Z_5 \tag{31}$$

which is approximately *randomly distributed about a fixed mean*, with standard deviation 0.067.

Taking antilogs this implies that an index of the form

$$H_p H_s/(R_p R_s)^{0.75} W^{0.50}$$

is approximately constant, having a percentage coefficient of variation given approximately by $100 \times \log_e 10 \times \sigma(Z_1) = 16\%$. The index is thus remarkably stable when it is remembered that over the time period studied the individual elements in the index were undergoing massive changes. For example, during this period hog prices increased tenfold. The numerator is obviously a measure of return to the farmer and the denominator a partial measure of his expenditure. The analysis points to the near constancy of this relation reminding us of the "economic law" that a viable business must operate so as to balance expenditure and income.

Again if we choose

$$\alpha = 46.51 \qquad \text{and} \qquad \beta = -137.8$$

we then obtain

$$\mathscr{Z}_2 = 1.00Z_1 - 1.02Z_2 + 0.99Z_3 - 0.26Z_4 + 0.21Z_5.$$

Upon taking antilogs and arguing as before, this implies that approximately

$$H_s R_p / H_p$$

is constant with coefficient of variation about 18%. This indicates a stable relationship between hog supply and the price ratio which is in accord with the well known findings of Wallace and Bressman [8], who showed that farmer's decision on hog production is heavily influenced by the ratio of hog to corn prices.

In some applications interest will be concentrated not on the smallest eigenvalues but the largest. The corresponding linear aggregate will maximize the ratio $\sigma_{\hat{Z}^*}^2/\sigma_{Z^*}^2$ and so will be the *most predictable* component, that is, the component which depends most on the past. Such components will be of value, for example, in representing the dynamic growth patterns of the original series.

It is shown in our paper that the eigenvalues close to unity can define a *nonstationary space*, the intermediate eigenvalues a *space of serial stationary relationships*, and the near-zero eigenvalues a space of *contemporaneous relationships*.

4. INTERVENTION ANALYSIS FOR DETECTING AND ESTIMATING CHANGES IN TIME SERIES

In the analysis of economic and environmental time series data, it is frequently of interest to determine the effects of exogeneous interventions such as a change in fiscal policy or the implementation of a certain pollution control measure that occurred at some known time point. Standard statistical procedures such as the t test of mean difference before and after the intervention are often not appropriate because of (i) the dynamic characteristics of the intervention and (ii) the existence of serial correlations in the observations. It is shown in our work [9] that a transfer function model of the form (6) can be employed to study the effect of interventions. Specifically, suppose we wish to estimate simultaneously the effects of, say, k interventions on an output series y_t, we may write the model as

$$y_t = \sum_{j=1}^{k} \frac{\omega_j(B)}{\delta_j(B)} x_{jt} + \phi^{-1}(B)\theta(B)a_t, \tag{32}$$

where x_{jt} are indicator variables taking the values 1 and 0 to denote the occurrence and nonoccurrence of exogeneous interventions and $\omega_j(B)/\delta_j(B)$ represent the dynamic effects on the output. The variables x_{jt} can assume the form of a step function $x_{jt} = S_t^{(T_j)}$, where

$$S_t^{(T_j)} = \begin{cases} 0 & t < T_j \\ 1 & t \geq T_j \end{cases}$$

or a pulse function $x_{jt} = P_t^{(T_j)}$, where

$$P_t^{(T_j)} = \begin{cases} 0 & t \neq T_j \\ 1 & t = T_j \end{cases}.$$

Figure 3 shows the responses to a step and a pulse input for various transfer function models.

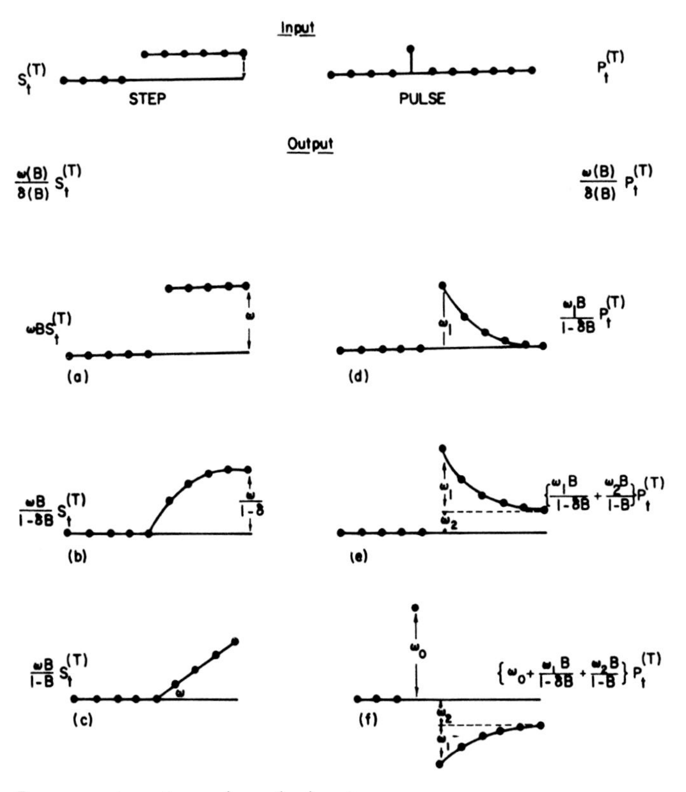

FIG. 3 Responses to a step and a pulse input
(a), (b), (c) show the response to a step input for various simple transfer function models
(d), (e), (f) show the response to a pulse for some models of interest [9].

Analysis of the Los Angeles Carbon Monoxide Data

To illustrate, data are used from a study made with the late Walter Hamming of air pollution in Los Angeles County [10]. Figure 4 shows monthly averages of CO from 1961 to 1972 for downtown Los Angeles. Because of meteorological effects the series are highly seasonal. An important problem is to estimate the effects of the control devices installed in all new model cars

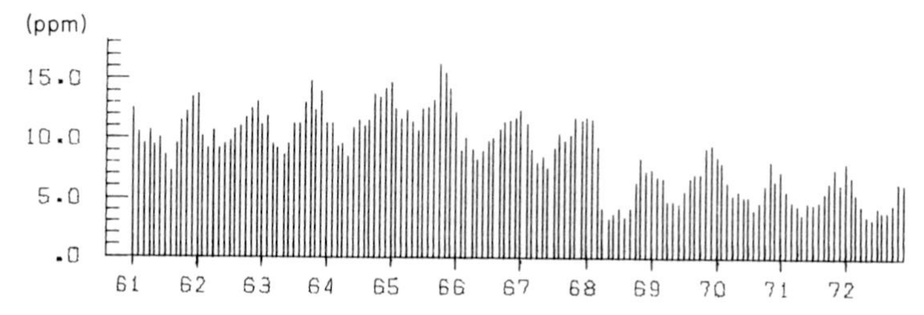

FIG. 4 Monthly averages of CO at downtown L.A. (Jan. 1961–Dec. 1972) [10].

from 1966 onwards. On casual inspection of the data it is easy to be persuaded that a dramatic improvement began with the initiation of the control measures at the beginning of 1966. However, it came to light on inquiry that the method of analysis for CO had been changed in April 1968. It is important, therefore, to separately estimate and distinguish (i) effects associated with change in the method of CO measurement from (ii) effects associated with the controls. In our work after some preliminary analysis of the data a model of the general form of (6) was postulated as follows:

$$y_t = \omega_0 x_{1t} + \omega_1 \frac{x_{2t}}{1 - B^{12}} + \omega_2 \frac{x_{3t}}{1 - B^{12}} + \frac{(1 - \theta_1 B)(1 - \theta_2 B^{12})}{(1 - \phi B)(1 - B^{12})} a_t, \tag{33}$$

where

$$x_{1t} = \begin{cases} 0 & t < \text{April 1968} \\ 1 & t \geq \text{April 1968} \end{cases}$$

$$x_{2t} = \begin{cases} 1 & t \geq \text{January 1966} \\ 0 & \text{otherwise} \end{cases}$$

$$x_{3t} = \begin{cases} 1 & t \leq \text{December 1965} \\ 0 & \text{otherwise.} \end{cases}$$

The indicator variable x_{1t} allows the possibility of a step change of size ω_0 in April 1968 due to the change in the analytical method, while x_{2t}, x_{3t} allow for yearly "staircase functions" with step size (trend) ω_1 before January 1966, and ω_2 after. The last term in (33) is a time series model capable of representing evolving seasonal and nonseasonal time dependent relationships in the data.

Table 4 shows the estimation results for seven different locations in Los Angeles County. Diagnostic checks of residuals failed to suggest model inadequacies, and we believe the fitted relationships provide a satisfactory basis for tentative conclusions. It appears that while there is evidence of some improvement due to control measures, the most dramatic change is an artifact arising from the change in analytical method.

TABLE 4

Location	Measurement effect April 1968	Trend before 1966	Trend since 1966	Trend difference	Noise model parameters		
	$\hat{\omega}_0$	$\hat{\omega}_1$	$\hat{\omega}_2$	$\hat{\omega}_2 - \hat{\omega}_1$	$\hat{\phi}$	$\hat{\theta}_1$	$\hat{\theta}_2$
Azusa	−6.50	−0.05	0.29	0.34	0.84	0.42	0.97
	(0.70)	(0.13)	(0.15)	(0.23)	(0.07)	(0.12)	(0.04)
Pasadena	−5.44	−0.29	−0.04	0.25	0.60	0.05	0.65
	(0.93)	(0.20)	(0.23)	(0.33)	(0.12)	(0.15)	(0.07)
Burbank	−5.72	0.40	−0.33	−0.73	0.78	0.22	0.79
	(1.10)	(0.25)	(0.26)	(0.40)	(0.08)	(0.12)	(0.05)
West	−5.17	0.11	−0.43	−0.54	0.79	0.28	0.82
Los Angeles	(0.87)	(0.19)	(0.20)	(0.31)	(0.07)	(0.11)	(0.05)
Downtown	−4.32	0.09	−0.28	−0.37	0.71	0.20	0.83
Los Angeles	(0.73)	(0.12)	(0.16)	(0.23)	(0.10)	(0.13)	(0.05)
Lennox	−5.08	0.51	−0.36	−0.87	0.79	0.19	0.59
	(0.85)	(0.55)	(0.25)	(0.61)	(0.08)	(0.13)	(0.08)
Long Beach	−5.29	0.40	−0.45	−0.85	0.77	0.27	0.81
	(0.84)	(0.17)	(0.19)	(0.29)	(0.09)	(0.13)	(0.06)

Intervention analysis in which dynamic effects of intervention are estimated against a background of noise which is usually highly serially dependent and often seasonal, have very many applications including the analysis of changes in ecological, economic, psychological, physiological, and business systems.

This paper ends with a brief summary of some very recent work at Madison, in the process of publication, which has been partially supported by the Air Force Office of Scientific Research and the Army Research Office, Durham.

Deterministic and Stochastic Models

Systems sometimes include a fixed deterministic component as well as an evolving stochastic one. Consideration is given as to how these elements may be detected and separated [11].

When is Exponential Smoothing Optimal?

Exponential smoothing (including higher order smoothing and the use of functions fitted by exponentially discounted least squares) has been a widely advocated method for forecasting. A recent study shows that these methods are optimal only for a very limited class of models of the general form

$$\phi(B)z_t = \phi(\beta B)a_t,$$

where β is the smoothing constant. There seems to be no reason why models occurring in nature should be of this particular form [12].

Analysis of Stochastic Models with Nonnormal Shocks

Recently it has been argued that, for stock price series, the generating shocks were nonnormal. This has led to speculation that the shocks have distributions belonging to an exotic class of functions known as stable distributions. The present study does not support this speculation and shows what effects nonnormality of a less bizarre form would have on forecasting and estimation [13].

Estimation with Closed Loop Data

When the system sampler is operating under feedback control special difficulties can occur in estimation of parameters. Methods of overcoming these difficulties are discussed [14].

Outliers in Time Series

Bayesian analyses are worked out providing methods for analyzing time series with outliers [15].

Temporal Aggregation of Time Series

Suppose that on a daily basis, a dynamic relationship of the form (6) exists between an input x and an output y. In many problems data are often available only in terms of some time aggregates such as a week or a month. It is then of interest to study the effect of aggregation on the relationship between x and y. General methods of determining the stochastic structure of the aggregates and of assessing the effects on estimation and forecasting are discussed [16].

Smoothing and Seasonal Adjustment of Time Series

It is often supposed that an observed economic series y_t can be written as

$$y_t = S_t + T_t + e_t,$$

where S_t is a seasonal component, T_t a trend component, and e_t a noise component. Specific moving average filters are then employed to estimate these components from the observed data $\mathbf{y}$—see e.g. the widely used X-11 procedure developed by the Census Bureau. In a recent paper [17], it has been shown that the X-11 procedure is optimal only when S_t and T_t follow particular models of the form (5). More recently, methods are developed to estimate S_t and T_t based on the overall model for y_t which can be built from the observed data [18].

Estimation of Parameters of Multiple Time Series

Suppose k time series $\mathbf{y}_t$ follow a multivariate autoregressive moving average model of the form (5). Various difficulties arise in estimating the parameters involved, especially when one or more of the zeros of det $\boldsymbol{\theta}(B)$ lie near or on the unit circle. An efficient algorithm has recently been developed to compute the exact likelihood function of the parameters. It is shown that this method will provide better estimates than a number of approximate methods currently in use [19].

REFERENCES

[1] Box, G. E. P. and Jenkins, G. M. (1962). Some statistical aspects of adaptive optimization and control. *J. Roy. Statist. Soc. Ser. B* **24** 297–343.

[2] Box, G. E. P. and Jenkins, G. M. (1963). Further contributions to adaptive quality control: simultaneous estimation of dynamics: non-zero cost. *Bull. Intl. Statist. Inst.* **34** 943–970.

[3] Box, G. E. P. and Jenkins, G. M. (1970). *Time Series Analysis Forecasting and Control.* Holden-Day, San Francisco.

[4] Quenouille, M. H. (1975). *Analysis of Multiple Time Series.* Hafner, New York.

[5] Coen, P. G., Gomme, E. D., and Kendall, M. G. (1969). Lagged relationships in economic forecasting. *J. Roy. Statist. Soc. Ser. A* **132** 133–163.

[6] Box, G. E. P. and Newbold, P. (1971). Some comments on a paper of Coen, Gomme, and Kendall. *J. Roy. Statist. Soc. Ser. A* **134** 229–240.

[7] Box, G. E. P. and Tiao, G. C. (1975). A canonical analysis of multiple time series. Tech. Rep. 428, Dept. of Statistics, University of Wisconsin-Madison. To appear in *Biometrika*, 1977.

[8] Wallace, H. A. and Bressman, E. N. (1949). *Corn and Corn Growing.* 5th ed. Wiley, New York.

[9] Box, G. E. P. and Tiao, G. C. (1975). Intervention analysis with applications to economic and environmental problems. *J. Amer. Statist. Assoc.* **70** 70–79.

[10] Tiao, G. C., Box, G. E. P., and Hamming, W. J. (1975). A statistical analysis of the Los Angeles ambient carbon monoxide data 1955–1972. *J. Air Pollution Control Assoc.* **25** 1129–1136.

[11] Box, G. E. P. and Abraham, B. (1975). Linear models, time series and outliers—3: stochastic difference equation models, Tech. Report 438, Dept. of Statistics, University of Wisconsin-Madison. Accepted for publication in *Appl. Statist.*

[12] Ledolter, J. and Box, G. E. P. (1975). Topics in time series analysis II. When are exponential smoothing forecast procedures optimal?, Tech. Report 447, Dept. of Statistics, University of Wisconsin-Madison.

[13] Ledolter, J. and Box, G. E. P. (1976). Topics in time series analysis III. ARIMA time series models with non-Normal shocks. Tech. Report 448, Dept. of Statistics, University of Wisconsin-Madison.

[14] Box, G. E. P. and MacGregor, J. F. (1976). Parameter estimation with closed-loop operating data. *Technometrics* **18** 371–380.

[15] Box, G. E. P. and Abraham, B. (1975). Linear models, time series and outliers—5: outliers in time series. Tech. Report 440, Dept. of Statistics, University of Wisconsin-Madison.

[16] Tiao, G. C., and Wei, W. S. (1976). Effect of temporal aggregation on the dynamic relationship of two time series variables. *Biometrika* **63** 513–524.
University of Wisconsin-Madison. To appear in *Biometrika*, 1977.

[17] Cleveland, W. P. and Tiao, G. C. (1976). Decomposition of seasonal time series: a model for the Census X-11 program, *J. Amer. Statist. Assoc.* **71** 581–587.

[18] Box, G. E. P., Hillmer, S. C., and Tiao, G. C. (1976). Analysis and modeling of seasonal time series. Tech. Report 473, Dept. of Statistics, University of Wisconsin-Madison.

[19] Hillmer, S. C. (1976). *Time Series: Estimation, Smoothing, and Seasonally Adjusting.* Ph.D. Thesis, University of Wisconsin-Madison.

Part IV

OUTLIERS, ROBUSTNESS, AND CENSORING

Testing for Outliers in Linear Regression

James E. Gentle

IOWA STATE UNIVERSITY

The problem of outliers in the regression model is considered. For the case of one outlier at most, the use of the maximum absolute studentized residual, R_n, for identification of the outlier has been suggested by a number of authors. Simulation studies of the power of a conservative test based on R_n for identifying single outliers in regression models with one, two, and three independent variables are reported. The case of multiple outliers is also considered and techniques for their identification are discussed. A simulation study of a sequential procedure for handling two outliers is reported.

1. INTRODUCTION

As the computational aspects of data analysis become more automatic through computer processing with canned software, the need for objective techniques for monitoring data and identifying outliers becomes ever greater. The analyst who sees just the original data and then only the summary statistics may be ill-prepared to form valid and meaningful conclusions about the problem at hand, unless the summary statistics also in some way indicate the validity of the data and the appropriateness of the model for the problem.

An important activity in data analysis, in addition to those of selecting a model form and estimating its parameters, is the identification of a set of

Key words and phrases Linear regression, outliers, studentized residuals.

data representing a sample from the population being modeled. The questions of valid data and correct model are intimately connected; suspicious data cast suspicion on the model. In the following, however, we consider the problem of data editing separately from the problem of model building. This is a reasonable approach if the number of suspicious data or outliers is quite small in relation to the total number of observations.

Consider the regression model

$$\mathbf{y} = \mathbf{X}\boldsymbol{\beta} + \boldsymbol{\varepsilon}, \tag{1.1}$$

where $\mathbf{y}$ is the $n \times 1$ vector of observations; $\mathbf{X}$, an $n \times m$ full-rank matrix of constants; $\boldsymbol{\beta}$, an $m \times 1$ vector of unknown parameters; and $\boldsymbol{\varepsilon}$, an $n \times 1$ vector of homoscedastic, independent, normally distributed errors with zero expectation. (It should be noted that the value of m includes the intercept if one is included in the model.)

Letting

$$\mathbf{A} = \mathbf{I} - \mathbf{X}(\mathbf{X}'\mathbf{X})^{-1}\mathbf{X}' = [a_{ij}] \tag{1.2}$$

and $\mathbf{e} = \mathbf{y} - \mathbf{X}\hat{\boldsymbol{\beta}}$, where $\hat{\boldsymbol{\beta}}$ is the usual least squares estimator, form the studentized residuals,

$$r_i = e_i/[a_{ii}\,\mathbf{e}'\mathbf{e}/(n-m)]^{1/2}. \tag{1.3}$$

A large absolute value of r_i may indicate that the ith observation is spurious. (Of course it may indicate the absence of an important independent variable from the model or the failure of a model assumption to be satisfied.) For identifying a single unspecified outlier, a test statistic would be

$$R_n = \max |r_i|. \tag{1.4}$$

Tests based on this criterion were suggested by Srikantan (1961) and a number of later authors.

2. ON THE DISTRIBUTION OF R_n

Since the covariances of the r_i depend on $\mathbf{X}$, the distribution of R_n likewise depends on $\mathbf{X}$. For this reason it is not practical to provide tables of the critical values of R_n. Hartley and Gentle (1975) give a numerical procedure, which could be implemented in a computer program for regression, for approximating the percentage points of R_n for a given $\mathbf{X}$.

Tietjen, Moore, and Beckman (1973) report on a Monte Carlo study of the distribution of R_n for $m = 2$, giving tables of the critical values averaged over several patterns of $\mathbf{X}$ matrices. Their study indicates, at least for $m = 2$, that the distribution of R_n changes only very slightly as $\mathbf{X}$ varies through moderately well-behaved patterns.

If the effect of $\mathbf{X}$ is ignored, percentage points of R_n can be approximated by means of Bonferroni inequalities and the relation of an arbitrary r_i to a t variable. Srikantan (1961), among others, shows this relationship (which is free of $\mathbf{X}$) to be given by

$$t^2 = [(n - m - 1)r_i^2/(n - m - r_i^2)], \tag{2.1}$$

where t has the Student's t distribution with $n - m - 1$ degrees of freedom (d.f.).

Since for any c, $P(r_i \geq c) \geq P(r_i = R_n)P(R_n \geq c)$, for given $\alpha = P(R_n \geq c)$, an over-approximation to c is given by c_0, where

$$P[t^2 \geq (n - m - 1)c_0^2/(n - m - c_0^2)] = \alpha/n, \tag{2.2}$$

and t is a t-variable with $n - m - 1$ d.f. The approximations are better in the upper tail of the distribution. In certain cases, depending on $\mathbf{X}$, the values obtained in this manner are exact. For example, from a result of Srikantan, for x_i uniformly spaced between 0 and 1 in a simple linear regression model, as studied by Tietjen *et al.* (1973, their pattern "D"), exact values are obtained when $\alpha = 0.05$ for n up to 9.

Lund (1975) gives a limited tabulation of over-approximations to the 0.10, 0.05, and 0.01 critical values of R_n in his Table 1. An indication of the extent to which the values of Lund's Table 1 differ from the true values for certain types of $\mathbf{X}$ may be obtained by comparing the points for $m = 2$ with the corresponding values from the simulation study of Tietjen *et al.* The values generally differ by less than 0.01.

For patterns of independent variables that induce large correlations among the r_i, such as in certain design models, the critical values of Lund's Table 1 are even more conservative (Srikantan, 1961). Large correlations among the r_i may be found in regression models when there are large correlations among the independent variables. To obtain an indication of the effect of collinearity on the distribution of R_n, the model $y_i = \beta_0 + \beta_1 x_i + \beta_2 x_i^2 + \varepsilon_i$, with $n = 20$, and $\varepsilon_i \sim N(0, 1)$ independently, was considered. Five sets of x's were drawn from the uniform distribution with limits 0 and 1, and each set was used in 200 simulations of the model (in the same manner as the Monte Carlo study described in Section 4). The empirical significance level of the 0.05 nominal critical value of Lund's Table 1, 2.76, was found to be 0.026. These samples had maximum absolute correlations between residuals of 0.300.

3. EQUIVALENT CRITERIA FOR SINGLE OUTLIERS

Andrews (1971) developed a test for outliers based on a projection of

$$\mathbf{d} = \mathbf{e}/(\mathbf{e}'\mathbf{e})^{1/2} \tag{3.1}$$

onto the columns of $\mathbf{A}$ (from 1.2). Letting $\mathbf{d}_{\pi i}$ denote the projection of $\mathbf{d}$ onto the ith column and $\mathbf{d}_{\perp i}$, the orthogonal complement, $\mathbf{d} - \mathbf{d}_{\pi i}$, Andrews' criterion for a single outlier is $\min \|\mathbf{d}_{\perp i}\|$. Since

$$\|\mathbf{d}_{\perp i}\|^2 = 1 - \|\mathbf{d}_{\pi i}\|^2 \qquad \text{and} \qquad \mathbf{d}_{\pi i} = [\mathbf{y}'\mathbf{A}\mathbf{J}_i / \mathbf{J}_i'\mathbf{J}_i(\mathbf{e}'\mathbf{e})^{1/2}]\mathbf{J}_i,$$

where $\mathbf{J}_i$ is the ith column of $\mathbf{A}$,

$$\|\mathbf{d}_{\perp i}\|^2 = 1 - (e_i^2/a_{ii}\,\mathbf{e}'\mathbf{e}) = 1 - [r_i^2/(n-m)]. \tag{3.2}$$

Hence, Andrews' single-outlier criterion is equivalent to Srikantan's R_n in (1.4).

Ellenberg (1976) discusses test statistics proposed by Mickey, Dunn, and Clark (1967) and by Snedecor and Cochran (1968) and shows that these procedures for single outliers are also equivalent to a procedure based on R_n.

Gentleman and Wilk (1975) employ a partitioning of the data set and regression statistics on various reduced data sets to identify outliers. For a single outlier Gentleman and Wilk use

$$Q_1^* = \max\{\mathbf{e}'\mathbf{e} - \tilde{\mathbf{e}}_i'\tilde{\mathbf{e}}_i\},$$

where $\mathbf{e}$ is the residual vector from a least squares fit of the full data set and $\tilde{\mathbf{e}}_i$ is the residual vector from a fit with the ith observation deleted. This is the numerator of the test statistic proposed by Mickey *et al.* (1967); and, hence, is essentially equivalent to R_n.

4. PERFORMANCE OF PROCEDURE FOR IDENTIFYING SINGLE OUTLIER

Tietjen *et al.* (1973) present a limited Monte Carlo study of the performance of a test based on their simulated critical values of R_n for a single outlier in a simple linear regression model. No details of their Monte Carlo power study are given. Again for a simple linear regression, Ellenberg (1976) describes a Monte Carlo study of the performance of the conservative test using the nominal critical values of R_n, as in (2.2). Ellenberg's results indicate that the power of this conservative test depends to some extent on the position of the outlying observation in relation to the values of the independent variable x.

In the present study, the performance of the conservative procedure was investigated by simulating regressions with models having a constant term and one, two, or three regressors, i.e., models with $m = 2, 3, 4$.

The independent variables (x's) were generated from a uniform (0, 1) population and the dependent variables were formed by a linear combination of the independent variables plus an error generated from a normal

(0, 1) population. The residuals are unaffected by the particular values of $\boldsymbol{\beta}$ used. The correlations of the residuals resulting from the values generated for the regressors were generally fairly small; hence, the performance of the test procedure would be expected to be slightly better in these cases than in the presence of higher degrees of multicollinearity, but slightly worse than in the case of orthogonal data. A multiplicative congruential generator with a large multiplier was used to generate the uniform variates; a modified Box–Muller transformation was used to form the normal variates. [The generator of the uniform random numbers used a multiplier of 16,807 and a modulus of $2^{31} - 1$ on an IBM 360/65. Tests on this particular generator are reported in Learmonth and Lewis (1974). The modification of the Box–Muller transformation that was used is described in Ahrens and Dieter (1972).] Five subsequences of 100 variates of the stream of errors were tested for normality by both a χ^2 test and a Kolmogorov–Smirnov test. In no case did either test indicate rejection of the hypothesis of normality at the 0.05 level.

Outliers were created by adding an amount λ to the dependent variable. Table 1 shows the proportion of correct identifications when only one outlier was present. Each entry in the table is the result of 4000 simulations, for

TABLE 1

Proportion of Correct Identifications of Single Outlier in
$Y_i = \beta_0 + \beta_1 x_{1i} + \cdots + \beta_p x_{pi} + \varepsilon_i$,
with λ Added to a Single Observation

p	1		2		3	
n	$\lambda = 2$	$\lambda = 5$	$\lambda = 2$	$\lambda = 5$	$\lambda = 2$	$\lambda = 5$
10	0.091	0.649	0.064	0.529	0.034	0.322
20	0.089	0.839	0.070	0.792	0.063	0.786
40	0.065	0.889	0.053	0.902	0.062	0.882
60	0.059	0.919	0.055	0.914	0.056	0.912

ten sets of independent variables, each set being used 400 times. The criterion is not very effective for identifying outliers whose means are only two standard deviations from the mean specified by the model; further, the power of the test for $\lambda = 2$ appears to decrease as the sample size increases. This same anomaly occurred in the Monte Carlo study for $m = 2$ by Tietjen *et al.* For $\lambda = 5$ the procedure is quite effective, and apparently increasing in power with increasing sample size, and generally decreasing in power with increasing number of independent variables.

In the presence of more than one outlier the effectiveness of R_n for identifying any of the outliers is degraded, due to the large contribution to **e**′**e** by the outliers when fitting a mean in a single sample.

Extreme values of the regressors may also degrade the performance of the outlier test, since the contribution to the least squares fit by the individual observations varies depending on the distances of the independent variables from their respective means. In Ellenberg's (1976) Monte Carlo study of a simple linear regression model, outliers were detected less often when they corresponded to x values far from the x mean. The inordinate effect on the least squares estimates by observations corresponding to x values far from the x mean can also result in suspiciously large residuals for valid observations. The artificial data of Table 2 illustrate this for the simple linear regression model. The presence of the outlier, observation 10, causes the studentized residual associated with observation 1 to exceed the 0.01 critical value (2.55 from Lund (1975), Table 1). Observation 1, however, appears quite valid upon removal of the true outlier. Since the maximum absolute studentized residual corresponds to that observation whose removal would result in maximum decrease in the residual sum of squares (see Section 3), use of R_n in situations similar to the example of Table 2 would not lead to identification of the wrong observation as the outlier. (A data set similar to that of Table 2 obviously would raise for the data analyst further questions, not to be explored here.)

TABLE 2

Artificial Data from $y_i = 1 + x_i + \varepsilon_i$, with Outlier (Observation 10) Causing Large Studentized Residual in Valid Datum (Observation 1)

Observation number	X	Y	Residual	Studentized residual
1	0.00000	1.00000	7.24452	2.60556
2	20.00000	21.00000	−0.43995	−0.11418
3	21.00000	22.00000	−0.82417	−0.21328
4	22.00000	23.00000	−1.20840	−0.31209
5	23.00000	24.00000	−1.59262	−0.41086
6	24.00000	25.00000	−1.97684	−0.50985
7	25.00000	26.00000	−2.36107	−0.60932
8	26.00000	27.00000	−2.74529	−0.70951
9	27.00000	28.00000	−3.12952	−0.81070
10	50.00000	70.00000	7.03334	2.82843

5. MULTIPLE OUTLIER PROCEDURES

Andrews (1971) shows how to apply his test for outliers to situations where possibly more than one outlier is present. If two outliers were suspected, for example, projections of the residual vector (3.1) onto hyperplanes generated by all combinations of pairs of columns of **A** (1.2) are considered. Andrews' test statistic is the minimum of the Euclidean norms of the orthogonal complements of the projections.

Mickey *et al.* (1967) recommend a forward selection procedure for the identification of multiple outliers. Observations are successively deleted (by adding dummy variables) in such a manner as to effect the maximum decrease in the residual sum of squares at each stage. The F distribution (as in forward selection regression) is used to indicate how far to proceed.

For identifying a maximum of k outliers, Gentleman and Wilk (1975) suggest performing regressions using all subsets of the original data with k observations removed. The criterion for k outliers is based on

$$Q_k^* = \max\{\mathbf{e}'\mathbf{e} - \tilde{\mathbf{e}}_i'\tilde{\mathbf{e}}_i\},$$

where $\tilde{\mathbf{e}}_i$ is the residual vector from the regression deleting the ith set of k observations. If Q_k^* does not appear aberrant, k is decreased by 1 and the procedure is repeated until $k = 1$.

A simple procedure can also be based on R_n. The method is to calculate R_n for the full data set and identify the corresponding observation, say the ith. If R_n exceeds the critical value for the appropriate significance level and sample size, remove the ith observation and repeat the calculations, obtaining R_{n-1} and the corresponding observation, say the jth. If R_{n-1} exceeds the critical value for the appropriate significance level and sample size $n - 1$, remove the jth observation from the data set. Repeat the calculations on the reduced data set of size $n - 2$. If a third outlier, say the kth observation, is identified remove it and then restore the ith observation to the data set. Perform the calculations on the data set of size $n - 2$ containing the ith observation but not the jth or kth. The R_{n-2} obtained at this stage is checked as above and the process continued. Effectively, all combinations of the suspicious data in and out of the sample are being explored in this procedure.

How far this procedure should be carried depends on several things, such as the initial sample size, the degree of belief in the validity of the model, etc. As more outliers are identified the model itself must undergo further inspection.

As noted previously, this procedure may not be powerful even at the first stage due to masking. A Monte Carlo study was carried out (as described in Section 4) to investigate the performance of this procedure. Table 3 shows

TABLE 3

Proportion of Correct Identifications of Outliers in
$y_i = \beta_0 + \beta_1 x_{1i} + \cdots + \beta_p x_{pi} + \varepsilon_i$,
with $\lambda = 5$ *Added to Two Observations*

	$p = 1$			$p = 2$		
n	O_1	O_2	Both	O_1	O_2	Both
20	0.539	0.541	0.284	0.405	0.413	0.218
40	0.839	0.838	0.743	0.813	0.816	0.699
60	0.889	0.892	0.822	0.870	0.878	0.806

the proportions of correct identification of one or both of two outliers. Each outlier was formed by adding five standard deviations to the dependent variable. The columns "O_1" and "O_2" are the proportions of times each outlier was identified. Since both outliers were formed in the same manner, differences in the columns "O_1" and "O_2" for given m and n are due to sampling variation. The column "Both" is the proportion of times both outliers were identified. The results are based on 4000 simulations as above.

From Table 3 it is seen that for sample sizes of 40 and 60 the performance of the procedure was almost as good for identifying each of the outliers as it was for identifying a single outlier, as in Table 1.

The performance of the procedure would not be expected to be as good in the presence of multicollinearity since the test is more conservative in this case.

6. EXAMPLE

A data set was chosen from Draper and Smith (1966, p. 8) and modified so as to illustrate the behavior of the studentized residuals, as well as the testing procedure. The second and third columns of Table 4 give the data used. The y values in observations numbers 7 and 11 have been perturbed. The equation $y_i = b_0 + b_1 x_i$ is fitted by least squares and the residuals for the full data set are shown in the fourth column. The residual greatest in absolute value is seen to occur for observation number 11; however, since x_{11} is closer to $\bar{x}$, 52.6, than is x_7, the variance associated with e_{11} is greater than that of e_7; and, consequently, the studentized residual (shown in column 5) for observation number 7 is greater in absolute value. The value of R_n is 2.886, which from Lund's Table 1 exceeds the conservative 0.05 critical value, 2.88.

TABLE 4

Data from Draper and Smith[a] Modified by Inserting Spurious Data for Observations 7 and 11

			Full data set		After removal of no. 7		After removal of nos. 7 and 11	
Obs. no.	X	Y	Residual	Studentized residual	Residual	Studentized residual	Residual	Studentized residual
1	35.3	10.98	0.274	0.198	0.331	0.292	0.09580	0.12634
2	29.7	11.13	−0.061	−0.045	0.070	0.063	−0.18394	−0.24706
3	30.8	12.51	1.414	1.032	1.531	1.370	1.28047	1.71291
4	58.8	8.40	0.266	−0.188	−0.526	−0.456	−0.68085	−0.88146
5	61.4	9.27	0.830	0.588	0.534	0.465	0.38867	0.50489
6	71.3	8.73	1.149	0.831	0.720	0.642	0.60837	0.80900
7	74.4	3.36	−3.952	−2.886*				
8	76.7	8.50	1.388	1.022	0.886	0.806	0.79276	1.07591
9	70.7	7.82	0.187	0.135	−0.234	−0.208	−0.34767	−0.46144
10	57.5	9.14	0.361	0.255	0.118	0.103	−0.04061	−0.05251
11	46.4	5.74	−4.002	−2.828	−4.096	−3.545*		
12	28.9	12.19	0.929	0.683	1.071	0.965	0.81467	1.09763
13	28.1	11.88	0.549	0.405	0.703	0.635	0.44328	0.59918
14	39.1	9.57	−0.806	−0.576	−0.800	−0.700	−1.02260	−1.33639
15	46.8	10.94	1.233	0.871	1.134	0.981	0.93829	1.21272
16	48.5	9.58	0.020	0.014	−0.102	−0.088	−0.29126	−0.37601
17	59.3	10.09	1.468	1.038	1.200	1.041	1.04752	1.35693
18	70.0	8.11	0.416	0.300	0.005	0.004	−0.11139	−0.14752
19	70.0	6.83	−0.863	−0.622	−1.275	−1.131	−1.39139	−1.84273
20	74.5	8.88	1.577	1.152	1.105	0.996	1.00394	1.35029
21	72.1	7.68	0.169	0.122	−0.271	−0.242	−0.38023	−0.50697
22	58.1	8.47	−0.257	−0.181	−0.508	−0.440	−0.66457	−0.85976
23	44.6	8.86	−1.038	−0.735	−1.108	−0.960	−1.31054	−1.69753
24	33.4	10.36	−0.511	−0.370	−0.429	−0.380	−0.67001	−0.88855
25	28.6	11.08	−0.207	−0.153	−0.061	−0.055	−0.31835	−0.42944

[a] Draper and Smith (1966, p. 8).

Observation number 7 is removed from the data set and the analysis is repeated. Column 7 shows that R_{24} is 3.545 and corresponds to observation number 11, which may then also be identified as an outlier. A final analysis on the 23 remaining data points completes the procedure.

This example illustrates the variable effect of studentization (the larger residual does not necessarily give the larger studentized residual) and the effect of masking (the studentized residual of observation 11 did not become significant until the removal of observation 7).

7. FURTHER COMMENTS

Outlier identification is only one aspect of data analysis; another very broad aspect is the treatment of those observations identified as outliers. The treatment of the data so classified need not necessarily be permanent removal from the data set. How the outliers are to be treated by the researcher after identification depends on his knowledge of the data gathering and transcribing process and on his own philosophy. Anscombe (1960), in an excellent discussion of the outlier problem, compared rules for treatment of outliers to insurance policies, with the fractional increase in variance of the estimator following a rejection of valid data being the "premium" paid for the "protection" measured by the fractional decrease in mean square error resulting from special treatment of invalid data.

Since observations can only be outliers in relation to a given model, the construction and validation of the model become related to identification of outliers.

Consideration must also be given to the pattern of the independent variables. The possibility of the appearance of valid data as outliers is enhanced by extreme patterns of the independent variables. The residuals of data points lying far from the centroid of the independent variables have relatively small variances since these points have a large influence on the orientation of the regression plane.

In outlier studies, as in any analysis, significance levels should not be slavishly observed; they should rather be viewed as guideposts. Although the previous discussion and examples have used 0.05 nominal levels, many workers in the field would recommend use of 0.01 nominal levels for outlier identification, particularly if the outliers are to be removed from the data set. In any event, the analyst must be cognizant of the broader aspects of data analysis and avail himself of a variety of analytic tools, such as, perhaps, procedures robust to the presence of outliers.

REFERENCES

AHRENS, J. H. and DIETER, V. (1972). Computer methods for sampling from the exponential and normal distributions. *Comm. ACM* **15** 873–882.

ANDREWS, D. F. (1971). Significance tests based on residuals. *Biometrika* **58** 139–148.

ANSCOMBE, F. J. (1960). Rejection of outliers. *Technometrics* **2** 123–147.

DRAPER, N. R. and SMITH, H. (1966). *Applied Regression Analysis*. Wiley, New York.

ELLENBERG, J. H. (1976). Testing for a single outlier from a general linear regression. *Biometrics* **32** 637–638.

GENTLEMAN, J. F. and WILK, M. B. (1975). Detecting outliers II. *Biometrics* **31** 387–410.

HARTLEY, H. O. and GENTLE, J. E. (1975). Data monitoring criteria for linear models. In *A Survey of Statistical Design and Linear Models*. 197–207 (J. N. Srivastava, ed.) North-Holland, Amsterdam.

Learmonth, G. P. and Lewis, P. A. W. (1974). Statistical tests of some widely used and recently proposed uniform random number generators. *Proc. Comput. Sci. and Statist.: Seventh Annual Symp. on the Interface, Iowa State University, Ames, Iowa, 163–171.*

LUND, R. E. (1975). Tables for an approximate test for outliers in linear regression. *Technometrics* **17** 473–476.

MICKEY, M. R., DUNN, O. J., and CLARK, V. (1967). Note on the use of stepwise regression in detecting outliers. *Comput. and Biomedical Res.* **1** 105–111.

SNEDECOR, G. W. and COCHRAN, W. G. (1968). *Statistical Methods.* 6th Ed. Iowa State University Press, Ames.

SRIKANTAN, K. S. (1961). Testing for the single outlier in a regression model. *Sankhya Ser. A* **23** 251–260.

TIETJEN, G. L., MOORE, R. H., and BECKMAN, R. J. (1973). Testing for a single outlier in simple linear regression. *Technometrics* **15** 717–721.

Robustness of Location Estimators in the Presence of an Outlier

H. A. David *V. S. Shu*

IOWA STATE UNIVERSITY

The bias and mean square error of various location estimators, expressible as linear functions of order statistics, are studied when an unidentified single outlier is present in a sample of size n. Specific attention is paid to the cases when the outlier comes from a population differing from the target population either in location or scale. When, in addition, the target population is normal, exact numerical results have been obtained for $n = 5, 10, 20$ and are presented here for $n = 10$. The estimators included are the mean, median, trimmed means, Winsorized means, linearly weighted means, and Gastwirth mean.

1. INTRODUCTION AND SUMMARY

There has been much interest in recent years in the robust estimation of the location parameter of symmetric distributions. Thus Crow and Siddiqui [2] examine, for various symmetric populations, the efficiency of estimators, such as the median, trimmed means, and Winsorized means, expressible as linear functions of order statistics. A much larger class of estimators, including many that are adaptive (i.e. adapt themselves to the special features of a particular sample) is considered in the valuable Princeton study [1]. While much other work on robustness has emphasized

Work supported by the U.S. Army Research Office.

Key words and phrases Order statistics, trimmed means, Winsorized means, slippage, mixtures, contaminated normal, recurrence relation for order statistics.

asymptotic results, the two references cited are very much concerned with the small-sample situation. Crow and Siddiqui base their work on the known means, variances, and covariances of order statistics for the populations considered. In [1] Monte Carlo methods are used to study the often complex estimators included. Attention is focused on symmetric populations, but some results are also obtained for an outlier situation with the observations coming from two normal distributions differing in location. Theoretical and numerical methods are used by Gastwirth and Cohen [4] to study a population which is a mixture of two normal populations with common means but different variances.

The primary aim of the present article is to examine the bias and mean square error of various location estimators expressible as linear functions of the order statistics when an unidentified single outlier is present in a sample of size n. More precisely, we represent the sample by n independent absolutely continuous random variables X_j $(j = 1, 2, \ldots, n-1)$ and Y, such that

$$\begin{aligned} &X_j \text{ has cdf } F(x) \text{ and pdf } f(x), \\ &Y \text{ has cdf } G(x) \text{ and pdf } g(x). \end{aligned} \tag{1}$$

The labeling is a matter of convenience; it is not known which is the Y observation. If these variates are arranged in combined ascending order, we obtain the order statistics

$$Z_{1:n} \leq Z_{2:n} \leq \cdots \leq Z_{n:n}. \tag{2}$$

The estimators considered are then of the form

$$M = \sum_{i=1}^{n} a_i Z_{i:n} \tag{3}$$

with

$$a_i \geq 0 \quad (i = 1, 2, \ldots, n), \qquad \sum_{i=1}^{n} a_i = 1. \tag{4}$$

We derive some general results on the properties of $Z_{r:n}$ $(r = 1, 2, \ldots, n)$ and M and deal in more detail with the important special cases when $G(x)$ differs from $F(x)$ either in location or scale. Numerical work concentrates on the normal cases:

$$X_j \sim N(0, 1), \qquad Y \sim N(\lambda, 1) \quad \text{for} \quad \lambda = 0(0.5)\,2, 3, 4 \tag{5a}$$

$$X_j \sim N(0, 1), \qquad Y \sim N(0, \tau^2) \quad \text{for} \quad \tau = 0.5, 2, 3, 4. \tag{5b}$$

X_j in (5) has been standardized which for our purposes can be done without loss of generality. In fact, we shall whenever convenient take $f(x)$ to be a standardized density with location parameter zero.

The means, variances, and covariances of the order statistics in the cases (5) have been tabulated by David *et al.* [3] for $n \leq 20$. With the help of these tables it is easy to obtain corresponding values of the bias and expected mean square error (EMS) of any linear function of the order statistics. On the basis of bias and EMS various location estimators are compared in the tables and charts of Section 5.

It may be noted here that Kale and Sinha [9] and Joshi [8] investigated the EMS of a class of estimators appropriate when the X_j and Y follow different *exponential* distributions.

2. BASIC THEORY

The cdf $H_{r:n}(x)$ of $Z_{r:n}$ defined in (2) may be obtained as follows ($r = 1, 2, \ldots, n-1$):

$$\begin{aligned} H_{r:n}(x) &= \Pr\{\text{at least } r \text{ of } X_1, \ldots, X_{n-1}, Y \leq x\} \\ &= \Pr\{\text{at least } r \text{ of } X_1, \ldots, X_{n-1} \leq x\} \\ &\quad + \Pr\{\text{exactly } r-1 \text{ of } X_1, \ldots, X_{n-1} \leq x \text{ and } Y \leq x\} \\ &= F_{r:n-1}(x) + \binom{n-1}{r-1} F^{r-1}(x)[1-F(x)]^{n-r}G(x), \end{aligned} \tag{6}$$

where $F_{r:n-1}(x)$ is the cdf of $X_{r:n-1}$, the rth order statistic among $X_1, \ldots, X_{n-1}$. For $r = n$ we have simply

$$H_{n:n}(x) = F^{n-1}(x)G(x). \tag{6'}$$

The null case $G(x) = F(x)$ is further discussed in the Appendix.

If F and G are absolutely continuous, a condition not required in the derivation of (6), differentiation gives

$$\begin{aligned} h_{r:n}(x) &= [(n-1)!/(r-2)!\,(n-r)!] \\ &\quad \cdot F^{r-2}(x)[1-F(x)]^{n-r}f(x)G(x) \\ &\quad + [(n-1)!/(r-1)!\,(n-r)!] \\ &\quad \cdot F^{r-1}(x)[1-F(x)]^{n-r}g(x) \\ &\quad + [(n-1)!/(r-1)!\,(n-r-1)!] \\ &\quad \cdot F^{r-1}(x)[1-F(x)]^{n-r-1}f(x)[1-G(x)]. \end{aligned} \tag{7}$$

The first term of (7) drops out if $r = 1$, the last if $r = n$. Equation (7) may be rewritten as

$$h_{r:n}(x) = f_{r-1:n-1}(x)G(x) + \binom{n-1}{r-1} F^{r-1}(x)[1 - F(x)]^{n-r}g(x)$$

$$+ f_{r:n-1}(x)[1 - G(x)]. \tag{7'}$$

Similarly, the joint pdf $h_{rs:n}(x, y)$ of $Z_{r:n}$ and $Z_{s:n}(r < s,\ x < y)$ may be expressed as (cf. [3]):

$$\begin{aligned} h_{rs:n}(x, y) = &\\ & f_{r-1,\,s-1:n-1}(x, y)G(x) + f_{rs:n-1}(x, y)[1 - G(y)] \\ & + \frac{(n-1)!}{(r-1)!\,(s-r-1)!\,(n-s)!} F^{r-1}(x) \\ & \cdot [F(y) - F(x)]^{s-r-1}[1 - F(y)]^{n-s} \\ & \cdot \left\{ f(x)g(y) + g(x)f(y) + (s-r-1)\frac{[G(y) - G(x)]}{[F(y) - F(x)]} \right\}. \end{aligned} \tag{8}$$

Here the first term drops out if $r = 1$, the second if $s = n$, and the last if $s = r + 1$.

LEMMA Let $G_1(x)$ and $G_2(x)$ be two continuous cdf's with $G_2(x) \leq G_1(x)$ for all x, the inequality being strict for some x. If Y_1 has cdf $G_1(x)$, then there exists a continuous function $\Delta(x) \geq 0$ such that the stochastically larger random variable (rv) $Y_2 = Y_1 + \Delta(Y_1)$ is distributed with cdf $G_2(x)$.

Proof Define $G_2^{-1}(t) = \inf\{x\colon G(x) \geq t\}$. Then it is easily seen that $\Delta(x) = G_2^{-1}[G_1(x)] - x$ is continuous with $\Delta(x) \geq 0$. Also

$$\Pr\{Y_1 + \Delta(Y_1) \leq x\} = \Pr\{G_2^{-1}[G_1(Y_1)] \leq x\} = \Pr\{G_1(Y_1) \leq G_2(x)\} = G_2(x),$$

showing that $\Delta(x)$ is the required function.

If $H_{r:n}^{(1)}(x)$, $H_{r:n}^{(2)}(x)$, denotes the cdf of $Z_{r:n}$ when Y_1, Y_2, respectively, is the outlier, then it follows at once from (6) that $H_{r:n}^{(2)}(x) \leq H_{r:n}^{(1)}(x)$, i.e. that $Z_{r:n}^{(2)}$ is stochastically larger than $Z_{r:n}^{(1)}$. Also since, for $1 \leq r_1 < \cdots < r_k \leq n$, at least r_j of $X_1, \ldots, X_{n-1}, Y_1 + \Delta(Y_1) \leq x_j$, $j = 1, 2, \ldots, k \Rightarrow$ at least r_j of $X_1, \ldots, X_{n-1}, Y_1 \leq x_j$, we have in obvious notation

$$H_{r_1,\ldots,r_k:n}^{(2)}(x_1, \ldots, x_k) \leq H_{r_1,\ldots,r_k:n}^{(1)}(x_1, \ldots, x_k). \tag{9}$$

Note that (9) can clearly be further generalized to the situation when more than one outlier is present.

3. OUTLYING POPULATION DIFFERING IN LOCATION

In the important special case of a location shift, $G(x) = F(x - \lambda)$ for all x, we may write $Y = X_n + \lambda$, where X_n is a rv with cdf $F(x)$ and independent of $X_1, \ldots, X_{n-1}$. Let $Z_{r:n}(\lambda)$ denote the rth order statistic in this case and $H_{r:n}(x; \lambda)$ its cdf. Evidently $H_{r:n}(x; \lambda)$ is a decreasing function of λ. If $E(X^k)$ exists, this ensures (but is not required for) the existence of $E[Z_{r:n}^k(\lambda)]$. The latter is seen to be an increasing function of λ for $k = 1$ (and indeed for all odd values of k).

It is also instructive to see how $Z_{r:n}(\lambda)$ itself behaves as a function of λ (cf. Hampel [7]). Lower case x, y, z will as usual represent realizations of X, Y, Z. Insert $y = x_n + \lambda$ into the ordered sample of size $n - 1$, viz. $x_{1:n-1}, \ldots, x_{n-1:n-1}$. Then for any fixed values of $x_1, \ldots, x_n$ we have

$$\begin{aligned} z_{1:n}(\lambda) &= x_n + \lambda \qquad && x_n + \lambda \leq x_{1:n-1} \\ &= x_{1:n-1} && x_n + \lambda > x_{1:n-1} \end{aligned}$$

and for $r = 2, \ldots, n - 1$

$$\begin{aligned} z_{r:n}(\lambda) &= x_{r-1:n-1} \qquad && x_n + \lambda \leq x_{r-1:n-1} \\ &= x_n + \lambda && x_{r-1:n-1} < x_n + \lambda \leq x_{r:n-1} \\ &= x_{r:n-1} && x_n + \lambda > x_{r:n-1} \end{aligned} \tag{10}$$

and

$$\begin{aligned} z_{n:n}(\lambda) &= x_{n-1:n-1} \qquad && x_n + \lambda \leq x_{n-1:n-1} \\ &= x_n + \lambda && x_n + \lambda > x_{n-1:n-1} \end{aligned}$$

Thus $z_{r:n}(\lambda)$ is a nondecreasing function of λ with $z_{n:n}(\infty) = \infty$, $z_{1:n}(-\infty) = -\infty$ and otherwise

$$z_{r:n}(\infty) = x_{r:n-1}, \qquad z_{r:n}(-\infty) = x_{r-1:n-1}. \tag{11}$$

Taking expectations we see from (10), for finite λ, that if $E(X)$ exists so does $\mu_{r:n}(\lambda) = EZ_{r:n}(\lambda)$, $r = 1, \ldots, n$. We write $\mu_{r:n}(0) = \mu_{r:n}$, etc. By the monotone convergence theorem it follows that, for $r = 1, \ldots, n - 1$,

$$\lim_{\lambda \to \infty} EZ_{r:n}(\lambda) = E \lim_{\lambda \to \infty} Z_{r:n}(\lambda),$$

$$\mu_{r:n}(\infty) = EX_{r:n-1} \equiv \mu_{r:n-1}. \tag{12}$$

Likewise, for $r = 2, \ldots, n$

$$\mu_{r:n}(-\infty) = \mu_{r-1:n-1}. \tag{12'}$$

Also

$$\mu_{1:n}(-\infty) = -\infty, \qquad \mu_{n:n}(\infty) = \infty. \tag{13}$$

Second moments, pure and mixed, are not in general increasing functions of λ. However, we may obtain their limiting values as $\lambda \to \pm\infty$ from the corresponding behavior of the cdf's or pdf's. Thus since, as $\lambda \to \infty$, we have for fixed x, y

$$f(x-\lambda) \to 0, \quad F(x-\lambda) \to 0, \quad F(y-\lambda) - F(x-\lambda) \to 0$$

it follows from (7′) and (8) that

$$h_{r:n}(x; \infty) = f_{r:n-1}(x) \qquad r = 1, 2, \ldots, n-1 \tag{14}$$

$$h_{rs:n}(x, y; \infty) = f_{rs:n-1}(x, y) \qquad 1 \le r < s \le n-1. \tag{15}$$

Likewise, as $\lambda \to -\infty$, we have

$$h_{r:n}(x; -\infty) = f_{r-1:n-1}(x) \qquad r = 2, \ldots, n, \tag{16}$$

$$h_{rs:n}(x, y; -\infty) = f_{r-1,\, s-1:n-1}(x, y) \qquad 2 \le r < s \le n. \tag{17}$$

Under suitable conditions we may invoke the Lebesgue dominated convergence theorem to obtain

$$\sigma_{rs:n}(\infty) = \sigma_{rs:n-1} \qquad r, s = 1, \ldots, n-1, \tag{18}$$

$$\sigma_{rs:n}(-\infty) = \sigma_{r-1,\, s-1:n-1} \qquad r, s = 2, \ldots, n. \tag{19}$$

Equations (18) and (19) have been established by Shu [13] specifically in the case where f is the normal density.

We turn now to the location estimator M of (3) which in the present context may be written as

$$M(\lambda) = \sum_{i=1}^{n} a_i Z_{i:n}(\lambda).$$

From (10) it follows that $\sum_{i=1}^{n} a_i z_{i:n}(\lambda)$ is a nondecreasing continuous function of λ made up of n line segments over the λ intervals $(-\infty, x_{1:n-1} - x_n]$, $(x_{1:n-1} - x_n, x_{2:n-1} - x_n], \ldots, (x_{n-1:n-1} - x_n, \infty)$, with respective slopes a_1, $a_2, \ldots, a_n$. Clearly, $\mathrm{EM}(\lambda)$ is an increasing function of λ, with $\mathrm{EM}(\infty) = \infty$ unless $a_n = 0$, and $\mathrm{EM}(-\infty) = -\infty$ unless $a_1 = 0$. With these respective conditions satisfied, we have from (11) and (12)

$$\mathrm{EM}(\infty) = \sum_{i=1}^{n-1} a_i \mu_{i:n-1}, \qquad \mathrm{EM}(-\infty) = \sum_{i=2}^{n} a_i \mu_{i-1:n-1}$$

Also, using integration by parts, we can write

$$\tfrac{1}{2}\mathrm{EM}^2(\lambda) = \int_0^\infty x \Pr\{|M(\lambda)| > x\}\, dx. \tag{20}$$

We further assume that f is symmetrical and accordingly take $a_i = a_{n+1-i}$ ($i = 1, \ldots, \frac{1}{2}[n + 1]$). If f is standardized, $M(0)$ is symmetrically distributed about zero, and $M(-\lambda)$ has the same distribution as $-M(\lambda)$. It follows that

$$\Pr\{|M(\lambda)| > x\} = \Pr\{M(\lambda) > x\} + \Pr\{M(-\lambda) > x\}$$

which may be expected to be an increasing function of $|\lambda|$ under broad conditions. Correspondingly $\mathrm{EM}^2(\lambda)$ is an increasing function of λ and for $a_1 = a_n = 0$

$$\begin{aligned} \mathrm{EM}^2(\pm\infty) &= \lim_{\lambda\to\infty} E[M(\lambda)]^2 \\ &= E\left(\lim_{\lambda\to\infty} \sum_{i=2}^{n-1} a_i Z_{i:n}(\lambda)\right)^2 = E\left(\sum_{i=2}^{n-1} a_i X_{i:n-1}\right)^2 \\ &= [\mathrm{EM}(\infty)]^2 + \sum_{i=2}^{n-1} \sum_{j=2}^{n-1} a_i a_j \sigma_{ij:n-1}. \end{aligned} \tag{21}$$

4. OUTLYING POPULATION DIFFERING IN SCALE

In this case we may write $G(x) = F(x/\tau)$ for $\tau > 0$ and all x. Also Y may be expressed as $Y = \tau X_n$. If the corresponding order statistics are denoted by $Z^*_{r:n}(\tau)$ we see that

$$z^*_{r:n}(\tau) \text{ is } \begin{matrix}\text{nondecreasing} \\ \text{nonincreasing}\end{matrix} \text{ in } \tau \qquad \text{if} \quad \begin{matrix} x_n > 0 \\ x_n < 0 \end{matrix}. \tag{22}$$

Thus, for $r = 2, \ldots, n - 1$,

$$\begin{aligned} z^*_{r:n}(\infty) &= x_{r:n-1} && \text{if} \quad x_n > 0 \\ &= x_{r-1:n-1} && \text{if} \quad x_n < 0. \end{aligned} \tag{23}$$

If X is symmetrically distributed about 0, it follows that for $r = 2, \ldots, n - 1$

$$EZ^*_{r:n}(\infty) \equiv \mu^*_{r:n}(\infty) = \tfrac{1}{2}(\mu_{r-1:n-1} + \mu_{r:n-1}). \tag{24}$$

Also

$$\mu^*_{1:n}(\infty) = -\infty, \qquad \mu^*_{n:n}(\infty) = \infty.$$

The location estimator now becomes

$$M^*(\tau) = \sum_{i=2}^{n-1} a_i Z^*_{i:n}(\tau)$$

and for $a_i - a_{n+1-i}$ $(i = 1, \ldots, \frac{1}{2}[n+1])$ is clearly symmetrically distributed about 0 for all τ. Thus $\mathrm{EM}^*(\tau) = 0$ for all τ. Also by comparing (23) with (11) we see that the limiting behavior of $M^*(\tau) \mid X_n > 0$ as $\tau \to \infty$ corresponds to that of $M(\lambda)$ as $\lambda \to \infty$, with a similar correspondence between $M^*(\tau) \mid X_n < 0$ and $M(\lambda)$ as $\lambda \to -\infty$. Thus

$$\lim_{\tau \to \infty} \mathrm{EM}^{*2}(\tau) = \tfrac{1}{2}[\mathrm{EM}^2(\infty) + \mathrm{EM}^2(-\infty)] = \mathrm{EM}^2(\infty). \qquad (25)$$

5. NUMERICAL RESULTS IN THE NORMAL CASES

In Fig. 1 we exhibit the bias in the rth order statistic for model (5a):

$$b_{r:n}(\lambda) = EZ_{r:n}(\lambda) - EX_{r:n} \qquad r = 1, 2, \ldots, n.$$

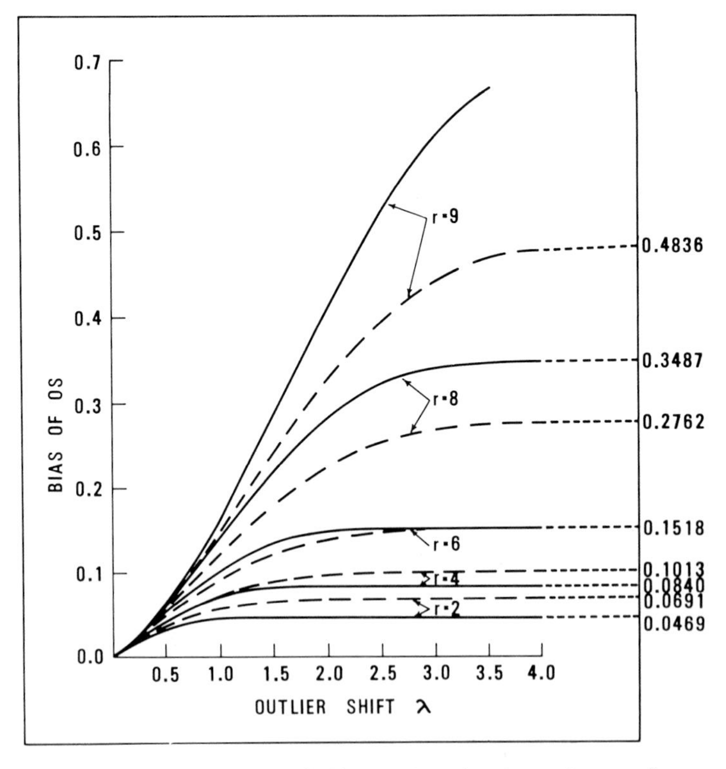

FIG. 1 Bias of the rth order statistic (OS) $Z_{r:10}$ when nine observations are from a normal (broken curves) or an extreme-value distribution (continuous curves) and one observation is from the corresponding distribution shifted to the right through a distance λ. The values on the right correspond to $\lambda = \infty$.

The results for $n = 10$ show that $b_{r:n}(\lambda)$ is not only increasing in λ but is also increasing in r.

Accompanying the results in the normal case which are based on the special tables in [3] the figure shows corresponding graphs for the more readily handled extreme-value population adjusted to have unit variance, viz. having cdf

$$F(x) = \exp(-e^{-\sigma x}) \qquad \sigma = \pi/\sqrt{6}.$$

Here

$$G(x) = F(x - \lambda) = F^k(x),$$

where $\lambda = (\ln k)/\sigma$. Thus $G(x)$ is a Lehmann alternative. When k is a positive integer $EZ_{r:n}(\lambda)$ can be expressed with the help of (7) as

$$\sigma EZ_{r:n}(\lambda) = \frac{(r + k - 1)^{(k)}}{(n + k - 1)^{(k)}} [EX_{r+k-1:n+k-1} - EX_{r+k:n+k-1}] + EX_{r:n-1}.$$

The expected values of the order statistics on the right can be obtained from Lieblein and Salzer [10].

We shall not pursue comparison with the extreme-value distribution any further but turn now to the location estimators included in this study (cf. [2]).

The forms appropriate when n is even are

(a) Sample mean:

$$\bar{X}_n = \frac{1}{n} \sum_{i=1}^{n} Z_{i:n};$$

(b) Trimmed means:

$$T_n(r) = \frac{1}{n - 2r} \sum_{i=r+1}^{n-r} Z_{i:n} \qquad 0 < r \leq n - 1;$$

(c) Winsorized means:

$$W_n(r) = \frac{1}{n} \left[(r + 1)(Z_{r+1:n} + Z_{n-r:n}) + \sum_{i=r+2}^{n-r-1} Z_{i:n} \right]$$

(d) Linearly weighted means:

$$L_n(r) = \frac{1}{2(\frac{1}{2}n - r)^2} \sum_{i=1}^{\frac{1}{2}n-r} (2i - 1)(Z_{r+i:n} + Z_{n-r-i+1:n})$$

(e) Gastwirth mean:

$$G_n = 0.3(Z_{[\frac{1}{3}n]+1:n} + Z_{n-[\frac{1}{3}n]:n}) + 0.2(Z_{\frac{1}{2}n:n} + Z_{\frac{1}{2}n+1:n}),$$

where $[\frac{1}{3}n]$ denotes the integral part of $\frac{1}{3}n$.

Table 1 gives the bias of these estimators under model (5a) as a function of $\lambda \geq 0$ when $n = 10$. Using $\prec$ to denote "inferior to" we have the ordering, for the entries in the table,

$$\bar{X}_{10} \prec W_{10}(1) \prec T_{10}(1) \prec W_{10}(2) \prec L_{10}(1) \prec T_{10}(2) \prec L_{10}(2) \prec G_{10} \prec \mathrm{Med}_{10}. \quad (26)$$

TABLE 1

Bias of Various Estimators of θ for Sample Size $n = 10$ When One Observation is from a $N(\theta + \lambda, 1)$ Population and the Others from $N(\theta, 1)$

λ:	0.0	0.5	1.0	1.5	2.0	3.0	4.0	∞
$\bar{X}_{10}$	0.0	0.05	0.10	0.15	0.20	0.30	0.40	∞
$T_{10}(1)$	0.0	0.04912	0.09325	0.12870	0.15400	0.17871	0.18470	0.18563
$T_{10}(2)$	0.0	0.04869	0.09023	0.12041	0.13904	0.15311	0.15521	0.15538
Med_{10}	0.0	0.04832	0.08768	0.11381	0.12795	0.13642	0.13723	0.13726
$W_{10}(1)$	0.0	0.04938	0.09506	0.13368	0.16298	0.19407	0.20239	0.20377
$W_{10}(2)$	0.0	0.04889	0.09156	0.12389	0.14497	0.16217	0.16504	0.16530
$L_{10}(1)$	0.0	0.04869	0.09024	0.12056	0.13954	0.15459	0.15727	0.15758
$L_{10}(2)$	0.0	0.04850	0.08892	0.11700	0.13328	0.14436	0.14576	0.14585
G_{10}	0.0	0.04847	0.08873	0.11649	0.13237	0.14285	0.14407	0.14414

Trimmed means are seen to be less subject to bias than the corresponding Winsorized means (except for $r = 4$ when both reduce to the median, Med_{10}). These results are as expected, as is the fact that the median is uniformly the least biased of all the estimators. The median is, however, more biased than one might naively have supposed.

In Fig. 2 graphs of the bias of $10T_{10}(1)$, $10W_{10}(2)$, and $10T_{10}(4)$ ($= 10\,\mathrm{Med}_{10}$) are shown, together with the respective sensitivity curves [1,14]. Note that for $\lambda > \mu_{n-r:n-1}$ the sensitivity curves for estimators of the form

$$M_n(r; \lambda) = \sum_{i=r+1}^{n-r} a_i Z_{i:n}(\lambda)$$

simply assume the constant value

$$\sum_{i=r+1}^{n-r} a_i \mu_{i:n-1} = M_n(r; \infty).$$

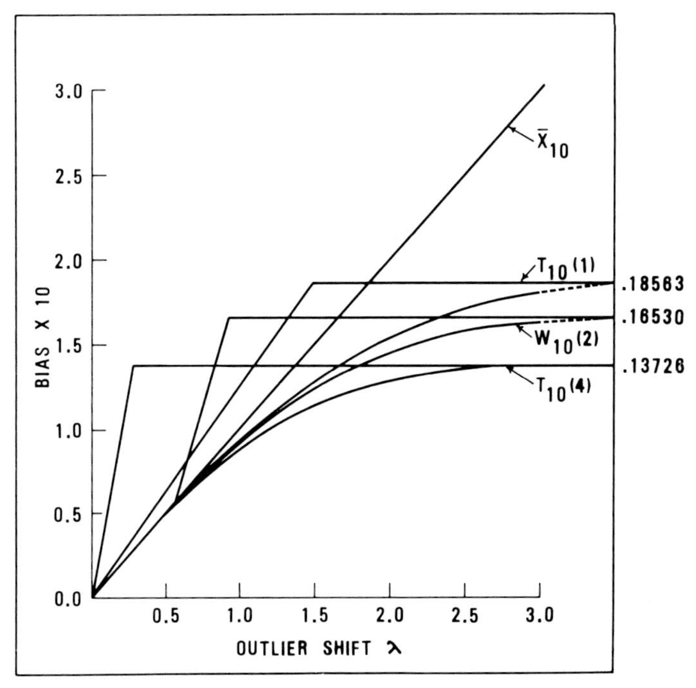

FIG. 2 Bias in the normal case of Fig. 1 for various location estimators, together with the corresponding sensitivity curves.

For $\lambda < 0$ recall that $\mathrm{EM}(-\lambda) = -\mathrm{EM}(\lambda)$.

Values under model (5a) of the mean square error (MSE) of the estimators are set out in Table 2. For $\lambda < 0$ use $\mathrm{MSE}[M(-\lambda)] = \mathrm{MSE}[M(\lambda)]$. The results are not as easily described as for Table 1. However, we can observe the partial ordering

$$\mathrm{Med}_{10} \prec G_{10} \prec L_{10}(2) \prec L_{10}(1) \prec T_{10}(2) \tag{27}$$

as well as $W_{10}(2) \prec T_{10}(1)$. Note the dilemma that (27) represents almost a reverse ranking of the best estimators in (26)! All the surviving MSE-admissible estimators are graphed in Fig. 3. Except for $T_{10}(2)$ they are inferior over part of the range to one or more of the omitted estimators; only the sample mean is ever inferior to the median.

Figure 3 shows the Winsorized means to do better than the corresponding trimmed means when λ is small. This is not surprising since the Winsorized means are roughly BLUEs based on the $n - 2r$ central order statistics. For large λ the trimmed means come to the fore with $T_{10}(2)$ optimal from $\lambda \doteq 3.3$ on.

TABLE 2

Mean Square Error of Estimators in Table 1

λ:	0.0	0.5	1.0	1.5	2.0	3.0	4.0	∞
$\bar{X}_{10}$	0.10	0.103	0.11	0.123	0.14	0.19	0.26	∞
$T_{10}(1)$	0.10534	0.10791	0.11471	0.12387	0.13285	0.14475	0.14865	0.14942
$T_{10}(2)$	0.11331	0.11603	0.12297	0.13132	0.13848	0.14580	0.14730	0.14745
Med_{10}	0.13833	0.14161	0.14964	0.15852	0.16524	0.17072	0.17146	0.17150
$W_{10}(1)$	0.10437	0.10693	0.11403	0.12405	0.13469	0.15039	0.15627	0.15755
$W_{10}(2)$	0.11133	0.11402	0.12106	0.12995	0.13805	0.14713	0.14926	0.14950
$L_{10}(1)$	0.11371	0.11644	0.12337	0.13169	0.13882	0.14626	0.14797	0.14820
$L_{10}(2)$	0.12097	0.12386	0.13105	0.13933	0.14598	0.15206	0.15310	0.15318
G_{10}	0.12256	0.12549	0.13276	0.14111	0.14777	0.15376	0.15472	0.15479

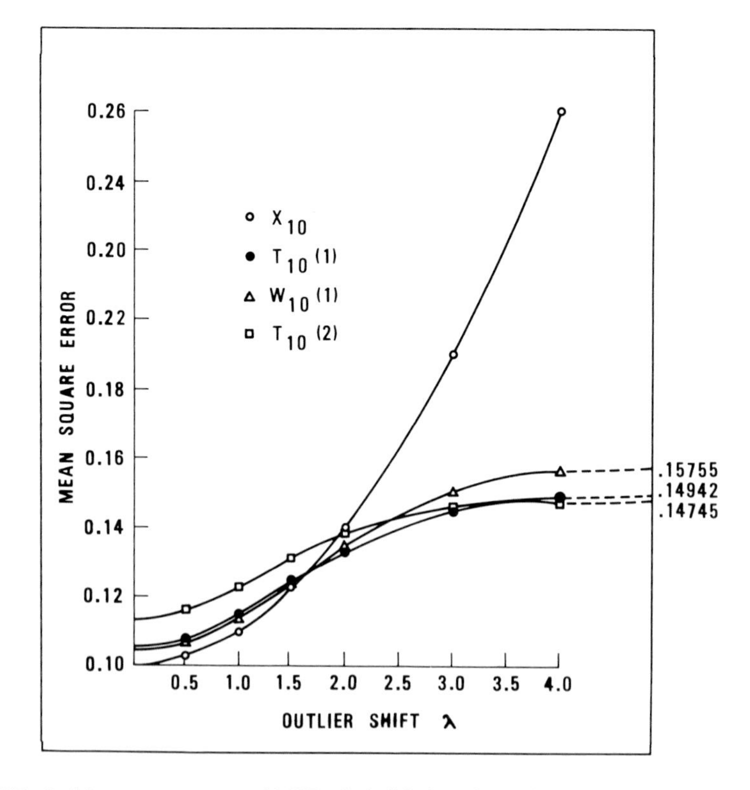

FIG. 3 Mean square error of MSE-admissible location estimators in the normal case of Fig. 1.

We have confined ourselves in this section to $n = 10$. Corresponding tables for $n = 5$ and $n = 20$ are also included in [13]. The results are qualitatively quite similar. The cross-over point between $T_{20}(1)$ and $T_{20}(2)$ occurs a little earlier, at $\lambda \doteq 2.6$.

In the case of model (5b) the estimators are, of course, unbiased for all τ. Table 3 gives their variances for $n = 10$. The trends are very similar to those in Table 2. Thus the ordering (27) still applies, except that $T_{10}(2)$ is now inferior to $L_{10}(1)$ for the "inlier" situation $\tau = 0.5$. Such general agreement is less surprising when one remembers that there is identity of results not only in the null case ($\lambda = 0$ or $\tau = 1$) but also for $\lambda = \infty = \tau$, the latter by (25).

TABLE 3

Mean Square Error of Various Estimators of θ for $n = 10$ When One Observation Is from $N(\theta, \tau^2)$ and the Others from $N(\theta, 1)$

τ:	0.5	1.0	2.0	3.0	4.0	∞
$\bar{X}_{10}$	0.09250	0.10000	0.13000	0.18000	0.25000	∞
$T_{10}(1)$	0.09491	0.10534	0.12133	0.12955	0.13417	0.14942
$T_{10}(2)$	0.09953	0.11331	0.12773	0.13389	0.13717	0.14745
Med_{10}	0.11728	0.13833	0.15375	0.15953	0.16249	0.17150
$W_{10}(1)$	0.09571	0.10436	0.12215	0.13221	0.13801	0.15754
$W_{10}(2)$	0.09972	0.11133	0.12664	0.13365	0.13745	0.14950
$L_{10}(1)$	0.09934	0.11371	0.12815	0.13436	0.13769	0.14820
$L_{10}(2)$	0.10432	0.12097	0.13531	0.14101	0.14398	0.15318
G_{10}	0.10573	0.12256	0.13703	0.14270	0.14565	0.15479

6. CONCLUDING REMARKS

The present article gives detailed answers in a limited setting. Besides obvious questions about other estimators and other parent populations the following remarks suggest themselves:

(a) Some related issues may be tackled with the help of tables in Milton [11]. For model (5a) these tables, which are more general, give the distribution of the rank of Y among the n variates for $n \leq 13$ and $\lambda = 0.2\ (.2)\ 1, 1.5, 2, 3$. Thus for $n = 10$ and $\lambda = 2$ we find the following values of $\pi_r = \Pr\{\text{rank } Y = r\}$:

r	1	2	3	4	5	6	7	8	9	10
π_r	0.001	0.003	0.005	0.009	0.014	0.023	0.040	0.072	0.159	0.674

This means that an estimator omitting the two (four) extreme observations excludes the outlier with probability 0.675 (0.837).

(b) The case for considering a single outlier in small samples has been well stated by Guttman [6]. In principle, the present work can be extended to more than one outlier but in practice this would be tedious. See, however, Prescott [12] for a generalization of sensitivity curves to two dimensions, corresponding to two outliers.

The most interesting extension of the present approach is to the case when $r(>1)$ outliers come from a common population. Here one may hope that results for mixtures, e.g., in case (5b) for a parent pdf (the contaminated normal)

$$f(x) = (1-\gamma)\phi(x) + (\gamma/\tau)\phi(x/\tau) \qquad 0 \le \gamma \le 1, \tag{28}$$

will give an increasingly good representation of the outlier situation for $r = \gamma n$

(i) as r gets closer to $\frac{1}{2}n$, with n fixed, and
(ii) as n increases, with r/n fixed.

Gastwirth and Cohen [4] have studied various location estimators for the case (28) and have obtained numerical values for $\tau = 3$, including the variances and covariances of the order statistics for $n \le 20$ and $\gamma = 0.01, 0.05, 0.1$. Comparison of the column $\tau = 3$ in Table 3 with the mixture case for $n = 10$, $\gamma = 0.1$ shows that the variances are higher in the latter case for all the estimators other than the mean, for which they are the same. Similar findings apply in a comparison for $n = 20$, with $\gamma = 0.05$. For $n = 10$ the maximal difference amounts to 10%, for $n = 20$ the agreement is even closer.

(c) In this article we have confined ourselves to estimators of location. Similar results for estimators of scale which are linear functions of the order statistics are readily obtained. A numerical illustration is given in [3].

APPENDIX

When $G(x) = F(x)$, (6) reduces to

$$F_{r:n}(x) = F_{r:n-1}(x) + \binom{n-1}{r-1} F^r(x)[1-F(x)]^{n-r}$$

$$r = 1, 2, \ldots, n-1. \tag{A1}$$

In spite of its simplicity this relation appears to be new. It provides a rather convenient way of building up tables of $F_{r:n}(x)$ for fixed r and $n = r, r+1, \ldots$ starting with $F_{r:r}(x) = F^r(x)$. Repeated application of (A1) gives

$$F_{r:n}(x) = F^r(x) \sum_{i=1}^{n-r+1} \binom{n-i}{r-1} [1-F(x)]^{n-r+1-i}, \tag{A2}$$

a result also valid for $r = n$.

Equation (A1) may be compared with

$$F_{r:n}(x) = F_{r+1:n}(x) + \binom{n}{r} F^r(x)[1 - F(x)]^{n-r} \tag{A3}$$

which follows from the standard formula

$$F_{r:n}(x) = \sum_{i=r}^{n} \binom{n}{i} F^i(x)[1 - F(x)]^{n-i}. \tag{A4}$$

In fact, (A1) may be obtained directly from (A4) or from (A3) combined with the basic recurrence relation

$$(n - r)F_{r:n}(x) + rF_{r+1:n}(x) = nF_{r:n-1} \qquad r = 1, 2, \ldots, n - 1.$$

From (A1) we have also, subject to the existence of the expectation involved,

$$E(X^k_{r:n-1} - X^k_{r:n}) = k\binom{n-1}{r-1} \int_{-\infty}^{\infty} x^{k-1} F^r(x)[1 - F(x)]^{n-r}\, dx \tag{A5}$$

for $k = 1, 2, \ldots$ and $r = 1, 2, \ldots, n - 1$. This type of integral has, of course, arisen before (e.g. Godwin [5]).

REFERENCES

[1] ANDREWS, D. F., BICKEL, P. J., HAMPEL, F. R., HUBER, P. J., ROGERS, W. H., and TUKEY, J. W. (1972). *Robust Estimates of Location.* Princeton Univ. Press, Princeton, New Jersey.

[2] CROW, E. L. and SIDDIQUI, M. M. (1967). Robust estimation of location. *J. Amer. Statist. Assoc.* **62** 353–389.

[3] DAVID, H. A., KENNEDY, W. J., and KNIGHT, R. D. (1977). Means, variances, and covariances of normal order statistics in the presence of an outlier. *Selected Tables in Mathematical Statistics*, Vol. V.

[4] GASTWIRTH, J. L. and COHEN, M. L. (1970). Small sample behavior of robust linear estimators of location. *J. Amer. Statist. Assoc.* **65** 946–973.

[5] GODWIN, H. J. (1949). Some low moments of order statistics. *Ann. Math. Statist.* **20** 279–285.

[6] GUTTMAN, I. (1973). Care and handling of univariate or multivariate outliers in detecting spuriosity—a Bayesian approach. *Technometrics* **15** 723–738.

[7] HAMPEL, F. R. (1974). The influence curve and its role in robust estimation. *J. Amer. Statist. Assoc.* **69** 383–393.

[8] JOSHI, P. C. (1972). Efficient estimation of the mean of an exponential distribution when an outlier is present. *Technometrics* **14** 137–144.

[9] KALE, B. K. and SINHA, S. K. (1971). Estimation of expected life in the presence of an outlier observation. *Technometrics* **13** 755–759.

[10] LIEBLEIN, J. and SALZER, H. E. (1957). Table of the first moment of ranked extremes. *J. Res., Nat. Bur. Stand.* **59** 203–206.

[11] Milton, R. C. (1970). *Rank Order Probability: Two-Sample Normal Shift Alternatives.* Wiley, New York.

[12] Prescott, P. (1976). Comparison of tests for normality using stylized sensitivity surfaces. *Biometrika* **63** 285–289.

[13] Shu, V. S. (1976). Robustness of the mean in the presence of a single observation from an outlying population. Unpublished Master's paper, Iowa State Univ.

[14] Tukey, J. W. (1970). *Exploratory Data Analysis*, Vol. 1, prelim. edition. Addison-Wesley, Reading, Mass.

The Ninther, a Technique for Low-Effort Robust (Resistant) Location in Large Samples

John W. Tukey

PRINCETON UNIVERSITY
and
BELL LABORATORIES

The ninther, a median of medians, has a number of desirable features as a constituent part of procedures to summarize a large number of values: it reduces the number of values by a factor of nine; it tends to shed the influence of "wild" values (unless they are clustered together fairly tightly); its Gaussian efficiency is 55% (its stretched-tail efficiency is higher); it can be computed with no arithmetic operations and an average of about 1.1 comparisons per input data value. Various specific uses are suggested.

1. INTRODUCTION

We sometimes want a robust estimate of location with low computational effort, particularly when we have a large amount of data. In such circumstances, we may

(1) not need extremely high efficiency
(2) not insist on the result being wholly independent of the order in which the data are made available to us.

Prepared in part in connection with research at Princeton University sponsored by the Army Research Office (Durham).

Key words and phrases Ninther, resistant summaries, large samples, quick computation, efficiency, robust estimates.

One possibility for a single-pass computation (in which each data value is calculated with only once) is the use of stochastic approximation techniques (Martin, 1972; Martin and Masreliez, 1974) using high efficiency robust/resistant estimators such as the biweight or the wave. However, several arithmetic operations are needed for each data value. (Conversely, only the earliest part of the sequence will provide any appreciable influence of order upon answer.)

Our purpose here is to show how to attain Gaussian efficiencies above 50% (and much higher stretched-tail efficiencies) with a very much smaller amount of computing (of the order of 1 to 1.5 comparisons—comparisons not arithmetic operations—per data value).

2. THE NINTHER

Let

$$y_A = \text{median}\{y_1, y_2, y_3\}$$

$$y_B = \text{median}\{y_4, y_5, y_6\}$$

$$y_C = \text{median}\{y_7, y_8, y_9\}$$

and then

$$y_E = \text{median}\{y_A, y_B, y_C\}$$

For brevity define

$$\text{9ther}(y_1, \ldots, y_9) = y_E$$

where "9ther" is the name of a function conveniently called "ninther."

3. DISTRIBUTION OF NINTHERS

If y_1 to y_9 are a random sample from some distribution, it will be shown subsequently that the distribution of

$$\text{9ther}(y_1, \ldots, y_9)$$

is a mixture of the distributions of medians of samples of 7 (27/35 weight) and 9 (8/35 weight) from the same distribution.

Should the nine y_i be monotonic increasing or decreasing, 9ther$(y_1, \ldots, y_9)$ will of course be the median of the y's.

Behavior in intermediate cases should, we may hopefully expect, be intermediate.

Gaussian case In the Gaussian case, which is on the unfavorable side for medians, we have

$$\text{var}\{\text{median of } 7\} = 0.2104$$

$$\text{var}\{\text{median of } 9\} = 0.1661$$

$$\text{var}\{\text{9ther of } 9\} = 0.2003$$

and the Gaussian efficiencies, compared to the mean, are

$$\text{Gaussian efficiency}\{\text{median of } 9\} = 67\%$$

$$\text{Gaussian efficiency}\{\text{9ther of } 9\} = 55\%$$

It is helpful to compare these, especially the last, with

$$\text{Gaussian efficiency}\{\text{median of } 3\} = 74\%$$

and

$$(\text{Gaussian efficiency}\{\text{median of } 3\})^2 = 55\%$$

and to note that efficiencies in stretch-tailed situations will be higher, often very much higher.

4. COMPUTING EFFORT

It requires either 2 or 3 comparisons to locate the median of 3 values. If these values are in random order, it requires an average of 2.5 comparisons.

If we anticipate either randomness or deviations toward order, we naturally choose a comparison pattern that takes only 2 comparisons when the values are monotonic.

Since there are 4 medians of 3 required for 9 observations, the needed number of comparisons per value for the ninther are:

$$\text{for randomness:} \quad \frac{10}{9} = 1.11$$

$$\text{for monotonicity:} \quad \frac{8}{9} = 0.89$$

$$\text{at worst:} \quad \frac{12}{9} = 1.33$$

We may expect performance in the range 1.11 to 1.00 under most natural circumstances.

5. THE NINTHER-MEDIAN COMBINATION

If we desire a robust location based on $9M$ observations, one natural approach is to:

ninther M sets of 9.
sort the M results.
take their median.

The computational effort required (assuming sorting for the median) can be estimated as

$$(1.0 \text{ to } 1.1)(9M) + M \log_2 M$$

comparisons, which is

$$(1.0 \text{ to } 1.1) + \tfrac{1}{9}(\log_2 M)$$

comparisons per value. This rises to

$$2.0 \quad \text{to} \quad 2.1$$

comparisons per value when $M = 512$ and $9M = 4608$.

The efficiency will be better than half that of a median of all the values, which can hardly fail to be adequate as a starter.

6. IMPRACTICALLY LARGE DATA SETS

If we have $81M$ values, for large M, we might:

ninther $9M$ sets of 9.
ninther M sets of 9 of these.
sort the M results.
take their median.

The computational effort required can be estimated as

$$(1.0 \text{ to } 1.1)(81M + 9M) + M \log_2 M$$

comparisons, which is

$$1.1 \quad \text{to} \quad 1.2 + \tfrac{1}{81}\log_2 M$$

comparisons per value. This rises to

$$1.4 \text{ to } 1.5 \quad \text{comparisons per value}$$

at M just less than 18 million, $81M$ about 1.4 billion.

Since a plausible Gaussian efficiency for the first two steps is about one quarter to one-third [note that $(55\%)^2 = 30\%$], this technique surely offers an adequate working summary for very large batches.

If we instead were to:
ninther $9M$ sets of 9.
mean M sets of 9 of these.
apply a more careful robust/resistant technique to the M results,

we would require about 1.1 to 1.2 comparisons, and 0.11 additions per data point followed by less than 1/81 of the effort required to apply the robust/resistant technique to the whole body of data.

7. THE NINTHER-MEAN COMBINATION

When data are only a little worse than usual, so far as wild and straggling values are concerned, we can do well enough by taking means of the results of ninthing. The computational cost per data value will be

1.0 to 1.1 comparisons plus 0.11 additions

surely not great, and even the Gaussian efficiency will be 55% for randomly ordered data (and much higher for almost ordered data or almost locally ordered data).

8. A COMMENT

A variety of suggestions for other summaries using the ninther as a constituent can be made. Any careful evaluation has to consider their performance against nearly least-favorable orderings of the data. This has only been done informally for the present account.

9. A PERMUTATION RESULT

Consider y_1 to y_9 in random order: what is the distribution of their ninther? We show now that it is

$$\tfrac{3}{14}\ y_4, \qquad \tfrac{4}{7}\ y_5, \qquad \tfrac{3}{14}\ y_6,$$

where we have now assumed the y_i to be, say, increasing in value when in subscript order, but to be met in a random order.

If we label the first three values met as A, the next three as B, and the last three as C, the labeling of y_1 to y_9 is a permutation of $AAABBBCCC$. Consider the labeling of y_1, y_2, y_3, whose structure may be AAA, AAB, ABA, BAA, or ABC (permutations of order among the three y's or among ABC do not matter here). In the first four cases y_A will be one of y_1 to y_3, while y_B and y_C will not. Hence y_A will not be y_E and y_E will not be y_1, y_2, or y_3 (ties aside). In the last case y_A, y_B, and y_C will not overlap y_1, y_2, y_3.

A symmetrical argument applies to y_7, y_8, y_9, so that we know that y_E is either y_4, y_5, or y_6.

In view of this we can relabel the y's as $y_L, y_L, y_L, y_4, y_5, y_6, y_H, y_H, y_H$.

How can y_5 be y_E? Let us suppose y_5 is an A, then the other two A's must be divided, one from y_L, y_L, y_L, y_4, and one from y_6, y_H, y_H, y_H. If this happens, there will remain 3 low and 3 high y's to be split into B's and C's. However this is done, one of y_B and y_C must be low and the other high. Thus if y_5 is y_A it is also y_E.

The chance of y_5 being a y_A (or y_B or y_C) is then the chance of the other two A's being split, namely:

$$\frac{\binom{4}{1}\binom{4}{1}}{\binom{8}{2}} = \frac{4 \cdot 4 \cdot 2}{8 \cdot 7} = \frac{4}{7}$$

The chance of y_4 being a y_A has to equal the chance of y_6 being a y_A; each is thus 3/14.

10. A SAMPLING RESULT

Suppose the y's are a sample from something. They are then of the form

$$y_i = R(u_i),$$

where the u_i are a sample from rectangular [0, 1], and

$$y_E = R(u_E).$$

What of u_E? The distributions of u_4, u_5, and u_6 are given by the densities:

$$\frac{9!}{3!5!}u^3(1-u)^5\,du, \qquad \frac{9!}{4!4!}u^4(1-u)^4\,du, \qquad \frac{9!}{5!3!}u^5(1-u)^3\,du$$

so that the density of their $\frac{3}{14}, \frac{4}{7}, \frac{3}{14}$ mixture is given by

$$\frac{9!}{4!5!}\left(4\cdot\frac{3}{14}u^3(1-u)^5 + 5\frac{4}{7}u^4(1-u)^4 + 4\cdot\frac{3}{14}u^5(1-u)^3\right) du$$

$$= \frac{9!}{4!5!14}u^3(1-u)^3[12(1-u)^2 + 40u(1-u) + 12u^2]\, du$$

$$= \frac{12}{14}\frac{9!}{4!5!}u^3(1-u)^3\, du + \frac{16}{14}\frac{9!}{4!5!}u^4(1-u)^4\, du$$

$$= \frac{27}{35}\frac{7!}{3!3!}u^3(1-u)^3\, du + \frac{8}{35}\frac{9!}{4!4!}u^4(1-u)^4\, du$$

which can be recognized as a 27 to 8 mixture of the distribution of the median of 7 and the median of 9.

Since this is so for the $\{u_i\}$ and since $y = R(u)$ for individual values, medians, and ninther alike, it also holds for the $\{y_i\}$. This is the result used above in Section 3.

REFERENCES

MARTIN, R. D. (1972). Robust estimation of signal amplitude. *IEEE Trans. Information Theory* **80** 596–606.

MARTIN, R. D. and MASRELIEZ, C. J. (1974). Robust estimation via stochastic approximation. *IEEE Trans. Information Theory* **20** 537–540.

Completeness Comparisons Among Sequences of Samples

Norman L. Johnson

UNIVERSITY OF NORTH CAROLINA
CHAPEL HILL

Two sets of sample values are available. It is known that one is a complete random sample while the other is the remainder after censoring of a complete random sample from the same population. It is required to decide which set was censored.

Two methods of approach are described—one requiring a knowledge of the population distribution and the other distribution free. The loss of power in the latter is shown to be quite small.

Extensions to more than two sets of sample values, and to sequences of sets are also discussed.

1. INTRODUCTION

In previous papers [1–3] we have considered methods of testing whether a set (or sequence of sets) of observed values represents a complete random sample (or a sequence of complete random samples) or is the consequence of some form of censoring having been applied to a complete random sample (or sequence of such samples). The methods required a knowledge of the population distribution of the measured character(s); some assessments of the effects of imperfect knowledge were discussed in [4] and [5].

This research was supported by the U.S. Army Research Office under Contract No. DAHCO4-74-C-0030.

Key words and phrases Censoring, distribution-free procedures, tests.

We now turn our attention to problems arising when there are two sets (or sequences of sets) of data, and we wish to compare the incidence of censoring them. We will suppose the appropriate population distribution to be the same for each set, and also (as in the earlier papers) that the distribution is continuous. Certain special problems will be selected for detailed discussion from among the very wide range of possible problems. A particularly interesting feature of the enquiry is that distribution-free procedures can be used (see Section 3), thus freeing us from the dependence on knowledge of population distributions which characterized the earlier work.

A considerable variety of problems can arise:

(i) We may have different types of censoring. We will restrict ourselves to symmetrical censoring of extreme values—omission of the s greatest and s least sample values.

(ii) We may have different hypotheses in mind. For example, we may know that one sample (or sequence of samples) is censored, and the other not, and wish to find which one.

(iii) We may, or may not, know the population distribution.

(iv) The sample sizes in each sequence may not stay the same.

2. "PARAMETRIC" TEST PROCEDURES

2.1 Fixed Numbers of Samples

We suppose that we have available two samples, one ($\mathscr{A}_{11}$) of size r_1 and one ($\mathscr{A}_{21}$) of size r_2. We denote the order statistics in $\mathscr{A}_{i1}$ $(i = 1, 2)$ by

$$X_{i1} \leq X_{i2} \leq \cdots \leq X_{ir_i}.$$

We further suppose that it is known that one of the two samples (it is not specified which one) is a complete random sample, while the other is a random sample which has been censored by removal of some extreme observations. We will specialize the problem by supposing that the censoring is symmetrical, i.e. the s greatest and s least values in the original random samples are removed. The value of s may, or may not, be known.

Finally, we suppose that the population cumulative distribution function $F(x)$ is the same for each of the two samples; and is absolutely continuous with $f(x) = dF/dx$.

Let H_i $(i = 1, 2)$ be the hypothesis that the censored sample is $\mathscr{A}_{i1}$. The likelihood function for H_i is

$$\mathscr{L}_i = \frac{r_{3-i}!\,(r_i + 2s)!}{(s!)^2}[F(X_{i,1})\{1 - F(X_{i,r_i})\}]^s \prod_{k=1}^{2} \prod_{j=1}^{r_k} f(X_{k,j}) \tag{1}$$

and so the likelihood ratio (H_1/H_2) is

$$\frac{\mathscr{L}_1}{\mathscr{L}_2} = \frac{(r_1+1)^{[2s]}}{(r_2+1)^{[2s]}} \left[\frac{F(X_{1,1})\{1 - F(X_{1,r_1})\}}{F(X_{2,1})\{1 - F(X_{2,r_2})\}} \right]^s, \tag{2}$$

where, e.g., $(r_1+1)^{[2s]} = (r_1+1)(r_1+2)\cdots(r_1+2s)$. (We also use (see (6)), $x^{(h)} = x(x-1)\cdots(x-h+1)$.)

Whatever the value of s, we see that discrimination between H_1 and H_2 will be based on the statistic

$$T = \frac{F(X_{1,1})\{1 - F(X_{1,r_1})\}}{F(X_{2,1})\{1 - F(X_{2,r_2})\}} \tag{3}$$

$$= \frac{Y_{11}(1 - Y_{1r_1})}{Y_{21}(1 - Y_{2r_2})}, \tag{3'}$$

where

$$Y_{ij} = F(X_{ij}). \tag{4}$$

(It is worth noting that if the X's have different distributions F_1 and F_2 in $\mathscr{A}_1$, and $\mathscr{A}_2$ respectively, we reach the same criterion (3′) with $Y_{ij} = F_i(X_{ij})$ and the Y's have the same joint distribution.) The hypothesis H_1 will be accepted if $T > K$, and the hypothesis H_2 will be accepted if $T < K$. (If $T = K$, an arbitrary decision can be made. Since $\Pr[T = K] = 0$, this will not affect the probability properties of the procedure.) Choice of K can depend on s, r_1, r_2 and relative costs of different incorrect decisions.

If $r_1 = r_2$, the likelihood ratio is just T^s and it is natural to take $K = 1$, in the absence of information on relative costs of errors (whatever be the value of s).

To evaluate probabilities of error, we need the distribution of T. A discussion of this distribution follows. We first evaluate the moments for purposes of record.

If both $\mathscr{A}_{11}$ and $\mathscr{A}_{21}$ have been censored with the s_i' least and s_i'' greatest observations removed from $\mathscr{A}_{i1}$ $(i = 1, 2)$ then the joint density function of Y_{11}, Y_{1r_1}, Y_{21}, Y_{2r_2} is

$$f(y_{11}, y_{1r_1}, y_{21}, y_{2r_2}) = \prod_{i=1}^{2} \left\{ \frac{(r_i + s_i' + s_i'')!}{s_i'!\,(r_i-2)!\,s_i''!} y_{i1}^{s_i'} (y_{ir_i} - y_{i1})^{r_i-2} (1 - y_{ir_i})^{s_i''} \right\}$$

$$0 \le y_{i1} \le y_{ir_i} \le 1, \quad i = 1, 2. \tag{5}$$

(Hypothesis H_i corresponds to $s_i' = s_i'' = s$; $s_{3-i}' = s_{3-i}'' = 0$.) and the hth moment of T (h an integer) is

$$\mu_h'(T) = E[T^h] = \prod_{i=1}^{2} \left\{\frac{(r_i + s_i' + s_i'')!}{s_i'!\, s_i''!}\right\} \cdot \frac{(s_1' + h)!\,(s_1'' + h)!\,(s_2' - h)!\,(s_2'' - h)!}{(r_1 + s_1' + s_1'' + 2h)!\,(r_2 + s_2' + s_2'' - 2h)!}$$

$$= \frac{(s_1' + 1)^{[h]}(s_1'' + 1)^{[h]}}{s_2'^{(h)} s_2''^{(h)}} \cdot \frac{(r_2 + s_2' + s_2'')^{(2h)}}{(r_1 + s_1' + s_1'' + 1)^{[2h]}}. \tag{6}$$

Moments of order $(\min(s_2', s_2'') + 1)$ or higher are infinite. We do not propose to use the moments in approximating the distribution of T.

The distribution of $Z_i = Y_{i1}(1 - Y_{ir_i})$ has been studied in [1] and [3]. It is quite a complicated distribution, and the distribution of $T = Z_1/Z_2$ is even more complicated. Here we describe two methods of obtaining an approximation.

As pointed out in [3] we can write

$$Z_i = U_i W_i (U_i + V_i + W_i)^{-2} \tag{7}$$

where U_i, V_i and W_i are independent random variables distributed as χ^2 with $2(s_i' + 1)$, $2(r_i - 1)$ and $2(s_i'' + 1)$ degrees of freedom, respectively. Hence

$$T = \frac{U_1 W_1 (U_2 + V_2 + W_2)^2}{U_2 W_2 (U_1 + V_1 + W_1)^2} \tag{8}$$

all six variables on the right hand side being mutually independent.

If r_1 and r_2 are large compared with s_1', s_1'', s_2' and s_2'' then writing

$$\frac{(r_1 - 1)^2}{(r_2 - 1)^2} T = \frac{U_1 W_1}{U_2 W_2} \cdot \left\{\frac{(r_2 - 1)^{-1}(U_2 + W_2) + (r_2 - 1)^{-1} V_2}{(r_1 - 1)^{-1}(U_1 + W_1) + (r_1 - 1)^{-1} V_1}\right\}^2, \tag{9}$$

and noting that $(r_i - 1)^{-1}(U_i + W_i) \sim 0$ and $(r_i - 1)^{-1} V_i \sim 2$ for r_i large, we would expect the large-sample distribution of T to be approximated by that of

$$\left(\frac{r_2 - 1}{r_1 - 1}\right)^2 \frac{U_1 W_1}{U_2 W_2}$$

i.e.

$$\left(\frac{r_2 - 1}{r_1 - 1}\right)^2 \frac{(s_1' + 1)(s_1'' + 1)}{(s_2' + 1)(s_2'' + 1)} \cdot F_{2(s_1' + 1),\, 2(s_2' + 1)} F_{2(s_1'' + 1),\, 2(s_2'' + 1)}. \tag{10}$$

The distribution of the product of two independent F-variables has been studied by Schumann and Bradley [6]. They give tables of upper percentage points of products of independent variables distributed as F_{ν_1, ν_2} and F_{ν_2, ν_1}, respectively. These will only apply to (10) for the special case $s_1' = s_2''$, $s_2' = s_1''$, while we are at present interested in $s_1' = s_1'' = s$; $s_2' = s_2'' = 0$.

In the case $r_1 = r_2 = r$ we can get some useful approximations when r is large. The rule is then to choose H_1 if $T > 1$, and

$$\Pr[T > 1 | H_1] \doteqdot \Pr[GG' < 1],$$

where the joint density of G and G' is

$$(s + 1)^2(1 + g)^{-(s+2)}(1 + g')^{-(s+2)} \qquad (0 < g, g').$$

Hence

$$\begin{aligned}\Pr[T > 1 | H_1] &\doteqdot 1 - (s + 1) \int_0^\infty (1 + g)^{-(s+2)}(1 + g^{-1})^{-(s+1)}\, dg \\ &= 1 - (s + 1) \int_0^\infty g^{s+1}(1 + g)^{-(2s+3)}\, dg \\ &= 1 - (s + 1)B(s + 1, s + 2) \\ &= 1 - \{\Gamma(s + 2)\}^2/\Gamma(2s + 3).\end{aligned}$$

In particular, when r is large the probability of a correct decision is approximately $\frac{5}{6}$ if $s = 1$, $\frac{19}{20}$ if $s = 2$.

Alternatively we can consider the distribution of $\ln T$. The characteristic function of $\ln T$ is (cf. (6))

$$E[T^{ih}] = \left[\prod_{j=1}^{2} \left\{\frac{(r_j + s_j' + s_j'')!}{s_j'!\, s_j''!}\right\}\right] \cdot \frac{\Gamma(s_1' + 1 + ih)\Gamma(s_1'' + 1 + ih)\Gamma(s_2' + 1 - ih)\Gamma(s_2'' + 1 - ih)}{\Gamma(r_1 + s_1' + s_1'' + 1 + 2ih)\Gamma(r_2 + s_2' + s_2'' + 1 - 2ih)}. \tag{11}$$

Taking derivatives of the cumulant generating function $\ln E[T^{ih}]$ we find

$$\begin{aligned}\kappa_t(\ln T) &= \psi^{(t-1)}(s_1' + 1) + \psi^{(t-1)}(s_1'' + 1) \\ &\quad - 2^t\psi^{(t-1)}(r_1 + s_1' + s_1'' + 1) \\ &\quad + (-1)^t\{\psi^{(t-1)}(s_2' + 1) + \psi^{(t-1)}(s_2'' + 1) \\ &\quad - 2^t\psi^{(t-1)}(r_2 + s_2' + s_2'' + 1)\},\end{aligned} \tag{12}$$

where

$$\psi^{(s)}(x) = \frac{d^{s+1} \ln \Gamma(x)}{dx^{s+1}}$$

is the $(s+2)$-gamma function when H_1 is true, and $r_1 = r_2 = r$, we have $s_1' = s_1'' = s$, and $s_2' = s_2'' = 0$, whence (12) becomes

$$\kappa_t(\ln T \mid H_1) = 2\psi^{(t-1)}(s+1) - 2^t\psi^{(t-1)}(r+2s+1) + (-1)^t\{2\psi^{(t-1)}(1) - 2^t\psi^{(t-1)}(r+1)\}. \tag{13}$$

Also

$$\kappa_t(\ln T \mid H_2) = 2\psi^{(t-1)}(1) - 2^t\psi^{(t-1)}(r+1) + (-1)^t\{2\psi^{(t-1)}(s+1) - 2^t\psi^{(t-1)}(r+2s+1)\} \tag{14}$$

In particular

$$\kappa_1(\ln T \mid H_1) = 2[\psi(s+1) - \psi(1) - \{\psi(r+2s+1) - \psi(r+1)\}] = -\kappa_1(\ln T \mid H_2) \tag{15}$$

and

$$\kappa_2(\ln T \mid H_1) = 2[\psi^{(1)}(s+1) + \psi^{(1)}(1) - 2\{\psi^{(1)}(r+1) + \psi^{(1)}(r+2s+1)\}] = \kappa_2(\ln T \mid H_2). \tag{16}$$

Some values of $\{\kappa_1(\ln T \mid H_1) - \kappa_1(\ln T \mid H_2)\}/\sqrt{\kappa_2(\ln T \mid H_i)} = \rho$ $(i = 1, 2)$ are shown in Table 1. (The values for $r = \infty$ are the limiting values

$$\frac{4\{\psi(s+1) - \psi(1)\}}{[2\{\psi^{(1)}(s+1) + \psi^{(1)}(1)\}]^{1/2}}.)$$

TABLE 1

Values of ρ

$r_1 = r_2 = r \backslash s$	1	2	$r \backslash s$	1	2
4	1.44	2.10	8	1.63	2.46
5	1.52	2.23	9	1.66	2.51
6	1.57	2.33	10	1.68	2.55
7	1.60	2.40	20	1.77	2.74
			∞	1.87	2.97

These values of ρ are of such a size as to indicate that even with a single pair of samples, fairly good discrimination is attainable, at least when $r_1 = r_2 = r$.

If we choose $H_1(H_2)$ when $\ln T > (<) 0$ then approximating the distribution of $\ln T$ by a normal distribution, the probability of correct decision would be $\Phi(\frac{1}{2}\rho)$.

If we have two *sequences* of samples

$$\mathscr{S}_1 : \mathscr{A}_{11}, \mathscr{A}_{12}, \ldots \mathscr{A}_{1m}$$

and

$$\mathscr{S}_2 : \mathscr{A}_{21}, \mathscr{A}_{22}, \ldots \mathscr{A}_{2m}$$

we would use as criterion

$$T_{(m)} = \prod_{j=1}^{m} T_j, \tag{17}$$

where T_j is the value of T [as defined in (3)] for the jth pair of samples $\mathscr{A}_{1j}$, $\mathscr{A}_{2j}$ $(j = 1, 2, \ldots m)$. Denoting the order statistics of sample $\mathscr{A}_{ij}$ by

$$X_{ij1} \leq X_{ij2} \leq \cdots \leq X_{ijr_i}$$

we have

$$T_j = \frac{Y_{1j1}(1 - Y_{1jr_1})}{Y_{2j1}(1 - Y_{2jr_2})}, \tag{18}$$

where $Y_{ijh} = F(X_{ijh})$. The second method of approximation (described above for T) is the more simply extended to $T_{(m)}$. Since $T_1, T_2, \ldots, T_m$ are mutually independent we have

$$\kappa_r(\ln T_{(m)}) = \sum_{j=1}^{m} \kappa_r(\ln T_j) = m\kappa_r(\ln T) \tag{19}$$

(on the assumption that $r_1, r_2, s_1', s_1'', s_2'$ and s_2'' remain constant throughout). It is not essential that the X's have the same distribution in each of the samples $\mathscr{A}_{ij}$, provided the appropriate transformations to the Y's are used (see the remarks following (4)). (Nor is it essential that there be the same numbers of samples in the sequences $\mathscr{S}_1, \mathscr{S}_2$; the modifications in such a case are obvious and will not be discussed here.)

From (19) we can see that there will be increasing power, i.e. probability of reaching a correct decision) as the lengths of the sequences $\mathscr{S}_1, \mathscr{S}_2$ increase. In fact, results obtained in the next section support the view, stated above, that the fixed sample procedure is likely to be quite effective, even with a single pair of samples.

2.2 A Sequential Procedure

In the circumstances described above if pairs of samples $\mathscr{A}_{1j}$, $\mathscr{A}_{2j}$ from the sequences $\mathscr{S}_1$, $\mathscr{S}_2$ become available at about the same time, while there is an appreciable interval between successive pairs—$(\mathscr{A}_{1j}, \mathscr{A}_{2j})$ and $(\mathscr{A}_{1,j+1}, \mathscr{A}_{2,j+1})$—it is natural to consider using a sequential test procedure. A sequential probability ratio test discriminating between H_1 and H_2, using the pairs of samples as they become available, is based on the continuation region

$$\frac{\alpha_1}{1-\alpha_2} < \left\{\frac{(r_1+1)^{[s]}}{(r_2+1)^{[s]}}\right\}^m \prod_{j=1}^{m} \left[\frac{Y_{1j1}(1-Y_{1jr_1})}{Y_{2j1}(1-Y_{2jr_2})}\right]^s < \frac{1-\alpha_1}{\alpha_2} \tag{20}$$

with acceptance of H_1 (H_2) if the right (left) hand inequality is violated. (α_1, α_2 are the nominal chances of error when the hypotheses H_1, H_2 respectively are valid.)

The remainder of this section will be devoted to the special case $r_1 = r_2 = r$, in which (15) becomes

$$\left(\frac{\alpha_1}{1-\alpha_2}\right)^{1/s} < \prod_{j=1}^{m} \left\{\frac{Y_{1j1}(1-Y_{1jr})}{Y_{2jr}(1-Y_{2jr})}\right\} < \left(\frac{1-\alpha_1}{\alpha_2}\right)^{1/s}. \tag{21}$$

The limits depend on s, as well as on α_1 and α_2. The larger s, the closer together are the limits and the earlier a decision is reached. If an incorrect value of s—$\hat{s}$, say—is used in (17), then the approximate actual chances of error $\alpha_1(\hat{s})$ and $\alpha_2(\hat{s})$ can be obtained from the formulae

$$\left\{\frac{\alpha_1(\hat{s})}{1-\alpha_2(\hat{s})}\right\}^{1/s} \doteqdot \left(\frac{\alpha_1}{1-\alpha_2}\right)^{1/\hat{s}}; \left\{\frac{1-\alpha_1(\hat{s})}{\alpha_2(\hat{s})}\right\}^{1/s} \doteqdot \left(\frac{1-\alpha_1}{\alpha_2}\right)^{1/\hat{s}}$$

whence

$$\alpha_i(\hat{s}) \doteqdot \left[\left(\frac{\alpha_i}{1-\alpha_{3-i}}\right)^{s/\hat{s}} - \left\{\frac{\alpha_1\alpha_2}{(1-\alpha_1)(1-\alpha_2)}\right\}^{s/\hat{s}}\right] \cdot \left[1 - \left\{\frac{\alpha_1\alpha_2}{(1-\alpha_1)(1-\alpha_2)}\right\}^{s/\hat{s}}\right]^{-1}. \tag{22}$$

From (15)

$$\begin{aligned} E_1 &= E\left[s \ln\left\{\frac{Y_{1j1}(1-Y_{1jr})}{Y_{2j1}(1-Y_{2jr})}\right\} \middle| H_1\right] \\ &= s[2\psi(s+1) - 2\psi(r+2s+1) - 2\psi(1) + 2\psi(r+1)] \\ &= 2s\left(\sum_{j=1}^{s} j^{-1} - \sum_{j=r+1}^{r+2s} j^{-1}\right) \end{aligned} \tag{23}$$

and

$$E_2 = -E_1 \tag{24}$$

in an obvious notation.

The standard approximate formula for the average sample number (ASN) of the procedure when H_1 is valid, is

$$E_1^{-1}\{\alpha_1 \ln[\alpha_1/(1-\alpha_2)] + (1-\alpha_1)\ln[(1-\alpha_1)/\alpha_2]\}. \tag{25}$$

Table 2 presents a few values of this quantity for the symmetrical case $\alpha_1 = \alpha_2 = \alpha$. They are so small that they clearly are only very rough approximations. (It is, of course, impossible for the ASN to be less than one, in reality.) They do indicate, however, that a decision can be expected very soon in the process.

TABLE 2

Approximate Average Sample Numbers

	$\alpha = 0.01$		$\alpha = 0.001$	
$r \backslash s$	1	2	1	2
4	3.6	1.3	5.4	2.0
5	3.3	1.2	5.0	1.8
6	3.1	1.1	4.7	1.7
7	2.9	1.0	4.5	1.6
8	2.9	1.0	4.4	1.5
9	2.8	1.0	4.3	1.5
10	2.7	1.0	4.2	1.5
20	2.5	(0.9)	3.8	1.3
∞[a]	2.3	(0.8)	3.4	1.1

[a] The $r = \infty$ values are calculated from the formula

$$\frac{(1-2\alpha)\ln\{(1-\alpha)/\alpha\}}{2s\sum_{j=1}^{s} j^{-1}}.$$

3. DISTRIBUTION-FREE TESTS

3.1 Single Pair of Samples

We suppose we have a single pair of samples $\mathscr{A}_{11}, \mathscr{A}_{21}$ as described at the beginning of Section 2.1. It is remarkable that it is possible to construct a test of H_1 (or H_2) in this case with a known significance level without knowing the population distribution function of the X's, provided only that it is continuous, and the same for both $\mathscr{A}_{11}$ and $\mathscr{A}_{21}$. Heuristically, one might "expect" this, on the grounds that one of the two samples (the one which is not censored) provides some information on the population distribution.

The only assumption we need (apart from continuity) is that the population distribution of X is the same for $\mathscr{A}_{11}$ and $\mathscr{A}_{21}$.

Suppose H_1 is valid. Then the original sample sizes were $(r_1 + 2s)$ for $\mathscr{A}_{11}$ and r_2 for $\mathscr{A}_{21}$. For the original set of $(r_1 + r_2 + 2s)$ sample values there would be

$$\binom{r_1 + r_2 + 2s}{r_2}$$

possible rankings of these values, according to origin, in order of ascending magnitude. Since the population distributions for $\mathscr{A}_{11}$ and $\mathscr{A}_{21}$ are identical, these orderings would be equally likely, each having probability

$$\binom{r_1 + r_2 + 2s}{r_2}^{-1}.$$

(Under H_2, the number of equally likely orderings would be

$$\binom{r_1 + r_2 + 2s}{r_1}.)$$

To calculate the likelihood, based on ranks, corresponding to the specific sets of observed ranks we have to evaluate the numbers of rankings of the original $(r_1 + r_2 + 2s)$ observations which could have produced the observed ranks [of $(r_1 + r_2)$ observed values] after censoring. Under H_1, we can recover the original ranking by adding to $\mathscr{A}_{11}$, s observations less than the least observed value in $\mathscr{A}_{11}$, and s greater than the greatest observed value in $\mathscr{A}_{11}$.

If, among the $(r_1 + r_2)$ observed values in $\mathscr{A}_{11}$ and $\mathscr{A}_{21}$ combined, the L_2 least and the G_2 greatest are from $\mathscr{A}_{21}$ then there are

$$\binom{L_2 + s}{s}\binom{G_2 + s}{s}$$

corresponding original rankings. Hence the likelihood function for H_1 is

$$\mathscr{L}'_1 = \binom{L_2 + s}{s}\binom{G_2 + s}{s}\Big/\binom{r_1 + r_2 + 2s}{r_2}. \tag{26}$$

Similarly, the likelihood function for H_2 is

$$\mathscr{L}'_2 = \binom{L_1 + s}{s}\binom{G_1 + s}{s}\Big/\binom{r_1 + r_2 + 2s}{r_1} \tag{27}$$

where L_1, G_1 are defined analogously to L_2, G_2. The likelihood ratio (H_1/H_2) is

$$\frac{\mathscr{L}'_1}{\mathscr{L}'_2} = \frac{(r_1+1)^{[2s]}}{(r_2+1)^{[2s]}} \cdot \frac{(L_2+1)^{[s]}(G_2+1)^{[s]}}{(L_1+1)^{[s]}(G_1+1)^{[s]}}. \tag{28}$$

In the special case of equal observed sample sizes $(r_1 = r_2 = r)$ we have

$$\frac{\mathscr{L}'_1}{\mathscr{L}'_2} = \frac{(L_2+1)^{[s]}(G_2+1)^{[s]}}{(L_1+1)^{[s]}(G_1+1)^{[s]}}. \tag{29}$$

If we construct a rule: Choose H_1 (H_2) if $\mathscr{L}'_1/\mathscr{L}'_2 > (<)\, 1$, then the decision appears to depend on s, as well as L_1, L_2, G_1 and G_2. This was not the case with the "parametric" test [see (3)].

However, we note that just one of L_1 and L_2 must be zero, and just one of G_1 and G_2 must be zero. If both L_1 and G_1, (or both L_2 and G_2) are zero then the test will accept H_1 (or H_2) whatever be the value of s. If L_1 and G_2 are zero then the test accepts H_1 (H_2) if $L_2 > (<)\, G_1$; and if L_2 and G_1 are zero then the test accepts H_1 (H_2) if $G_2 > (<)\, L_1$. In all these cases the decision is in fact not dependent on s. So we can present the test in the form: accept H_1 (H_2) if

$$L_2 > G_1 \quad \text{or} \quad G_2 > L_1 \; (L_2 < G_1 \quad \text{or} \quad G_2 < L_1) \tag{30}$$

or even more simply: accept H_1 (H_2) if

$$L_2 + G_2 > (<)\, L_1 + G_1 \tag{30$'$}$$

whatever be the value of s. (Note that $L_2 > G_1$ implies $L_2 > 0$ and so also $L_1 = 0$.)

If $L_2 = G_1 > 0$ or $G_2 = L_1 > 0$, we are unable to reach a decision.

In order to obtain the distribution of $\mathscr{L}'_1/\mathscr{L}'_2$ we have to take into account not only the probabilities (26) or (27) but also the number of possible orderings of the values between the $\max(L_1+1,\ L_2+1)$ and $\min(r_1+r_2-G_1,\ r_1+r_2-G_2)$ order statistics of the combined $(\mathscr{A}_1 \cup \mathscr{A}_2)$ sample.

We have to multiply by

$$\left.\begin{array}{ll} 1\begin{cases} \text{if} & L_1 = r_1 \quad (\text{and so} \quad G_2 = r_2) \\ \text{or} & G_1 = r_1 \quad (\text{and so} \quad L_2 = r_2) \end{cases} & \\ \dbinom{r_1+r_2-L_1-L_2-G_1-G_2-2}{r_1-L_1-G_1-\phi} & \text{otherwise} \end{array}\right\} \tag{31}$$

where ϕ $(=0, 1, 2)$ is the number of zeroes in (L_1, G_1).

Note that (31) equals 1 if either L_i or G_i equals r_i or $(r_i - 1)$ for either $i = 1$ or $i = 2$.

Table 3 sets out the probabilities for the case H_1 valid. A similar table can easily be constructed for H_2 valid. [In calculating (31) we take $\binom{0}{0} = 1$.] (For $\max(l_1, g_1) < r_1$; $\max(l_2, g_2) < r_2$. If $\max(l_1, g_1) = r_1$, then $\max(l_2, g_2) = r_2$ and the entry in the fifth column is $(r_2 + 1)^{[s]}/s!$.)

TABLE 3

Distribution of $\mathscr{L}'_1/\mathscr{L}'_2$ under H_1

$L_1 =$	$L_2 =$	$G_1 =$	$G_2 =$	Probability $\times \binom{r_1 + r_2 + 2s}{r_2}$	$(\mathscr{L}'_1/\mathscr{L}'_2) \cdot (r_2 + 1)^{[2s]}/(r_1 + 1)^{[2s]}$
0	l_2	0	g_2	$(s!)^{-2}(l_2 + 1)^{[s]}(g_2 + 1)^{[s]} \times (31)$	$(l_2 + 1)^{[s]}(g_2 + 1)^{[s]}/(s!)^2$
0	l_2	g_1	0	$(s!)^{-1}(l_2 + 1)^{[s]} \times (31)$	$(l_2 + 1)^{[s]}/(g_1 + 1)^{[s]}$
l_1	0	0	g_2	$(s!)^{-1}(g_2 + 1)^{[s]} \times (31)$	$(g_2 + 1)^{[s]}/(l_1 + 1)^{[s]}$
l_1	0	g_1	0	$1 \times (31)$	$(s!)^2/\{(l_1 + 1)^{[s]}(g_1 + 1)^{[s]}\}$

When $r_1 = r_2$, the last column gives the values of $\mathscr{L}'_1/\mathscr{L}'_2$. If acceptance of H_1 follows when $(L_2 > G_1)$ or $(G_2 > L_1)$, then

$$\begin{aligned}\Pr[\text{correct decision} \mid H_1] &= \Pr[(L_2 > G_1) \cup (G_2 > L_1) \mid H_1] \\ &= \Pr[L_2 > G_1 > 0 \mid H_1] \\ &\quad + \Pr[G_2 > L_1 > 0 \mid H_1] \\ &\quad + \Pr[L_1 = G_1 = 0 \mid H_1] \qquad (32)\end{aligned}$$

(remembering that $L_2 > 0$, $G_2 > 0$ imply $L_1 = 0$, $G_1 = 0$ and conversely).

TABLE 4

$\binom{2r+2s}{r} = \binom{8+2s}{4} = 210 (s = 1);\ 495 (s = 2)$

l_1	l_2	g_1	g_2	(31)	$(l_2 + 1)^{[s]}(g_2 + 1)^{[s]}/(s!)^2$ $s = 1$	$s = 2$			
4	0	0	4	1	5	15			$(\alpha)''$
3	0	0	3	1	4	10			$(\alpha)''$
2	0	0	3	1	4	10		$(\alpha)'$	
1	0	0	3	1	4	10		$(\alpha)'$	
0	1	0	3	1	8	30	(α)		
2	0	0	2	2	3	6			$(\alpha)''$
1	0	0	2	3	3	6		$(\alpha)'$	
0	1	0	2	3	6	18	(α)		
0	2	0	2	1	9	36	(α)		
1	0	0	1	6	2	3			$(\alpha)''$
0	1	0	1	6	4	9	(α)		

		$s = 1$	$s = 2$
(α)	$p = \Pr[L_1 = G_1 = 0 \mid H_1] = \dfrac{(2 \times 8) + (2 \times 18) + 9 + 24}{210}$	$= \dfrac{85}{210}$	$= \dfrac{258}{495}$
$(\alpha)'$	$p' = \Pr[G_2 > L_1 > 0 \mid H_1] = \dfrac{4 + 4 + 9}{210}$	$= \dfrac{17}{210}$	$= \dfrac{38}{495}$
$(\alpha)''$	$p'' = \Pr[G_2 - L_1 > 0 \mid H_1] = \dfrac{5 + 4 + 6 + 12}{210}$	$= \dfrac{27}{210}$	$= \dfrac{55}{495}$
Probability of correct decision $(p + 2p')$:		$\dfrac{119}{210} = 0.567$	$\dfrac{334}{495} = 0.675$
Probability of no decision $(2p'')$:		$\dfrac{54}{210} = 0.257$	$\dfrac{110}{495} = 0.222$
Probability of incorrect decision:		$\dfrac{37}{210} = 0.176$	$\dfrac{51}{495} = 0.103$

From (32) and Table 3 it is possible to evaluate the probability of a correct decision (and the probability of not reaching a decision) when H_1 is valid. In the symmetrical case $(r_1 = r_2 = r)$ these probabilities have the same values when H_2 is valid.

The calculations for $r_1 = r_2 = 4$ are set out in detail in Table 4.

Table 5 presents the results of similar calculations.

The figures in Table 5 are fairly encouraging, when it is realized that we are trying to reach a conclusion on the basis of a single pair of samples, without knowing anything of the population distribution, except that it is continuous.

By analogy with results in [3] it is to be expected that as $r \to \infty$ the probability correct will tend to limits depending on s.

3.2 Sequence of Pairs of Samples

If we have two sequences, $\mathscr{S}_1$ and $\mathscr{S}_2$, as described in Section 2 we can expect considerable increases in power. The likelihood ratio test criterion, based on two sequences of m samples each, would be

$$\prod_{j=1}^{m} (\mathscr{L}'_{1j}/\mathscr{L}'_{2j}) \tag{33}$$

in an obvious notation. If $r_1 = r_2 = r$, the criterion would be

$$\prod_{j=1}^{m} \left[\frac{(L_{2j} + 1)^{[s]}(G_{2j} + 1)^{[s]}}{(L_{1j} + 1)^{[s]}(G_{1j} + 1)^{[s]}}\right] \tag{34}$$

TABLE 5

Properties of Distribution-free Test Based on a Single Pair of Samples Each Containing r Observed Values

	1			2		
$r \backslash s$	*Prob. correct*	*Prob. no dec.*	*Prob. incorrect*	*Prob. correct*	*Prob. no dec.*	*Prob. incorrect*
4	0.567	0.257	0.176	0.675	0.222	0.103
5	0.619	0.194	0.187	0.742	0.152	0.107
6	0.647	0.165	0.188	0.778	0.117	0.104
7	0.664	0.149	0.187	0.801	0.099	0.100
8	0.676	0.139	0.184	0.815	0.088	0.097
9	0.684	0.133	0.182	0.827	0.081	0.092
10	0.690	0.129	0.181	0.835	0.076	0.089
11	0.695	0.125	0.179	0.842	0.072	0.086
12	0.700	0.123	0.178	0.847	0.069	0.084
15	0.709	0.117	0.174	0.858	0.063	0.079
20	0.717	0.112	0.171	0.869	0.058	0.074
25	0.723	0.109	0.169	0.875	0.055	0.071
30	0.726	0.107	0.167	0.879	0.052	0.069
60[a]	0.735	0.102	0.164	0.889	0.047	0.064
∞[a]	0.743	0.097	0.160	0.899	0.043	0.058

[a] Values for $r = 60$ and $r = \infty$ were obtained by harmonic extrapolation. This table was calculated with the assistance of Anna Colosi, to whom I express my thanks.

and it would be reasonable to assign to H_1 (H_2) if its value is $> (<) 1$, reaching no conclusion if it equals 1.

Probabilities of correct and incorrect decision with this criterion, for the case $m = 2$, are shown in Table 6 for $r = 4(1)6$.

TABLE 6

Properties of Distribution-free Test Based on Two Pairs of Samples Each Containing r Observed Values

	1			2		
$r \backslash s$	*Prob. correct*	*Prob. no dec.*	*Prob. incorrect*	*Prob. correct*	*Prob. no dec.*	*Prob. incorrect*
4	0.791	0.117	0.093	0.897	0.076	0.027
5	0.832	0.084	0.085	0.936	0.044	0.020
6	0.851	0.066	0.083	0.954	0.029	0.017

These figures do, indeed, show considerable improvement over those in Table 5. They also confirm the conclusions in Section 2 on the power of discrimination when the population distribution(s) is (are) known, since the latter will be more powerful than the present procedure.

If it be supposed that the population distribution common to the two members of a pair of samples is the *same* for each pair, then an even better distribution free procedure should be feasible, taking account of this additional knowledge.

To construct such a procedure, it is necessary to enumerate all arrangements of the original $m(r_1 + r_2 + 2s)$ observations, which could result in the observed sequences $\mathscr{S}_1$ ($\equiv \mathscr{A}_{11}, \ldots, \mathscr{A}_{1m}$) and $\mathscr{S}_2$ ($\equiv \mathscr{A}_{21}, \ldots, \mathscr{A}_{2m}$) after removal of the s greatest and s least observations from each of the members of $\mathscr{S}_j$ (for H_j valid; $j = 1, 2$). Under H_j, the mr_{3-j} observations in $\mathscr{S}_{3-j}$ constitute a random sample of that size, but the mr_j observations in $\mathscr{S}_j$ have been censored in a special way.

Evaluation of the required probabilities appears to be rather complex, though not impossible. In view of the good power attainable with the procedure already discussed in this section (which can also be applied when variation in population distribution from pair to pair is suspected), we will not investigate further possibilities here.

A distribution free sequential probability ratio test has the continuation region

$$\frac{\alpha_1}{1-\alpha_2} < \left\{\frac{(r_1+1)^{[2s]}}{(r_2+1)^{[2s]}}\right\}^m \prod_{j=1}^{m} \left\{\frac{(L_{2j}+1)^{[s]}(G_{2j}+1)^{[s]}}{(L_{1j}+1)^{[s]}(G_{1j}+1)^{[s]}}\right\} < \frac{1-\alpha_1}{\alpha_2} \tag{35}$$

with $H_1(H_2)$ accepted if the left (right) inequality is the first to be violated.

4. SOME OTHER PROBLEMS

In the situation described in Section 3.1, if we just want to test the hypothesis: "Is $\mathscr{A}_1$ uncensored (assuming $\mathscr{A}_2$ to be uncensored)?", the likelihood criterion is simply

$$W = (L_2 + 1)^{[s]}(G_2 + 1)^{[s]} \tag{36}$$

with high values of W leading to rejection of the hypothesis.

(Note that $L_2 + 1 =$ rank order of least member of $\mathscr{A}$, and $r_1 + r_2 - G_2 =$ rank order of greatest member of $\mathscr{A}$ in the combined set of $r_1 + r_2$ observed sample values.)

Under $H_{s,s}$ (using the notation of [3] for symmetrical censoring of the s

extreme values from each end of the sample), the joint distribution of L_2 and G_2 is

$$\Pr[(L_2 = l_2) \cap (G_2 = g_2)] = \frac{\binom{r_1 + r_2 - l_2 - g_2 - 2}{r_2 - 2}\binom{l_2 + s}{s}\binom{g_2 + s}{s}}{\binom{r_1 + r_2 + 2s}{r_2}}$$

$$(0 < l_2, g_2; l_2 + g_2 \leq r_2). \quad (37)$$

It is interesting to note that if r_2 is large, so that we have a lot of information on the population distribution provided by a large random sample, known to be uncensored, then

$$Wr_2^{-2s} \doteqdot Y_{11}(1 - Y_{1r_1}) \quad (38)$$

which is the criterion used to test for symmetrical censoring (see (2) of [3]) when the population distribution is known.

We now, for a moment, consider the problem of detecting which one, out of k sequences $\mathscr{S}_1, \mathscr{S}_2, \ldots, \mathscr{S}_k$ of samples, has been symmetrically censored. We identify the hypothesis H_i with the statement "$\mathscr{S}_i$ is the censored sequence." For simplicity we will suppose that each available sample contains r observed values.

The method, of course, is to consider each set in turn as the censored one, the other $k - 1$ sets being uncensored. The appropriate likelihoods are those appropriate to the case of two samples (as in Sections 2 and 3) of sizes r and $(k - 1)r$ respectively, the smaller sample being the residue after removing the s greatest and s least values from a complete random sample of size $r + 2s$.

If the population distribution function(s) $F_j(\cdot)$ are known, the sequence $\mathscr{S}_i$ indicated as the censored one by this method is that for which

$$T_i = \prod_{j=1}^{m} [F_i(X_{j1})\{1 - F_i(X_{jr})\}] \quad (39)$$

is maximum among $T_1, T_2, \ldots, T_k$. The distribution-free approach leads to selection of that $\mathscr{S}_i$ for which

$$V_i = \prod_{j=1}^{m} (L_{ij} + 1)(G_{ij} + 1) \quad (40)$$

is maximum among $V_1, V_2, \ldots, V_k$ where $L_{ij}(G_{ij})$ is the number of observations less (greater) than the least (greatest) observation in $\mathscr{A}_{ij}$ among $\mathscr{A}_{1j}, \mathscr{A}_{2j}, \ldots, \mathscr{A}_{kj}$ for $j = 1, 2, \ldots, i - 1, i + 1, \ldots, m$.

Since we have a relatively large uncensored sample available (though we are not certain where it is) we should be able to construct good distribution-free procedures in this case. When H_i is valid and we are considering $\mathscr{S}_i$ versus the remaining sequences we are in effect in the situation, described earlier in this section, wherein we have a relatively large uncensored random sample providing information on the population distribution. When any other sequence—$\mathscr{S}_h$, say—is the one compared with the remainder, such bias as is introduced by the presence of $\mathscr{S}_i$ among the remainder tends to reduce apparent significance. It therefore seems that a procedure in which each sequence in turn is tested for symmetrical censoring against the remainder, in the way described at the beginning of this section, will provide useful information. The situation becomes more complicated if there can be more than one censored sequence, especially if the number of censored sequences is not known precisely.

Sequential procedures, choosing among the hypotheses $H_1, H_2, \ldots, H_k$ may be constructed in a straightforward way.

The discussion in this paper has been restricted to symmetrical censoring of extreme values. Analysis for other cases (e.g., censoring from above or below (i.e., on right or left)) follows parallel lines.

REFERENCES

[1] Johnson, N. L. (1966). Sample censoring. *Proc. Twelfth Conf. Des. Exp. Army Res. Dev. Testing* 403–424.

[2] Johnson, N. L. (1970). A general purpose test of censoring of sample extreme values. In *Essays in Probability and Statistics.* (S. N. Roy Memorial Volume) 379–384 Univ. of North Carolina Press, Chapel Hill.

[3] Johnson, N. L. (1971). Comparison of some tests of sample censoring of extreme values. *Austral. J. Statist.* **13** 1–6.

[4] Johnson, N. L. (1973–4). Robustness of certain tests of censoring of extreme sample values; I, II. *Univ. of North Carolina Mimeo Series* Nos. 866, 940.

[5] Johnson, N. L. (1974). Study of possibly incomplete samples. *Proc. Fifth Intern. Conf. Prob., Brasov, Romania.*

[6] Schumann, D. E. W. and Bradley, R. A. (1959). The comparison of the sensitivities of similar experiments: Model II of the analysis of variance. *Biometrics* **15** 405–416.

Part V

MATHEMATICAL PROGRAMMING AND COMPUTING

Absolute Deviations Curve Fitting: An Alternative to Least Squares

Roger C. Pfaffenberger *John J. Dinkel*

UNIVERSITY OF MARYLAND THE PENNSYLVANIA STATE UNIVERSITY

The determination of minimum absolute deviation estimators of regression coefficients by linear programming is reviewed. The linear program is shown to be equivalent to a geometric program. As an illustration of the small sample properties of the estimators, a Monte Carlo simulation experiment is conducted using Normal, Cauchy, and Laplace error distributions.

1. INTRODUCTION

In the past few years, there has been growing interest, especially in the Business and Economic disciplines, in the use of nonnormal error distributions in fitting linear models to certain economic processes, such as changes in stock prices, changes in other speculative prices, and interest rate changes (e.g., see Fama [14] and Mandelbrot [18, 19]). In particular, the Cauchy distribution has drawn considerable attention in modeling error distributions of highly speculative investment return rates due to its "fat tails." Other symmetric distributions used in these applications are the Laplace, logistic (see Chew [11]) and members of the symmetric stable family of distributions to which both the normal and Cauchy distributions belong (see Feller [15] for a discussion of the theoretical properties of the stable distributions).

When the errors in a linear model are nonnormally distributed, alternative estimators to least squares have been proposed. One such alternative is

Key words and phrases Minimum absolute deviation, geometric programming, curve fitting, normed smoothing.

M.A.D.—the Minimum Absolute Deviations estimator. In this paper, we review the determination of the M.A.D. estimates of regression coefficients by linear programming, develop the equivalence of the LP program to a geometric program, and compare the least squares and the M.A.D. estimators in "small" samples for Cauchy and Laplace distributed errors by Monte Carlo simulation experiments.

2. M.A.D. ESTIMATION AND GEOMETRIC PROGRAMMING

Let $x_1, x_2, \ldots, x_m$ be m independent variables and let y be dependent on these. Let $(y_1; x_{11}, x_{12}, \ldots, x_{1m}), \ldots, (y_n; x_{n1}, x_{n2}, \ldots, x_{nm})$ be n $(n > m)$ observations on these variables. The M.A.D. estimation problem is

$$\underset{(b_0, b_1, \ldots, b_m)}{\text{minimize}} \sum_{i=1}^{n} \left| y_i - \left(b_0 + \sum_{j=1}^{m} b_j x_{ij} \right) \right|, \tag{1}$$

where $b_0, b_1, \ldots, b_m$ are unrestricted in sign. Let X_i denote the ith deviational value, given by:

$$X_i = y_i + \sum_{j=0}^{m} a_{ij} b_j \qquad i = 1, 2, \ldots, n, \tag{2}$$

where $a_{ij} = -x_{ij}, j \neq 0$, and $a_{i0} = -1$, for all i. Now, define the l_p norm:

$$\|X\|_p = \left\{ \sum_{i=1}^{n} |X_i|^p \right\}^{1/p} \tag{3}$$

From (3), it is clear that the M.A.D. estimator is the minimum over $b = (b_0, b_1, \ldots, b_m)$ of the l_1 norm; that is,

$$\min_b \|X\|_1 \qquad b = (b_0, b_1, \ldots, b_m). \tag{4}$$

Linear programming was proposed as a means of solving (4) by Wagner [28, 29], among others, who showed that the LP program,

$$\min_{b, d^+, d^-} \sum_{i=1}^{n} (d_i^+ + d_i^-) \tag{5}$$

$$\text{subject to} \quad b_0 + \sum_{j=1}^{m} b_j x_{ij} + d_i^+ - d_i^- = y_i \qquad i = 1, 2, \ldots, n$$

$$d_i^+, d_i^- \geq 0, \ b_j \text{ unrestricted in sign}, \tag{6}$$

where d_i^+ and d_i^- represent the positive and negative deviation, respectively, for the ith observation, produces the solution to (4). As Wagner [28], and others have pointed out, the dual program to (5) and (6) is more easily solved than the primal program. Recently, a number of papers have appeared in the literature which are related to the efficient numerical solution to (5) and (6), or its dual. The research by Barrodale [4], Barrodale and Young [7], and Barrodale and Roberts [5] has resulted in a modified version of the simplex algorithm for solving (5) and (6). The complete details of the algorithm may be found in [6], where the authors present some numerical results showing the efficiency of this algorithm as compared to the standard simplex algorithm. Robers and Ben-Israel [23] and Appa and Smith [2] have developed some specialized algorithms for dealing with the dual program to (5) and (6). The dual is computationally more attractive to solve due to the reduced number of constraints, especially when GUB (Generalized Upper Bound) simplex algorithms are used. Several additional computational procedures have been proposed: a descent method by Usow [27], a determination of l_1 as a limit of l_p approximation as $p \to 1^+$ by Abdelmalek [1] and a similar approach using l_2 by Schlossmacher [24]. Sielken and Hartley [25] have developed an algorithm which guarantees unbiased estimates of the regression coefficients, while other algorithms will typically produce biased estimates when the solution to (5) and (6) is not unique.

In order to develop a unified presentation of duality for unconstrained and constrained l_1 estimation, we present the notion of extended geometric programming. The theory of ordinary geometric programming is based on the arithmetic–geometric inequality. Extended geometric programming is based on more general inequalities; for example, the harmonic mean inequality. The general properties of such inequalities are given in Chapter 7 of [13].

Duffin *et al.* [13] show that the l_p norms satisfy the conditions of a geometric inequality except for the nondifferentiability of l_1 and l_∞. To illustrate that geometric programming can be extended by relaxing the differentiability requirement and, thus, can be used to provide an approach to l_1 estimation, we give the following version of Lemma 7.5.1 in [13]:

If X, δ are arbitrary vectors in E^n then

$$\sum_{i=1}^{n} X_i \delta_i \leq \|\delta\|_\infty \|X\|_1, \tag{7}$$

where $\|\delta\|_\infty = \max_i |\delta_i|$, $\|X\|_1 = \sum_{i=1}^{n} |X_i|$. Moreover, (7) is an equality if and only if

$$X_i \delta_i > 0 \qquad \text{for} \quad i \in I(X) \cap I(\delta) \tag{8}$$

and

$$|\delta_i| = \max_j |\delta_j| \qquad i \in I(X), \tag{9}$$

where $I(v) = \{i\colon v_i \neq 0\}$.

Proof From elementary inequalities, we have

$$\sum_{i=1}^{n} X_i \delta_i \leq \sum_{i=1}^{n} |X_i| \, |\delta_i| \leq \left\{ \sum_{i=1}^{n} |X_i| \right\} \max_i |\delta_i|$$

which is (7). The equality conditions are:

The first inequality is an equality if and only if the algebraic sign of $X_i \delta_i$ is positive; that is,

$$X_i \delta_i > 0 \qquad \text{for} \qquad X_i, \delta_i \neq 0,$$

which is (8).

The second inequality is an equality if and only if

$$|\delta_i| = \max_j |\delta_j|, \qquad \delta_j \neq 0$$

which is (9).

Thus, equality in (7) is achieved if and only if *both* of these conditions hold.

Based on the inequality of Lemma 7.5.1, we can develop the geometric programming dual to (4) and use the equality conditions (8) and (9) of the inequality in (7) to characterize optimal solutions. The geometric programming dual program for (4) is

$$\text{maximize} \quad \sum_{i=1}^{n} y_i \delta_i \tag{10}$$

$$\text{subject to} \quad \|\delta\|_\infty = 1 \tag{11}$$

$$\sum_{i=1}^{n} a_{ij} \delta_i = 0 \qquad j = 0, 1, \ldots, m. \tag{12}$$

To develop this relationship, we can assume that rank $(a_{ij}) = m$ and $y \notin$ column space (a_{ij}). With these assumptions, the consistency of the constraints (11) and (12) may be established as follows:

The general solution to (10)–(12) can be written as

$$z_i = \sum_{j=1}^{n-m} r_j d_i^{(j)} \qquad i = 1, 2, \ldots, n \tag{13}$$

and $\delta_i = z_i / \|z\|_\infty$, $i = 1, 2, \ldots, n$ satisfies (11) since $z \neq 0$ due to $y \notin$ column space of (a_{ij}). It also follows from these assumptions that $\sum_{i=1}^{n} y_i \delta_i$ must be positive. From Lemma 7.5.1 we infer

$$\sum_{i=1}^{n} y_i \delta_i \leq \|X\|_1$$

and hence the maximum of the dual problem is less than or equal to the minimum of the primal problem. To show they are equal we note by the compactness of the dual constraints and the continuity of (10) that there exists a δ^* which maximizes the dual problem. Using (13), we write

$$\sum_{i=1}^{n} y_i\delta_i = \sum_{j=1}^{n-m} L_j r_j \qquad \text{where} \quad L_j = \sum_{i=1}^{n} y_i d_i^{(j)}.$$

Now, define the index sets

$$\begin{aligned} J^+ &= \{i\colon\ \delta_i^* = 1\} \\ J^- &= \{i\colon\ \delta_i^* = -1\} \end{aligned} \tag{14}$$

and consider the linear system

$$\begin{aligned} \sum_j (-d_i^{(j)}) r_j \geq 0 \qquad & i \in J^+ \\ \sum_j d_i^{(j)} r_j \geq 0 \qquad & i \in J^- \\ \sum_j L_j r_j \geq 0 \qquad & \end{aligned} \tag{15}$$

and its "dual system"

$$\sum_{J^+} (-d_i^{(j)}) X_i + \sum_{J^-} d_i^{(j)} X_i + L_j X_{n+1} = 0 \qquad j = 1, \ldots, n-m$$

$$X_i \geq 0 \qquad i = 1, 2, \ldots, n+1. \tag{16}$$

By Tucker's theorem [13] there exist solutions to (15) and (16), r_j^* and X_i^* respectively, such that

$$\begin{aligned} X_i^* + \sum_j (-d_i^{(j)}) r_j^* > 0 \qquad & i \in J^+ \\ X_i^* + \sum_j d_i^{(j)} r_j^* > 0 \qquad & i \in J^- \\ X_{n+1}^* + \sum_j L_j r_j^* > 0. \qquad & \end{aligned}$$

Since $y \notin$ column space of (a_{ij}) then not all $X_i^* = 0$, $i \in J^+ \cup J^-$. However,

$$\sum_{j=1}^{n-m} L_j r_j^* = \sum_{i=1}^{n} y_i \delta_i^* > 0$$

implies by Tucker's theorem that $X_{n+1}^* = 0$. Thus, define

$$X_i' = \begin{cases} X_i^* & i \in J^+ \\ -X_i^* & i \in J^- \\ 0 & i = n+1 \end{cases} \tag{17}$$

and the equality conditions of Lemma 7.5.1. follow from (14) and (17).

The equality conditions (8) and (9) of the Lemma serve to characterize the optimal solutions to the primal and dual l_1 programs. Since the value of $\|X^*\|_1$ is assumed to be greater than zero, then according to (9) there is at least one δ_i^* such that $|\delta_i^*| = 1$ and thus from (8) it follows that

if $\delta_i^* = 1$ and $X_i^* \neq 0$, then $X_i^* > 0$—the observation lies *above* the optimal hyperplane;
if $\delta_i^* = -1$ and $X_i^* \neq 0$, then $X_i^* < 0$—the observation lies *below* the optimal hyperplane.

We also note from (8) and (9) that we may have $|\delta_i^*| = 1$ for $X_i^* = 0$; that is, the optimal hyperplane passes through the observation X_i, but the value of the dual variable is ± 1. However, from (8) and (9), we note that

if $\delta_i^* = 1$, then $X_i^* \geq 0$ is optimal for any such X_i;
if $\delta_i^* = -1$, then $X_i^* \leq 0$ os optimal for any such X_i.

Thus, $|\delta_i^*| = 1$, $X_i^* = 0$ characterizes those programs with alternative optimal solutions.

The equality conditions of Lemma 7.5.1 can be used to show the effect of parametrization of the data (changes in y_i) and thus show the insensitivity of the l_1 norm to "wild data points". Suppose $X_i^* = y_i + \sum_{j=1}^m a_{ij} b_j^*$ is optimal for the set of data points given by y_i, $i = 1, 2, \ldots, n$, and we want to study the effect on the optimal solution of parametrizing the values of the y_i. That is, consider

$$X_i = (y_i + \rho_i) + \sum_{j=1}^{m} a_{ij} b_j \qquad i = 1, 2, \ldots, n, \tag{18}$$

where ρ_i is a scalar and define

$$\hat{X}_i \equiv X_i - \rho_i. \tag{19}$$

Now, if X_i^*, δ_i^* are optimal for the primal and dual programs and if $X_i^* > 0$, then X_i^* maintains the optimality of the current solution as long as

$$\hat{X}_i^* \delta_i^* = (X_i^* - \rho_i)\delta_i^* \geq 0.$$

Thus, since the observation (X_i^*) lies *above* the optimal hyperplane, any reduction is allowable until the observation lies on the optimal hyperplane. On the other hand, any increase in the value of the observation away from the optimal hyperplane will not affect the solution; that is, from (18) and (19), ρ_i can be made arbitrarily large which is a restatement of the condition that the l_1 criterion is insensitive to "wild data points".

In addition to this insensitivity to wild data points, the previous result also shows that the original data can be reparametrized while maintaining the current optimal hyperplane.

To illustrate these results, consider the example presented by Appa and Smith [2] with the data:

x_i:	0	1	2	3	4	5	6	7	8	9
y_i:	5	5.8	4	5.6	3	5.4	2	5.2	1	5

The optimal hyperplane according to the l_1 criterion is $b_0 = 5$ (intercept) and $b_1 = 0$ (slope) and as seen in Fig. 1, it is clear that the line does not reflect the data trend. While somewhat surprising initially, this result is not unexpected in light of the previous discussion on parametrizing the y_i values. For example, the hyperplane described previously is also optimal for the data set (among others)

x_i:	0	1	2	3	4	5	6	7	8	9
y_i:	5	5.2	4	5.2	4	5.2	4	5.2	4	5

for which the optimal hyperplane does reflect the data trend (see Fig. 1). In addition, we could "move" points away from the optimal hyperplane without affecting the solution. For example, we could replace the data point (3, 5.6) by (3,100) and the hyperplane (line) in Fig. 1 remains optimal.

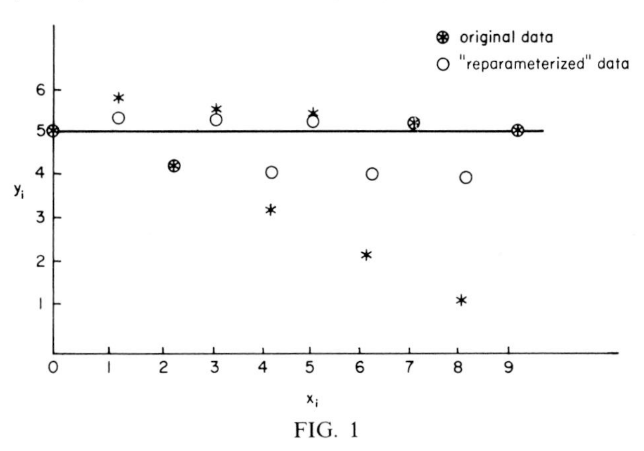

FIG. 1

The robustness of l_1 norm estimators to "outliers" is its most attractive property in linear model applications to certain economic variables. The bothersome attribute, indicated in the previous example, is the potential for nonunique solutions. When the solution to (5) and (6) is nonunique, Harter [17] suggests determining the limiting l_1 lines and forming as an estimate the line which is the locus of points whose vertical distances from the limiting lines are equal. Sielken and Hartley [25] propose an algorithm which produces unbiased estimates when the solution is nonunique (or unique). The obvious appeal of this method is the selection of a line, among all optimal lines, which has the presumably desirable property of unbiasedness.

We have used geometric programming here specifically to modify and to extend results first due to Appa and Smith [2]. Since the Appa and Smith paper, these properties, as well as others, have been reported by a number of researchers using LP techniques (see, in particular, Sposito *et al.* [26]). The geometric programming approach offers these advantages to LP in our view:

(1) by using the extended inequality, special algorithms may be developed for solving the l_1 programming problem and
(2) some properties are more easily understood from the GP structure.

In particular, we report [12] on properties of the constrained l_1 (and l_p) norm problem by using GP, where the form of the constraints are

(i) boundary constraints on the parameters in b of the form, $L_j \leq b_j \leq U_j, j = 0, 1, 2, \ldots, m$; L_j, U_j constant and
(ii) a constraint on the largest absolute deviation, of the form $\max_i |X_i| \leq M$, where M is a predetermined constant.

3. PROPERTIES OF THE l_1 NORM ESTIMATORS IN REGRESSION FOR SMALL SAMPLES

Since small-sample properties of the M.A.D. estimator of the $(m + 1)$-vector **b** are unknown due to mathematical problems in deriving its sampling distribution, simulation experiments have been performed to assess the behavior of the estimator (e.g., [3], [8], [16], [20], and [22]). To illustrate important small-sample properties of the M.A.D. estimator and to verify conclusions reached by others through simulation experiments, we conducted a $3 \times 6 \times 4$ factorial experiment, with the following factors:

(1) Error distribution
 (i) Normal: $N(0, 1)$
 (ii) Cauchy: $f(x) = [\pi(1 + x^2)]^{-1} \quad -\infty < x < +\infty$
 (iii) Laplace: $f(x) = (\sqrt{2}/2) \exp(-\sqrt{2}|x|) \quad -\infty < x < +\infty$
(2) Sample Size
 $N = 10, 20, 40, 80, 100$, and 120
(3) Regression Model Form
 (i) $y = b_0 + b_1 x + e, \quad x \sim N(10, 4)$
 (ii) $y = b_0 + b_1 x + e, \quad x_i = i/N, \quad i = 1, 2, \ldots, N$ (The "Trend line" model)
 (iii) $y = b_0 + b_1 x_1 + b_2 x_2 + e; \quad x_1, x_2 \sim N(10, \sqrt{8})$
 (iv) $y = b_0 + \sum_{j=1}^{5} b_j x_j + e, x_j \sim N(10, \sqrt{16/5}) \quad j = 1, \ldots, 5$

For each treatment (combination of factor levels), 100 Monte Carlo samples were drawn and the means and standard deviations of the l_1 and l_2 estimates were computed.

The error distribution deviates for the Laplace and Cauchy distributions were produced by inversion of the CDF functions and the use of RAND uniform (0, 1) variables produced from the IBM 370-168 computer system and software package. The normal error deviates were generated by using the Box–Muller [9] method. Double precision was used throughout the experiment.

Only a portion of the simulation results are reported here. For a complete discussion of the simulation experiment and the results, see [21]. Our results do concur with those of others, most notably the results reported in [16] and in [22]. The simulation experiments conducted to date on comparing l_1 and l_2 estimators of linear model coefficients in small samples confirm the robustness of l_1 to outliers and fat-tailed error distributions. Due to the increased computing power available today, our experiment is more extensive than the one conducted by Rice and White [22] on the Laplace and Cauchy distributed errors. Interestingly, our results on the robustness of l_1 are similar to those experienced by Rice and White, yet they appear to have used an improperly defined Laplacian function (see [22], p. 248) and incorrect inversion formulas for both the Cauchy and the Laplace distributions (see [22], p. 254).

To summarize our results, l_1 performs much better than does l_2 when the error distribution is Cauchy. This result is not surprising, since l_2 (least squares) is very sensitive to "outliers"—values of the variate in the sample that are atypically large or small. The l_2 estimator is, of course, optimal when the error distribution is normal and the simulation experiments reflect this fact. Since the Laplace distribution is similar to the normal distribution in its characteristics, the l_1 and l_2 estimators were much alike for most treatments, though l_1 tended to outperform the l_2 estimator slightly in the majority of cases.

In Table 1 are given the ratios of mean square errors† ($\mathrm{MSE}(l_1)/\mathrm{MSE}(l_2)$) for the two estimators in the regression model $y = b_0 + b_1 x_1 + b_2 x_2 + e$, where e follows each of the error distributions and for sample sizes $n = 20$, 40, and 100. Notice that l_2 is more efficient (the ratio $\mathrm{MSE}(l_1)/\mathrm{MSE}(l_2) > 1$) when the error distribution is normal, but that l_1 is more efficient for Laplace errors and *greatly* more efficient for Cauchy errors. The experience with these treatments typified the results for all treatments in the experiment.

† The mean square error of an estimator of the kth regression coefficient, b_k, is given by $\mathrm{MSE} = \sum_{i=1}^{M} (\hat{b}_k - b_k)^2/M$, where $\hat{b}_k$ estimates b_k, and M is the number of Monte Carlo samples for the specified treatment.

TABLE 1

Ratio of Mean Square Errors (l_1/l_2)

	$n = 20$			$n = 40$			$n = 100$		
Parameter	Normal	Laplace	Cauchy	Normal	Laplace	Cauchy	Normal	Laplace	Cauchy
b_0	1.24	0.92	0.003	1.38	0.82	0.005	1.49	0.96	0.003
b_1	1.45	0.81	0.006	1.62	0.88	0.001	1.18	0.83	0.009
b_2	1.42	0.98	0.002	1.60	0.92	0.001	1.24	0.78	0.003

The Appendix gives comparison of the standard deviations of the l_1 and l_2 estimators in the model, $y = b_0 + b_1 x$ for the three error distributions. As described previously, l_1 is more efficient (in MSE ratio) than l_2 for the Laplace and dramatically so for the Cauchy.

4. SUMMARY

The difficulty in calculating l_1 (M.A.D.) estimates of regression coefficients should no longer present a reason for selecting the l_2 (least squares) estimator over the l_1 estimator. By using LP or specific l_1 norm algorithms, the computational tasks of producing l_1 estimates is simplified immensely. More importantly, due to the current efficiency in computing l_1 estimates, it is now possible to study the small sample properties of the l_1 estimators by simulation techniques.

The family of symmetric stable distributions is attracting much attention in the economic literatures and with the recent article by Chambers *et al.* [9], considerable simulation research on the small-sample properties of l_1 norm estimators in linear models can be anticipated. Additionally, it can be expected that mathematical programming techniques will be productively used to identify properties of l_p norm estimators, particularly in constrained linear and nonlinear systems.

APPENDIX

Standard Deviations of Estimators Plotted Against Sample Size; l_1 is the M.A.D. Estimator; l_2 is the Least Squares Estimator.

Normal errors $Y = b_0 + b_1 x + e$; $x \sim N(10, 4)$, $(l_i, 1)$ denotes the ith estimate standard deviation of the slope, $(l_i, 2)$ is the standard deviation of the intercept.

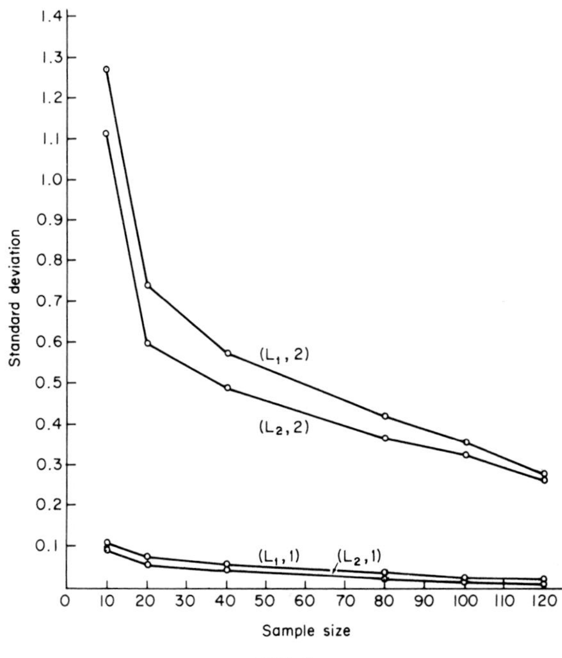

FIG. 2

Standard Deviations of the l_1 and l_2 Estimators Plotted Against Sample Size.

Laplace errors $Y = b_0 + b_1 x + e; x \sim N(10, 4)$.

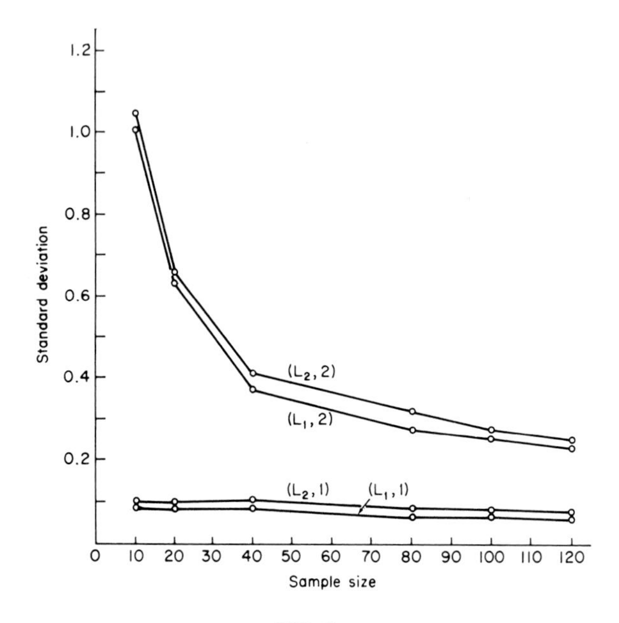

FIG. 3

Standard Deviations of the l_1 Estimator Plotted Against Sample Size.

Cauchy errors $Y = b_0 + b_1 x + e;\ x \sim N(10, 4)$.

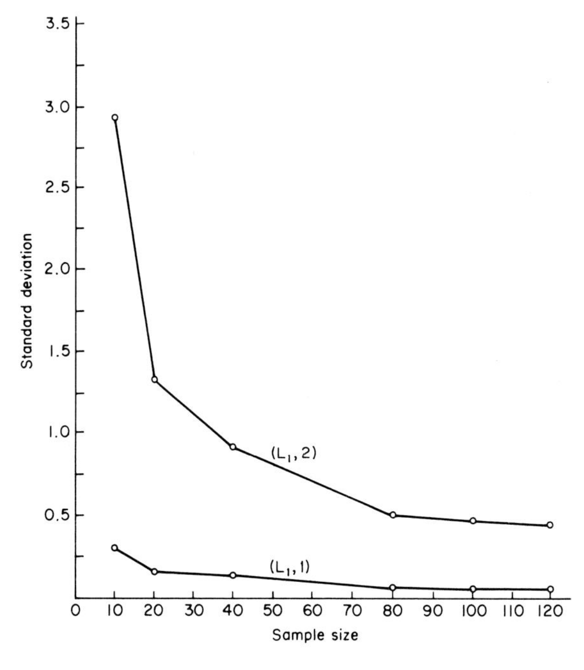

FIG. 4

Standard Deviations of the l_2 Estimator Plotted Against Sample Size.

Cauchy errors $Y = b_0 + b_1 x + e;\ x \sim N(10, 4)$.

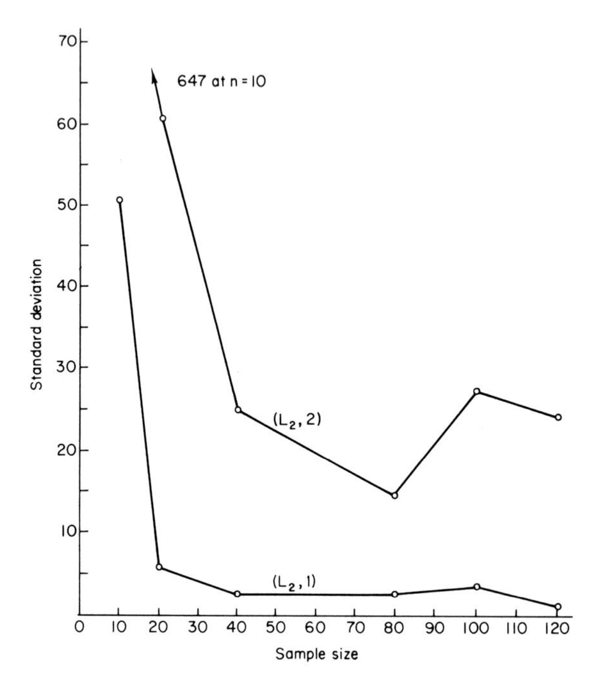

FIG. 5

REFERENCES

[1] ABDELMALEK, N. N. (1971). Linear l_1 approximation for a discrete point set and l_1 solutions of overdetermined linear equations. *J. Assoc. Comput. Mach.* **18** 41–47.

[2] APPA, G. and SMITH, C. (1973). On l_1 and Chebyshev estimation. *Math. Programming* **5** 73–87.

[3] ASHER, V. G. and WALLACE, T. D. (1963). A sampling study of minimum absolute deviations estimators. *Operations Res.* **2** 747–758.

[4] BARRODALE, I. (1968). l_1 approximation and the analysis of data. *Appl. Statist.* **17** 51–57.

[5] BARRODALE, I. and ROBERTS, F. D. K. (1970). Application of mathematical programming to l_p approximation. In *Nonlinear Programming*. 447–464. J. B. Rosen, O. L. Mangasarian, and K. Ritter (eds.). Academic Press, New York.

[6] BARRODALE, I. and ROBERTS, F. D. K. (1973). An improved algorithm for discrete l_1 linear approximation. *SIAM J. Numer. Anal.* **10** 839–848.

[7] BARRODALE, I. and YOUNG, A. (1966). Algorithms for best l_1 and linear approximations on a discrete set. *Numer. Math.* **8** 295–306.

[8] BLATTBERG, R. and SARGENT, T. (1971). Regression with non-Gaussian disturbances. *Econometrica* **39** 501–510.

[9] BOX, G. E. P. and MULLER, M. E. (1958). A note on the generation of normal deviates. *Ann. Math. Statist.* **29** 610–611.

[10] CHAMBERS, J. M., MALLOWS, C. L., and STUCK, B. W. (1976). A method for simulating stable random variables. *J. Amer. Statist. Assoc.* **71** 340–344.

[11] CHEW, V. (1968). Some useful alternatives to the normal distribution. *Amer. Statist.* **22**(3) 22–24.

[12] DINKEL, J. and PFAFFENBERGER, R. (1977). Constrained l_1 estimation via geometric programming. Submitted to the *J. Optimization Theory Appl.*

[13] DUFFIN, R. J., PETERSON, E. L., and ZENER, C. (1967). *Geometric Programming.* Wiley, New York.

[14] FAMA, E. F. (1965). The behavior of stock market prices. *J. Business* **38** 34–105.

[15] FELLER, W. (1966). *An Introduction to Probability Theory and Its Applications.* Vol. II. Wiley, New York.

[16] GLAHE, F. R. and HUNT, J. G. (1970). The small sample properties of simultaneous least absolute estimators vis-a-vis least squares estimators. *Econometrica* **38** 742–753.

[17] HARTER, H. L. (1976). Nonuniqueness of least absolute values regression. Paper presented at the Conference on Robust Procedures, Iowa City, Iowa, July 19–23.

[18] MANDELBROT, B. (1963). The stable Paretian income distribution when the apparent exponent is near two. *Internat. Econom. Rev.* **4** 111–114.

[19] MANDELBROT, B. (1963). The variation of certain speculative prices. *J. Business* **36** 394–419.

[20] MEYER, J. R. and GLAUBER, R. R. (1964). *Investment Decisions, Economic Forecasting and Public Policy.* Div. of Res., Grad. School Bus. Admin., Harvard Univ., Cambridge, Massachusetts.

[21] PFAFFENBERGER, R. C. and DINKEL, J. (197). A Monte Carlo experiment on l_1 and l_2 norm estimation in linear models with Cauchy- and Laplace-distributed errors. *Working Paper Series*, #14. College of Business and Management, Univ. of Maryland, College Park, Maryland.

[22] RICE, J. R. and WHITE, J. S. (1964). Norms for smoothing and estimation. *SIAM Rev.* **6** 243–256.

[23] ROBERS, P. D. and BEN-ISRAEL, A. (1969). An interval programming algorithm for discrete linear l_1 approximation problems. *J. Approximation Theory* **2** 323–336.

[24] SCHLOSSMACHER, E. J. (1973). An iterative technique for absolute deviations curve fitting. *J. Amer. Statist. Sssoc.* **68** 857–859.

[25] SIELKEN, R. L. and HARTLEY, H. O. (1973). Two linear programming algorithms for unbiased estimation of linear models. *J. Amer. Statist. Assoc.* **68** 639–641.

[26] SPOSITO, V. A., KENNEDY, W. J., and GENTLE, J. E. (1976). Generalized properties of l_1 estimators. Technical Report, Iowa State University.

[27] USOW, K. H. (1967). On l_1 approximation II: Computation for discrete functions and discretization effects. *SIAM J. Numer. Anal.* **4** 233–244.

[28] WAGNER, H. M. (1969). Linear programming techniques for regression analysis. *J. Amer. Statist. Assoc.* **54** 206–212.

[29] WAGNER, H. M. (1962). Nonlinear regression with minimal assumptions. *J. Amer. Statist. Assoc.* **57** 572–578.

Tables of the Normal Conditioned on t-Distribution

D. B. Owen *R. W. Haas*

SOUTHERN METHODIST UNIVERSITY

Various representations of the joint distribution of Y and T are obtained when (X, Y) has a joint bivariate normal distribution and $T = (x - \mu_x)/s_x$, where s_x is the usual estimator of σ_x. Percentiles of the normal conditioned on t-distribution are then tabulated. An application is also discussed.

1. INTRODUCTION

In many situations a buyer of an article will want to ensure that the lifetime, Y, of an article is greater than some value, L, say, but he cannot obtain assurance of this without destroying the article. Here we consider the bivariate normal model of the situation where there is a variable, X, correlated with Y which is more easily observed than Y. The variable X is called a screening variable.

In Owen *et al.* (1975) this model was explained in more detail and tables were given for the case where all of the parameters are known. Specifically the procedure given there is to select all articles with $X \geq \mu_x - K_\beta \sigma_x$, where μ_x and σ_x are the mean and variance of X, and K_β is read from a table of the normal conditioned on normal distribution given in Owen *et al.* (1975). This

Research for this paper was supported in part by Office of Naval Research Contract N00014-76-C-0613.

Key words and phrases Screening variables, bivariate normal model, selection procedures using correlated variates, correlated variates, acceptance sampling.

procedure guarantees that $Y \geq L$ with probability $\delta = 0.95$, 0.99, or 0.999 depending on which of the tables given in Owen *et al.* (1975) is used.

In order to enter those tables it is necessary to know the correlation coefficient, ρ, and also the proportion of $Y \geq L$ before selection is started. This latter proportion is designated γ. For a thorough discussion of this model and the various implications the reader is referred to the paper by Owen *et al.*

This approach is extended in Owen and Boddie (1976) and in Owen and Su (1977) to some cases where parameters are unknown. In this paper we will consider only the case where μ_x and σ_x are unknown, but ρ and γ are known. Extensions to unknown ρ or γ are considered by Owen and Su (1977), but if μ_x and σ_x are also unknown it is necessary to have the tables presented here in order to implement those procedures.

2. MODEL DEVELOPMENT

If μ_x is known and σ_x is unknown we take a preliminary sample of size n to estimate this parameter in the usual way, i.e. we compute $s^{*2} = \sum_{i=1}^{n} (X_i - \mu)^2/n$. Then for each additional article we select only those for which $X \geq \mu_x - K_\beta s^*$.

Hence we want to find K_β so that

$$P\{Y \geq \mu_y - K_\gamma \sigma_y | X \geq \mu_x - K_\beta s^*\} = \delta,$$

where K_γ is a $100\gamma\%$ point of the normal distribution, i.e.,

$$\frac{1}{\sqrt{2\pi}} \int_{-\infty}^{K_\gamma} e^{-1/2t^2}\, dt = \gamma.$$

If this probability is standardized, we obtain

$$P\{Z \geq -K_\gamma | T_f \geq -K_\beta\} = \delta$$

or equivalently

$$P\{Z \leq K_\gamma | T_f \leq K_\beta\} = \delta, \tag{1}$$

where $Z = (Y - \mu_y)/\sigma_y$ and $T_f = (X - \mu)/s^*$ has the student t-distribution with $f = n$ degrees of freedom.

The value of K_β which satisfies (1) is the quantity given in Table 1 at the end of the chapter.

We note in passing that if the mean, μ_x, is also unknown, we estimate μ_x and σ_x^2 by $\bar{X} = \sum_{i=1}^{n} X_i/n$ and $s^2 = \sum_{i=1}^{n} (X_i - \bar{X})^2/(n-1)$, respectively. In this case the procedure is to select all articles with $X \geq \bar{X} - K_\beta s[(n+1)/n]^{1/2}$ where K_β is read from the tables given here, and where the tables are entered with $(\rho[n/(n+1)]^{1/2}, \gamma, \delta, f)$ instead of $(\rho, \gamma, \delta, f)$.

3. MATHEMATICAL DEVELOPMENT

We give several representations of $P_f(K_\gamma, K_\beta) = P\{Z \le K_\gamma,\ T_f \le K_\beta\}$. Hsiao (1976) derived two of these, (2) and (3) below. We start with the joint bivariate (V, Z) normal distribution and square root of a chi-square divided by degrees of freedom, U, distribution with f degrees of freedom. Since these two densities are independent we can write the joint distribution of the three variables as

$$\frac{1}{(1-\rho^2)^{1/2}}\, G'\left(\frac{Z-\rho V}{(1-\rho^2)^{1/2}}\right) G'(V) \frac{f^{f/2}\sqrt{2\pi}}{\Gamma\left(\frac{f}{2}\right) 2^{(f-2)/2}}\, U^{f-1} G'(U\sqrt{f}),$$

where $G'(Z) = (2\pi)^{-1/2} e^{-Z^2/2}$.

We then make the change of variables $t = V/U$, $Z = Z$, and $V = V$ and integrate V from $-\infty$ to $+\infty$, Z from $-\infty$ to K_γ, and t from $-\infty$ to K_β. If this is done in different orders the following representations are found:

$$P_f(K_\gamma, K_\beta) = \int_{-\infty}^{K_\gamma} P\left\{T_f \le \frac{K_\beta}{(1-\rho^2)^{1/2}} \,\middle|\, \lambda\right\} G'(X)\, dX, \tag{2}$$

where T_f has a noncentral t-distribution with f degrees of freedom and noncentrality parameter $\lambda = \rho X/(1-\rho^2)^{1/2}$.

$$P_f(K_\gamma, K_\beta) = \frac{\Gamma\left(\frac{f+1}{2}\right)}{\Gamma\left(\frac{f}{2}\right)\sqrt{\pi f}} \int_{-\infty}^{K_\beta} \left(1 + \frac{X^2}{f}\right)^{-(f+1)/2}$$

$$\cdot P\left\{T_{f+1} \le \frac{-\rho X\sqrt{f+1}}{(1-\rho^2)^{1/2}(f+X^2)^{1/2}} \,\middle|\, \lambda\right\} dX, \tag{3}$$

where T_{f+1} has a noncentral t-distribution with $f+1$ degrees of freedom and noncentrality parameter $\lambda = -K_\gamma/(1-\rho^2)^{1/2}$.

$$P_f(K_\gamma, K_\beta) = \int_{-\infty}^{+\infty} G'(X) G\left(\frac{K_\gamma - c_1 X}{(1-c_1^2)^{1/2}}\right) P\left\{T_f \le \frac{K_\beta}{(1-c_2^2)^{1/2}} \,\middle|\, \lambda\right\} dX, \tag{4}$$

where $G(X) = \int_{-\infty}^{X} G'(t)\, dt$ and T_f has a noncentral t-distribution with f degrees of freedom and noncentrality parameter $\lambda = c_2 X/(1-c_2^2)^{1/2}$, and $c_1 c_2 = \rho$ with $0 \le c_i^2 \le 1$.

$$P_f(K_\gamma, K_\beta) = \frac{\sqrt{2\pi}}{\Gamma\left(\frac{f}{2}\right) 2^{(f-2)/2}} \int_0^{\infty} X^{f-1} G'(X) B\left(K_\gamma, \frac{K_\beta X}{\sqrt{f}}; \rho\right) dX, \tag{5}$$

where $B(h, k; \rho)$ is the standardized bivariate normal cumulative to (h, k) with correlation ρ.

$$P_f(K_\gamma, K_\beta) = \int_{-\infty}^{+\infty} G'(X)G\left(\frac{K_\gamma - \rho X}{(1-\rho^2)^{1/2}}\right)P\left\{\chi_f^2 \geq \frac{fX^2}{K_\beta^2}\right\} dX, \tag{6}$$

where χ_f^2 has the chi square distribution with f degrees of freedom.

For one degree of freedom, we have

$$P_1(K_\gamma, K_\beta)$$
$$= 2C\left(0, K_\gamma, 0; \rho_{12} = \frac{\rho}{(1+K_\beta^2)^{1/2}}, \rho_{13} = \frac{K_\beta}{(1+K_\beta^2)^{1/2}}, \rho_{23} = 0\right) \tag{7}$$

where $C(h, k, m; \rho_{12}, \rho_{13}, \rho_{23})$ is the cumulative trivariate normal distribution to (h, k, m) with correlations $(\rho_{12}, \rho_{13}, \rho_{23})$.

For two degrees of freedom, we have

$$P_2(K_\gamma, K_\beta)$$
$$= B(0, K_\gamma; \rho)$$
$$+ \frac{K_\beta}{(K_\beta^2+2)^{1/2}} B\left(0, \frac{K_\gamma(K_\beta^2+2)^{1/2}}{[2(1-\rho^2)+K_\beta^2]^{1/2}}; -\frac{\rho K_\beta}{[2(1-\rho^2)+K_\beta^2]^{1/2}}\right), \tag{8}$$

where $B(h, k; \rho)$ is the standardized bivariate normal cumulative as used above.

As checks for special cases we have for $\rho = 0$

$$P_f(K_\gamma, K_\beta \mid \rho = 0) = G(K_\gamma)P\{T_f \leq K_\beta \mid \lambda = 0\}$$

and for $K_\beta = 0$

$$P_f(K_\gamma, 0) = B(K_\gamma, 0; \rho).$$

In order to compute this function for any number of degrees of freedom we derived the following recursion formula by integrating (2) by parts and making use of the fact that for the noncentral t-distribution

$$P\{T_f \leq t \mid \lambda\} = \frac{\sqrt{2\pi}}{\Gamma\left(\frac{f}{2}\right)2^{(f-2)/2}} \int_0^\infty G\left(\frac{tX}{\sqrt{f}} - \lambda\right)X^{f-1}G'(X)\, dX.$$

The result can be expressed as

$$P_f(K_\gamma, K_\beta) = P_{f-2}\left(K_\gamma, K_\beta\left(\frac{f-2}{f}\right)^{1/2}\right) + \frac{K_\beta\Gamma\left(\frac{f-1}{2}\right)}{2\Gamma\left(\frac{f}{2}\right)\sqrt{\pi f}}\left(\frac{f}{K_\beta^2+f}\right)^{(f-1)/2}$$
$$\cdot P\left\{T_{f-1} \leq \frac{-\rho K_\beta\sqrt{f-1}}{(1-\rho^2)^{1/2}(K_\beta^2+f)^{1/2}} \,\middle|\, \lambda\right\},$$

where $\lambda = -K_\gamma/(1-\rho^2)^{1/2}$ for $f > 2$.

The computation process was then to use the false position method to find the value of K_β which satisfies

$$P\{Z \le K_\gamma, T_f \le K_\beta\} - \delta P\{T_f \le K_\beta\} = 0,$$

by using the recursion formula and either $P_1(K_\gamma, K_\beta)$ or $P_2(K_\gamma, K_\beta)$, depending on whether f is odd or even. The partial derivative of $P\{Z \le K_\gamma, T_f \le K_\beta\}$ with respect to K_β shows that there is exactly one maximum for the function and clearly the function is zero for $K_\beta = -\infty$. As K_β approaches $+\infty$, the function approaches $\gamma - \delta$ from above, and hence there is only one finite value of K_β which satisfies this equation.

As a check on the accuracy of Table 1, a 20-point Gauss–Hermite integration was performed on (4) with $c_1 = c_2 = \sqrt{\rho}$ using the value of K_β obtained from the false position process applied to the recursion formula. If the value of $P\{K_\gamma, K_\beta)/P\{T_f \le K_\beta \mid \lambda = 0\}$ was within 0.0005 of δ, the value was accepted and put into the table.

REFERENCES

Owen, D. B., McIntire, D., and Seymour, E. (1975). Tables using one or two screening variables to increase acceptable product under one-sided specifications. *J. Quality Tech.* **7** 127–138.

Owen, D. B. and Boddie, J. W. (1976). A screening method for increasing acceptable product with some parameters unknown. *Technometrics* **18** 195–199.

Owen, D. B. and Su, Yueh-ling Hsiao (1977). Screening based on normal variables. *Technometrics* **19** 65–68.

Hsiao, Yueh-ling (1976). A selection problem using a screening variable with some parameters unknown. Ph.D. Dissertation, Southern Methodist University.

TABLE 1[a]

Values of K_β Satisfying Equation (1)

D.F. = 3 D = .95

G \ R	.70	.75	.80	.85	.90	.95	.99
.80	.136	.300	.450	.590	.724	.855	.959
.81	.208	.365	.511	.648	.779	.908	1.010
.82	.282	.434	.575	.709	.838	.964	1.064
.83	.359	.506	.644	.775	.901	1.025	1.124
.84	.440	.583	.717	.845	.970	1.092	1.189
.85	.527	.666	.797	.922	1.045	1.165	1.261
.86	.621	.756	.884	1.008	1.128	1.247	1.341
.87	.724	.855	.981	1.103	1.222	1.340	1.433
.88	.839	.968	1.092	1.212	1.330	1.447	1.540
.89	.972	1.099	1.221	1.340	1.458	1.574	1.666
.90	1.129	1.255	1.377	1.496	1.613	1.729	1.822
.91	1.325	1.451	1.573	1.693	1.811	1.929	2.023
.92	1.587	1.714	1.838	1.961	2.082	2.203	2.299
.93	1.979	2.112	2.243	2.372	2.500	2.627	2.729

D.F. = 3 D = .99

G \ R	.70	.75	.80	.85	.90	.95	.99
.85	-.600	-.255	.016	.245	.452	.643	.788
.86	-.481	-.167	.086	.305	.503	.688	.829
.87	-.367	-.080	.157	.366	.557	.736	.873
.88	-.255	.008	.231	.431	.614	.788	.921
.89	-.145	.098	.308	.499	.676	.844	.974
.90	-.034	.191	.390	.572	.744	.907	1.032
.91	.080	.289	.478	.653	.818	.976	1.099
.92	.199	.396	.575	.742	.902	1.056	1.175
.93	.329	.513	.683	.845	.999	1.149	1.265
.94	.474	.647	.810	.965	1.115	1.261	1.375
.95	.645	.809	.964	1.115	1.261	1.403	1.516
.96	.860	1.017	1.167	1.313	1.457	1.597	1.708
.97	1.168	1.319	1.467	1.612	1.755	1.896	2.008

[a] f, δ, ρ, γ, and ∞ are represented in this table by D.F., D, R, G, and INF.

TABLE 1—(*continued*)

D.F. = 4 D = .95

G \ R	.70	.75	.80	.85	.90	.95	.99
.80	.134	.297	.448	.590	.726	.859	.965
.81	.205	.363	.510	.648	.781	.911	1.015
.82	.279	.432	.574	.709	.840	.967	1.069
.83	.356	.504	.643	.775	.902	1.027	1.127
.84	.438	.581	.716	.845	.970	1.092	1.190
.85	.524	.663	.795	.921	1.043	1.163	1.259
.86	.618	.753	.881	1.005	1.124	1.242	1.336
.87	.720	.852	.977	1.098	1.215	1.331	1.423
.88	.834	.962	1.085	1.203	1.319	1.432	1.522
.89	.965	1.089	1.209	1.325	1.439	1.551	1.640
.90	1.117	1.239	1.357	1.472	1.584	1.694	1.781
.91	1.304	1.424	1.541	1.654	1.765	1.874	1.960
.92	1.548	1.667	1.783	1.896	2.006	2.115	2.202
.93	1.903	2.023	2.140	2.254	2.366	2.477	2.564

D.F. = 4 D = .99

G \ R	.70	.75	.80	.85	.90	.95	.99
.85	-.557	-.244	.015	.245	.457	.659	.817
.86	-.453	-.161	.085	.305	.510	.706	.859
.87	-.349	-.078	.156	.367	.566	.756	.904
.88	-.245	.008	.230	.433	.625	.809	.953
.89	-.141	.097	.308	.503	.688	.867	1.007
.90	-.033	.189	.391	.579	.757	.931	1.066
.91	.078	.288	.480	.661	.834	1.001	1.133
.92	.198	.396	.579	.752	.919	1.082	1.210
.93	.328	.515	.689	.856	1.017	1.175	1.299
.94	.474	.651	.818	.978	1.134	1.287	1.407
.95	.647	.814	.973	1.127	1.278	1.427	1.544
.96	.863	1.022	1.174	1.323	1.470	1.614	1.729
.97	1.167	1.319	1.466	1.611	1.754	1.896	2.009

TABLE 1—(*continued*)

D.F. = 5 P = .95

G \ P	.70	.75	.80	.85	.90	.95	.99
.80	.133	.296	.447	.590	.728	.862	.970
.81	.204	.362	.509	.648	.783	.914	1.019
.82	.278	.431	.574	.710	.841	.970	1.072
.83	.355	.503	.642	.775	.903	1.029	1.129
.84	.436	.580	.715	.845	.970	1.093	1.191
.85	.523	.662	.794	.921	1.043	1.163	1.258
.86	.616	.752	.880	1.003	1.123	1.240	1.333
.87	.718	.850	.975	1.095	1.212	1.326	1.416
.88	.832	.959	1.081	1.198	1.312	1.424	1.512
.89	.960	1.084	1.203	1.317	1.428	1.537	1.623
.90	1.110	1.231	1.346	1.458	1.566	1.672	1.756
.91	1.292	1.409	1.522	1.631	1.737	1.841	1.922
.92	1.526	1.641	1.751	1.858	1.962	2.063	2.142
.93	1.859	1.972	2.081	2.186	2.288	2.388	2.465

D.F. = 5 P = .99

G \ P	.70	.75	.80	.85	.90	.95	.99
.85	-.534	-.238	.015	.245	.461	.670	.838
.86	-.437	-.158	.084	.305	.514	.718	.881
.87	-.339	-.076	.155	.368	.571	.769	.927
.88	-.240	.008	.229	.435	.632	.824	.977
.89	-.138	.096	.308	.506	.696	.883	1.031
.90	-.033	.189	.391	.583	.767	.947	1.091
.91	.078	.288	.482	.666	.844	1.019	1.158
.92	.197	.396	.581	.759	.931	1.100	1.235
.93	.327	.516	.693	.864	1.030	1.194	1.324
.94	.474	.653	.823	.987	1.147	1.305	1.431
.95	.648	.817	.979	1.136	1.291	1.443	1.565
.96	.865	1.025	1.180	1.331	1.479	1.627	1.744
.97	1.167	1.319	1.467	1.612	1.755	1.898	2.011

TABLE 1—(*continued*)

D.F. = 6 D = .95

G \ D	.70	.75	.80	.85	.90	.95	.99
.80	.132	.295	.447	.590	.729	.865	.973
.81	.203	.361	.508	.649	.784	.917	1.022
.82	.276	.430	.573	.710	.842	.972	1.075
.83	.354	.502	.642	.775	.904	1.030	1.131
.84	.435	.579	.715	.845	.971	1.094	1.192
.85	.522	.662	.794	.920	1.043	1.163	1.258
.86	.615	.751	.879	1.003	1.122	1.239	1.331
.87	.717	.849	.973	1.093	1.209	1.323	1.412
.88	.830	.957	1.079	1.195	1.308	1.418	1.505
.89	.958	1.081	1.198	1.312	1.421	1.528	1.612
.90	1.106	1.225	1.339	1.449	1.555	1.658	1.739
.91	1.284	1.399	1.510	1.616	1.719	1.819	1.896
.92	1.512	1.623	1.730	1.833	1.932	2.028	2.102
.93	1.831	1.939	2.042	2.141	2.237	2.329	2.400

D.F. = 6 D = .99

G \ D	.70	.75	.80	.85	.90	.95	.99
.85	-.520	-.235	.015	.244	.463	.679	.854
.86	-.427	-.156	.083	.305	.518	.727	.898
.87	-.332	-.076	.154	.369	.575	.779	.944
.88	-.236	.008	.229	.436	.637	.835	.995
.89	-.136	.095	.308	.508	.702	.894	1.049
.90	-.033	.188	.392	.586	.774	.960	1.110
.91	.077	.287	.483	.670	.852	1.032	1.177
.92	.196	.396	.583	.763	.939	1.114	1.254
.93	.326	.516	.696	.869	1.039	1.207	1.343
.94	.474	.655	.827	.993	1.156	1.319	1.449
.95	.649	.819	.983	1.143	1.300	1.456	1.581
.96	.867	1.028	1.184	1.336	1.487	1.636	1.756
.97	1.168	1.320	1.468	1.613	1.757	1.900	2.014

TABLE 1—(*continued*)

D.F. = 7 D = .95

G \ R	.70	.75	.80	.85	.90	.95	.99
.80	.132	.294	.446	.591	.730	.867	.976
.81	.202	.360	.508	.649	.785	.918	1.025
.82	.276	.429	.573	.711	.843	.973	1.077
.83	.353	.502	.642	.776	.905	1.032	1.132
.84	.434	.579	.715	.845	.972	1.095	1.192
.85	.521	.661	.794	.921	1.043	1.163	1.258
.86	.615	.750	.879	1.002	1.122	1.238	1.329
.87	.716	.848	.973	1.092	1.208	1.321	1.409
.88	.829	.956	1.077	1.193	1.305	1.414	1.499
.89	.956	1.079	1.196	1.308	1.416	1.521	1.603
.90	1.103	1.221	1.334	1.442	1.547	1.648	1.726
.91	1.279	1.393	1.501	1.606	1.706	1.803	1.877
.92	1.502	1.611	1.715	1.815	1.911	2.003	2.073
.93	1.811	1.915	2.015	2.110	2.201	2.287	2.352

D.F. = 7 D = .99

G \ R	.70	.75	.80	.85	.90	.95	.99
.85	-.510	-.232	.015	.244	.465	.685	.867
.86	-.420	-.155	.083	.305	.520	.734	.911
.87	-.328	-.075	.154	.370	.578	.787	.958
.88	-.233	.008	.229	.437	.640	.843	1.009
.89	-.135	.095	.308	.510	.707	.903	1.064
.90	-.032	.187	.392	.588	.779	.969	1.125
.91	.077	.287	.484	.673	.858	1.042	1.192
.92	.195	.396	.585	.767	.946	1.124	1.269
.93	.326	.517	.698	.874	1.046	1.218	1.358
.94	.474	.656	.829	.998	1.164	1.329	1.463
.95	.649	.821	.987	1.148	1.307	1.466	1.594
.96	.868	1.030	1.187	1.341	1.492	1.644	1.765
.97	1.168	1.321	1.469	1.614	1.758	1.902	2.016

TABLE 1—(*continued*)

D.F. = 8 D = .95

G \ R	.70	.75	.80	.85	.90	.95	.99
.80	.131	.294	.446	.591	.731	.868	.978
.81	.202	.360	.508	.649	.786	.920	1.026
.82	.275	.429	.573	.711	.844	.974	1.078
.83	.352	.501	.642	.776	.906	1.033	1.133
.84	.434	.578	.715	.846	.972	1.095	1.193
.85	.521	.661	.794	.921	1.043	1.163	1.257
.86	.614	.750	.879	1.002	1.121	1.237	1.328
.87	.716	.847	.972	1.092	1.207	1.319	1.407
.88	.828	.955	1.076	1.191	1.303	1.411	1.495
.89	.955	1.077	1.193	1.305	1.413	1.516	1.597
.90	1.101	1.218	1.330	1.438	1.541	1.640	1.716
.91	1.275	1.388	1.495	1.598	1.696	1.790	1.862
.92	1.494	1.602	1.704	1.802	1.895	1.984	2.050
.93	1.796	1.898	1.995	2.087	2.174	2.255	2.316

D.F. = 8 D = .99

G \ R	.70	.75	.80	.85	.90	.95	.99
.85	-.503	-.230	.015	.244	.466	.690	.877
.86	-.415	-.154	.083	.305	.522	.740	.921
.87	-.325	-.075	.154	.370	.581	.793	.969
.88	-.232	.008	.228	.438	.643	.849	1.020
.89	-.134	.095	.308	.511	.710	.910	1.076
.90	-.032	.187	.392	.589	.783	.977	1.137
.91	.077	.287	.484	.675	.862	1.050	1.205
.92	.195	.396	.586	.770	.951	1.133	1.281
.93	.326	.517	.700	.877	1.052	1.227	1.370
.94	.475	.657	.832	1.002	1.170	1.338	1.475
.95	.650	.823	.989	1.152	1.313	1.474	1.604
.96	.869	1.032	1.190	1.344	1.497	1.650	1.773
.97	1.169	1.321	1.470	1.615	1.760	1.903	2.019

TABLE 1—(*continued*)

D.F. = 9 D = .95

G \ R	.70	.75	.80	.85	.90	.95	.99
.80	.131	.293	.446	.591	.731	.869	.979
.81	.201	.359	.508	.649	.786	.921	1.028
.82	.275	.428	.573	.711	.845	.976	1.079
.83	.352	.501	.642	.776	.906	1.034	1.134
.84	.433	.578	.715	.846	.972	1.096	1.193
.85	.520	.661	.793	.921	1.044	1.163	1.257
.86	.614	.750	.878	1.002	1.121	1.237	1.327
.87	.715	.847	.971	1.091	1.206	1.318	1.404
.88	.828	.954	1.075	1.190	1.302	1.409	1.492
.89	.954	1.076	1.192	1.303	1.410	1.513	1.591
.90	1.099	1.216	1.327	1.434	1.536	1.634	1.708
.91	1.272	1.384	1.490	1.592	1.689	1.781	1.850
.92	1.488	1.595	1.696	1.792	1.883	1.969	2.032
.93	1.785	1.885	1.979	2.069	2.153	2.231	2.288

D.F. = 9 D = .99

G \ R	.70	.75	.80	.85	.90	.95	.99
.85	-.498	-.228	.015	.244	.468	.694	.885
.86	-.411	-.153	.083	.306	.524	.744	.930
.87	-.322	-.074	.153	.370	.583	.798	.978
.88	-.230	.008	.228	.439	.646	.855	1.030
.89	-.134	.094	.307	.512	.713	.916	1.086
.90	-.032	.187	.393	.591	.786	.983	1.147
.91	.076	.286	.485	.677	.866	1.057	1.215
.92	.195	.396	.587	.772	.955	1.140	1.292
.93	.326	.518	.701	.880	1.056	1.234	1.380
.94	.475	.658	.833	1.005	1.174	1.345	1.484
.95	.650	.824	.991	1.155	1.317	1.480	1.613
.96	.869	1.033	1.192	1.347	1.501	1.655	1.780
.97	1.169	1.322	1.470	1.616	1.761	1.905	2.020

TABLE 1—(*continued*)

D.F. = 10 D = .95

G \ R	.70	.75	.80	.85	.90	.95	.99
.80	.131	.293	.446	.591	.732	.870	.981
.81	.201	.359	.508	.650	.787	.922	1.029
.82	.274	.428	.573	.711	.845	.976	1.080
.83	.351	.501	.642	.777	.907	1.034	1.135
.84	.433	.578	.715	.846	.973	1.097	1.193
.85	.520	.660	.793	.921	1.044	1.164	1.257
.86	.613	.749	.878	1.002	1.121	1.237	1.326
.87	.715	.846	.971	1.091	1.206	1.317	1.403
.88	.827	.954	1.074	1.189	1.300	1.407	1.489
.89	.953	1.075	1.191	1.301	1.408	1.509	1.587
.90	1.097	1.214	1.325	1.431	1.533	1.629	1.702
.91	1.269	1.380	1.486	1.587	1.683	1.773	1.841
.92	1.484	1.589	1.689	1.783	1.873	1.956	2.018
.93	1.776	1.874	1.967	2.054	2.136	2.211	2.265

D.F. = 10 D = .99

G \ R	.70	.75	.80	.85	.90	.95	.99
.85	-.493	-.227	.015	.244	.468	.697	.893
.86	-.408	-.152	.082	.306	.525	.748	.938
.87	-.320	-.074	.153	.370	.584	.802	.986
.88	-.229	.008	.228	.439	.648	.859	1.038
.89	-.133	.094	.307	.513	.715	.921	1.094
.90	-.032	.186	.393	.592	.789	.989	1.156
.91	.076	.286	.485	.678	.869	1.063	1.224
.92	.194	.396	.588	.774	.959	1.146	1.301
.93	.325	.518	.702	.882	1.060	1.240	1.389
.94	.475	.658	.835	1.007	1.178	1.351	1.493
.95	.651	.825	.993	1.158	1.321	1.486	1.620
.96	.870	1.034	1.193	1.349	1.504	1.660	1.785
.97	1.169	1.322	1.471	1.617	1.762	1.906	2.022

TABLE 1—(*continued*)

D.F. = 11 D = .95

G \ R	.70	.75	.80	.85	.90	.95	.99
.80	.130	.293	.445	.591	.732	.871	.982
.81	.201	.359	.508	.650	.788	.923	1.030
.82	.274	.428	.573	.711	.846	.977	1.081
.83	.351	.501	.642	.777	.907	1.035	1.135
.84	.433	.578	.715	.846	.973	1.097	1.194
.85	.520	.660	.793	.921	1.044	1.164	1.257
.86	.613	.749	.878	1.002	1.121	1.236	1.325
.87	.715	.846	.971	1.090	1.205	1.316	1.401
.88	.827	.953	1.074	1.189	1.299	1.405	1.486
.89	.952	1.074	1.190	1.300	1.406	1.507	1.583
.90	1.096	1.213	1.323	1.429	1.529	1.625	1.696
.91	1.267	1.378	1.483	1.583	1.678	1.767	1.832
.92	1.480	1.584	1.683	1.777	1.865	1.946	2.006
.93	1.769	1.866	1.957	2.042	2.122	2.195	2.246

D.F. = 11 D = .99

G \ R	.70	.75	.80	.85	.90	.95	.99
.85	-.490	-.226	.015	.244	.469	.700	.899
.86	-.406	-.152	.082	.306	.526	.751	.944
.87	-.319	-.074	.153	.371	.586	.805	.993
.88	-.228	.008	.228	.440	.649	.863	1.045
.89	-.133	.094	.307	.513	.717	.925	1.101
.90	-.032	.186	.393	.593	.791	.993	1.163
.91	.076	.286	.486	.679	.872	1.067	1.231
.92	.194	.396	.588	.775	.961	1.150	1.308
.93	.325	.518	.703	.884	1.063	1.245	1.396
.94	.475	.659	.836	1.009	1.182	1.356	1.500
.95	.651	.826	.995	1.160	1.325	1.491	1.626
.96	.871	1.035	1.195	1.352	1.507	1.663	1.790
.97	1.170	1.323	1.472	1.618	1.763	1.907	2.024

TABLE 1—(*continued*)

D.F. = 12 D = .95

G \ R	.70	.75	.80	.85	.90	.95	.99
.80	.130	.293	.445	.591	.733	.872	.983
.81	.200	.359	.507	.650	.788	.923	1.031
.82	.274	.428	.573	.712	.846	.978	1.082
.83	.351	.500	.642	.777	.908	1.036	1.136
.84	.433	.578	.715	.846	.974	1.097	1.194
.85	.519	.660	.793	.921	1.044	1.164	1.256
.86	.613	.749	.878	1.002	1.121	1.236	1.325
.87	.714	.846	.971	1.090	1.205	1.316	1.400
.88	.826	.953	1.073	1.188	1.299	1.404	1.484
.89	.952	1.073	1.189	1.299	1.404	1.505	1.580
.90	1.095	1.212	1.322	1.427	1.527	1.621	1.691
.91	1.266	1.376	1.480	1.580	1.674	1.761	1.825
.92	1.477	1.581	1.679	1.771	1.858	1.938	1.995
.93	1.763	1.858	1.948	2.033	2.111	2.181	2.230

D.F. = 12 D = .99

G \ R	.70	.75	.80	.85	.90	.95	.99
.85	-.487	-.226	.015	.244	.470	.702	.904
.86	-.404	-.151	.082	.306	.527	.754	.950
.87	-.317	-.074	.153	.371	.587	.808	.999
.88	-.227	.008	.228	.440	.650	.866	1.051
.89	-.132	.094	.307	.514	.719	.929	1.108
.90	-.032	.186	.393	.594	.793	.997	1.170
.91	.076	.286	.486	.680	.874	1.071	1.238
.92	.194	.396	.589	.777	.964	1.155	1.315
.93	.325	.518	.704	.885	1.066	1.250	1.403
.94	.475	.659	.837	1.011	1.185	1.360	1.506
.95	.651	.826	.996	1.162	1.328	1.495	1.632
.96	.871	1.036	1.196	1.353	1.509	1.667	1.794
.97	1.170	1.323	1.472	1.619	1.764	1.908	2.025

TABLE 1—(*continued*)

D.F. = 15 D = .95

G \ R	.70	.75	.80	.85	.90	.95	.99
.80	.130	.292	.445	.591	.734	.874	.985
.81	.200	.358	.507	.650	.789	.925	1.032
.82	.273	.427	.573	.712	.847	.979	1.083
.83	.350	.500	.642	.777	.909	1.037	1.137
.84	.432	.577	.715	.847	.974	1.098	1.194
.85	.519	.660	.793	.921	1.045	1.164	1.256
.86	.612	.749	.878	1.002	1.121	1.236	1.323
.87	.714	.845	.970	1.090	1.204	1.314	1.397
.88	.826	.952	1.072	1.187	1.297	1.401	1.479
.89	.951	1.072	1.187	1.297	1.401	1.500	1.572
.90	1.093	1.209	1.319	1.423	1.521	1.614	1.680
.91	1.262	1.371	1.475	1.573	1.665	1.749	1.809
.92	1.470	1.572	1.669	1.759	1.843	1.919	1.972
.93	1.749	1.843	1.930	2.011	2.086	2.151	2.194

D.F. = 15 D = .99

G \ R	.70	.75	.80	.85	.90	.95	.99
.85	-.481	-.224	.015	.244	.471	.708	.917
.86	-.399	-.150	.082	.306	.529	.760	.964
.87	-.314	-.073	.153	.371	.589	.815	1.013
.88	-.225	.008	.227	.441	.654	.874	1.066
.89	-.132	.094	.307	.515	.723	.937	1.123
.90	-.032	.186	.393	.595	.797	1.006	1.185
.91	.076	.286	.487	.683	.879	1.081	1.253
.92	.194	.396	.590	.780	.970	1.165	1.330
.93	.325	.519	.706	.889	1.072	1.260	1.418
.94	.475	.660	.839	1.015	1.191	1.371	1.520
.95	.652	.828	.999	1.167	1.334	1.504	1.645
.96	.872	1.038	1.199	1.357	1.515	1.674	1.804
.97	1.171	1.324	1.474	1.620	1.766	1.911	2.028

TABLE 1—(*continued*)

D.F. = 18 D = .95

G \ R	.70	.75	.80	.85	.90	.95	.99
.80	.130	.292	.445	.592	.734	.875	.986
.81	.200	.358	.507	.650	.790	.926	1.034
.82	.273	.427	.573	.712	.848	.980	1.084
.83	.350	.500	.642	.778	.909	1.038	1.137
.84	.432	.577	.715	.847	.975	1.099	1.194
.85	.519	.660	.793	.921	1.045	1.165	1.255
.86	.612	.749	.878	1.002	1.121	1.236	1.322
.87	.714	.845	.970	1.089	1.204	1.313	1.394
.88	.825	.952	1.072	1.186	1.296	1.400	1.475
.89	.950	1.071	1.186	1.295	1.399	1.497	1.567
.90	1.092	1.207	1.316	1.420	1.518	1.608	1.672
.91	1.260	1.368	1.471	1.568	1.659	1.741	1.798
.92	1.466	1.567	1.662	1.751	1.833	1.907	1.955
.93	1.740	1.832	1.918	1.997	2.069	2.131	2.169

D.F. = 18 D = .99

G \ R	.70	.75	.80	.85	.90	.95	.99
.85	-.477	-.223	.015	.244	.472	.712	.926
.86	-.397	-.149	.082	.306	.530	.764	.973
.87	-.313	-.073	.152	.372	.591	.820	1.023
.88	-.224	.008	.227	.441	.656	.879	1.076
.89	-.131	.093	.307	.516	.725	.943	1.133
.90	-.032	.186	.393	.596	.800	1.012	1.196
.91	.076	.286	.487	.684	.882	1.087	1.265
.92	.193	.396	.591	.782	.974	1.172	1.341
.93	.325	.519	.707	.892	1.077	1.267	1.429
.94	.475	.661	.841	1.018	1.196	1.378	1.531
.95	.652	.829	1.001	1.170	1.339	1.511	1.654
.96	.873	1.039	1.201	1.360	1.519	1.680	1.812
.97	1.171	1.325	1.474	1.622	1.767	1.913	2.031

TABLE 1—(*continued*)

D.F. = 21 P = .95

G \ R	.70	.75	.80	.85	.90	.95	.99
.80	.129	.292	.445	.592	.735	.876	.987
.81	.199	.358	.507	.651	.790	.927	1.034
.82	.273	.427	.573	.712	.848	.981	1.084
.83	.350	.500	.642	.778	.910	1.039	1.137
.84	.431	.577	.715	.847	.975	1.100	1.194
.85	.518	.660	.793	.922	1.045	1.165	1.255
.86	.612	.748	.878	1.002	1.121	1.236	1.321
.87	.713	.845	.970	1.089	1.204	1.313	1.393
.88	.825	.951	1.071	1.186	1.295	1.398	1.472
.89	.949	1.070	1.185	1.294	1.398	1.494	1.562
.90	1.091	1.206	1.315	1.418	1.515	1.605	1.666
.91	1.258	1.366	1.468	1.564	1.654	1.735	1.790
.92	1.463	1.563	1.657	1.745	1.826	1.898	1.943
.93	1.734	1.825	1.909	1.987	2.057	2.116	2.151

D.F. = 21 P = .99

G \ R	.70	.75	.80	.85	.90	.95	.99
.85	-.475	-.222	.015	.244	.473	.714	.933
.86	-.395	-.149	.082	.306	.531	.767	.981
.87	-.311	-.073	.152	.372	.592	.823	1.031
.88	-.224	.008	.227	.442	.657	.883	1.084
.89	-.131	.093	.307	.516	.727	.947	1.142
.90	-.032	.185	.393	.597	.802	1.016	1.204
.91	.076	.286	.487	.686	.885	1.092	1.273
.92	.193	.396	.591	.783	.976	1.177	1.350
.93	.325	.519	.708	.893	1.080	1.272	1.437
.94	.475	.661	.842	1.020	1.199	1.383	1.539
.95	.653	.830	1.002	1.172	1.342	1.516	1.662
.96	.873	1.040	1.202	1.362	1.522	1.684	1.817
.97	1.172	1.325	1.475	1.623	1.769	1.915	2.032

TABLE 1—(*continued*)

D.F. = 24 D = .95

G \ R	.70	.75	.80	.85	.90	.95	.99
.80	.129	.291	.445	.592	.735	.876	.987
.81	.199	.357	.507	.651	.790	.928	1.035
.82	.273	.427	.573	.713	.849	.982	1.085
.83	.350	.500	.642	.778	.910	1.039	1.138
.84	.431	.577	.715	.847	.976	1.100	1.194
.85	.518	.659	.793	.922	1.046	1.165	1.254
.86	.612	.748	.878	1.002	1.121	1.236	1.320
.87	.713	.845	.970	1.089	1.204	1.312	1.391
.88	.825	.951	1.071	1.185	1.295	1.397	1.470
.89	.949	1.070	1.184	1.293	1.396	1.493	1.559
.90	1.091	1.205	1.314	1.417	1.513	1.602	1.662
.91	1.257	1.365	1.466	1.562	1.651	1.731	1.783
.92	1.460	1.560	1.654	1.741	1.821	1.891	1.934
.93	1.730	1.819	1.903	1.979	2.048	2.105	2.137

D.F. = 24 D = .99

G \ R	.70	.75	.80	.85	.90	.95	.99
.85	-.473	-.221	.015	.243	.474	.717	.939
.86	-.393	-.149	.082	.306	.532	.770	.986
.87	-.310	-.073	.152	.372	.593	.826	1.037
.88	-.223	.008	.227	.442	.658	.886	1.090
.89	-.130	.093	.307	.517	.728	.950	1.148
.90	-.032	.185	.393	.598	.804	1.020	1.211
.91	.075	.285	.488	.686	.887	1.096	1.280
.92	.193	.396	.592	.784	.979	1.181	1.357
.93	.324	.520	.709	.895	1.082	1.277	1.444
.94	.475	.662	.843	1.022	1.202	1.387	1.545
.95	.653	.830	1.003	1.174	1.345	1.520	1.667
.96	.874	1.041	1.203	1.364	1.524	1.687	1.821
.97	1.172	1.326	1.476	1.623	1.770	1.916	2.034

TABLE 1—(*continued*)

D.F. = 30 D = .95

G \ R	.70	.75	.80	.85	.90	.95	.99
.80	.129	.291	.445	.592	.735	.877	.988
.81	.199	.357	.507	.651	.791	.929	1.036
.82	.272	.426	.573	.713	.849	.983	1.085
.83	.349	.499	.642	.778	.911	1.040	1.138
.84	.431	.577	.715	.848	.976	1.101	1.194
.85	.518	.659	.793	.922	1.046	1.165	1.253
.86	.611	.748	.878	1.002	1.121	1.235	1.318
.87	.713	.845	.970	1.089	1.203	1.312	1.389
.88	.824	.951	1.071	1.185	1.294	1.396	1.467
.89	.948	1.069	1.184	1.292	1.395	1.490	1.554
.90	1.090	1.204	1.312	1.415	1.511	1.598	1.655
.91	1.255	1.362	1.464	1.558	1.646	1.725	1.773
.92	1.457	1.556	1.649	1.735	1.813	1.881	1.920
.93	1.723	1.812	1.894	1.969	2.035	2.089	2.117

D.F. = 30 D = .99

G \ R	.70	.75	.80	.85	.90	.95	.99
.85	-.470	-.220	.015	.243	.474	.720	.947
.86	-.391	-.148	.081	.306	.533	.773	.995
.87	-.309	-.072	.152	.372	.594	.830	1.046
.88	-.222	.008	.227	.442	.660	.890	1.100
.89	-.130	.093	.307	.518	.730	.955	1.158
.90	-.031	.185	.394	.599	.806	1.025	1.221
.91	.075	.285	.488	.688	.890	1.102	1.290
.92	.193	.396	.592	.786	.982	1.187	1.367
.93	.324	.520	.709	.897	1.086	1.283	1.454
.94	.475	.662	.844	1.024	1.206	1.393	1.555
.95	.653	.831	1.005	1.176	1.349	1.526	1.676
.96	.874	1.042	1.205	1.366	1.528	1.692	1.828
.97	1.172	1.326	1.477	1.624	1.771	1.918	2.036

TABLE 1—(*continued*)

D.F. = 40 D = .95

G \ R	.70	.75	.80	.85	.90	.95	.99
.80	.129	.291	.444	.592	.736	.878	.989
.81	.199	.357	.507	.651	.791	.930	1.036
.82	.272	.426	.573	.713	.850	.984	1.086
.83	.349	.499	.642	.778	.911	1.041	1.138
.84	.431	.577	.715	.848	.977	1.101	1.193
.85	.518	.659	.793	.922	1.046	1.166	1.252
.86	.611	.748	.878	1.002	1.121	1.235	1.316
.87	.713	.844	.970	1.089	1.203	1.311	1.386
.88	.824	.950	1.070	1.184	1.293	1.394	1.463
.89	.948	1.069	1.183	1.291	1.393	1.487	1.549
.90	1.089	1.203	1.311	1.413	1.508	1.594	1.647
.91	1.253	1.360	1.461	1.555	1.642	1.718	1.763
.92	1.454	1.552	1.644	1.729	1.806	1.871	1.906
.93	1.717	1.804	1.885	1.958	2.022	2.073	2.097

D.F. = 40 D = .99

G \ R	.70	.75	.80	.85	.90	.95	.99
.85	-.467	-.220	.015	.243	.475	.723	.956
.86	-.389	-.148	.081	.306	.534	.777	1.004
.87	-.307	-.072	.152	.372	.596	.834	1.055
.88	-.221	.008	.227	.443	.662	.895	1.110
.89	-.130	.093	.307	.518	.732	.960	1.168
.90	-.031	.185	.394	.600	.809	1.030	1.232
.91	.075	.285	.488	.689	.892	1.107	1.301
.92	.193	.396	.593	.788	.985	1.193	1.378
.93	.324	.520	.710	.899	1.089	1.289	1.465
.94	.475	.663	.846	1.027	1.209	1.400	1.565
.95	.653	.832	1.006	1.179	1.352	1.532	1.684
.96	.875	1.043	1.207	1.369	1.531	1.697	1.834
.97	1.173	1.327	1.478	1.626	1.773	1.920	2.037

TABLE 1—(*continued*)

D.F. = 60 D = .95

G \ R	.70	.75	.80	.85	.90	.95	.99
.80	.129	.291	.444	.592	.737	.879	.990
.81	.199	.357	.507	.651	.792	.931	1.037
.82	.272	.426	.572	.713	.850	.985	1.086
.83	.349	.499	.642	.779	.912	1.042	1.138
.84	.430	.577	.715	.848	.977	1.102	1.192
.85	.518	.659	.794	.922	1.047	1.166	1.251
.86	.611	.748	.878	1.002	1.122	1.235	1.314
.87	.712	.844	.969	1.089	1.203	1.310	1.383
.88	.824	.950	1.070	1.184	1.292	1.393	1.458
.89	.947	1.068	1.182	1.290	1.392	1.485	1.543
.90	1.088	1.202	1.309	1.411	1.505	1.589	1.639
.91	1.251	1.358	1.458	1.551	1.637	1.711	1.752
.92	1.450	1.548	1.639	1.723	1.798	1.861	1.891
.93	1.710	1.797	1.876	1.948	2.009	2.057	2.076

D.F. = 60 D = .99

G \ R	.70	.75	.80	.85	.90	.95	.99
.85	-.464	-.219	.015	.243	.476	.726	.965
.86	-.387	-.147	.081	.306	.535	.780	1.014
.87	-.306	-.072	.152	.372	.597	.838	1.066
.88	-.220	.008	.227	.443	.663	.899	1.121
.89	-.129	.093	.307	.519	.735	.965	1.179
.90	-.031	.185	.394	.601	.811	1.036	1.243
.91	.075	.285	.489	.690	.895	1.113	1.313
.92	.192	.396	.593	.789	.988	1.199	1.390
.93	.324	.520	.711	.901	1.093	1.296	1.476
.94	.475	.663	.847	1.029	1.214	1.407	1.576
.95	.654	.833	1.008	1.181	1.357	1.538	1.694
.96	.875	1.044	1.208	1.371	1.535	1.702	1.841
.97	1.173	1.328	1.478	1.627	1.775	1.922	2.039

TABLE 1—(*continued*)

D.F. = 120 D = .95

G \ R	.70	.75	.80	.85	.90	.95	.99
.80	.129	.290	.444	.592	.737	.881	.991
.81	.198	.357	.507	.651	.793	.932	1.037
.82	.271	.426	.572	.714	.851	.986	1.086
.83	.349	.499	.642	.779	.912	1.042	1.137
.84	.430	.576	.715	.849	.978	1.103	1.191
.85	.517	.659	.794	.923	1.047	1.167	1.249
.86	.611	.748	.878	1.002	1.122	1.235	1.312
.87	.712	.844	.969	1.089	1.203	1.310	1.379
.88	.823	.950	1.070	1.184	1.292	1.391	1.453
.89	.947	1.067	1.181	1.289	1.390	1.482	1.536
.90	1.087	1.201	1.308	1.409	1.502	1.585	1.630
.91	1.250	1.356	1.455	1.548	1.633	1.705	1.740
.92	1.447	1.544	1.634	1.717	1.791	1.850	1.875
.93	1.704	1.789	1.867	1.937	1.997	2.040	2.054

D.F. = 120 D = .99

G \ R	.70	.75	.80	.85	.90	.95	.99
.85	-.461	-.218	.015	.243	.477	.729	.976
.86	-.385	-.147	.081	.306	.536	.784	1.025
.87	-.305	-.072	.152	.373	.599	.842	1.077
.88	-.220	.008	.227	.444	.665	.904	1.132
.89	-.129	.093	.307	.520	.737	.970	1.192
.90	-.031	.185	.394	.602	.814	1.042	1.256
.91	.075	.285	.489	.692	.898	1.120	1.326
.92	.192	.396	.594	.791	.992	1.206	1.403
.93	.324	.521	.712	.903	1.097	1.303	1.489
.94	.475	.664	.848	1.031	1.218	1.414	1.588
.95	.654	.834	1.010	1.184	1.361	1.545	1.704
.96	.876	1.045	1.210	1.374	1.538	1.708	1.848
.97	1.174	1.328	1.480	1.628	1.777	1.925	2.040

TABLE 1—(*continued*)

D.F. = INF D = .95

G \ R	.70	.75	.80	.85	.90	.95	.99
.80	.128	.290	.444	.592	.738	.882	.991
.81	.198	.356	.507	.652	.793	.933	1.037
.82	.271	.426	.572	.714	.852	.987	1.086
.83	.348	.499	.642	.779	.913	1.043	1.136
.84	.430	.576	.715	.849	.978	1.103	1.190
.85	.517	.659	.794	.923	1.048	1.167	1.247
.86	.611	.748	.878	1.002	1.122	1.235	1.308
.87	.712	.844	.969	1.089	1.203	1.309	1.374
.88	.823	.949	1.069	1.183	1.291	1.390	1.447
.89	.946	1.067	1.180	1.288	1.389	1.479	1.528
.90	1.086	1.199	1.306	1.407	1.500	1.580	1.619
.91	1.248	1.354	1.453	1.545	1.628	1.698	1.727
.92	1.444	1.540	1.629	1.711	1.783	1.839	1.858
.93	1.697	1.782	1.858	1.926	1.984	2.023	2.032

D.F. = INF D = .99

G \ R	.70	.75	.80	.85	.90	.95	.99
.85	-.459	-.217	.015	.243	.478	.733	.987
.86	-.383	-.146	.081	.306	.537	.788	1.037
.87	-.303	-.072	.151	.373	.600	.847	1.090
.88	-.219	.008	.226	.444	.667	.909	1.146
.89	-.129	.093	.307	.520	.739	.976	1.206
.90	-.031	.184	.394	.603	.817	1.048	1.270
.91	.075	.285	.489	.693	.901	1.126	1.340
.92	.192	.396	.595	.793	.995	1.213	1.417
.93	.324	.521	.713	.905	1.101	1.310	1.503
.94	.475	.665	.850	1.034	1.222	1.421	1.601
.95	.654	.835	1.011	1.187	1.365	1.552	1.715
.96	.877	1.046	1.212	1.376	1.543	1.714	1.855
.97	1.174	1.329	1.481	1.630	1.779	1.927	2.038

A B C 8 D 9 E 0 F 1 G 2 H 3 I 4 J 5